CO9324147

Procurement Strategies

Procurement Strategies

A Relationship-based Approach

Edited by

Derek Walker and Keith Hampson

Blackwell
Science

First published 2003 by Blackwell Science Ltd

Library of Congress
Cataloging-in-Publication Data
Procurement strategies : a relationship-based approach/
   edited by Derek Walker & Keith Hampson.
      p. cm.
   Includes bibliographical references and index.
   ISBN 0-632-05886-2
   1. Construction industry—Management.
2. Building trades—Management.   3. Industrial
procurement.   I. Walker, Derek H.H.   II. Hampson,
Keith.

HD9715.A2 P727 2002
624′.068′7—dc21
                                        2002074742

ISBN 0-632-05886-2

A catalogue record for this title is available from the
British Library

Set in 10/12pt Times
by DP Photosetting, Aylesbury, Bucks
Printed and bound in Great Britain by
TJ International, Padstow, Cornwall

For further information on
Blackwell Publishing, visit our website:
www.blackwellpublishing.com

Contents

Chapter 9   Developing an Innovative Culture Through Relationship-based Procurement Systems   **236**

Derek Walker, Keith Hampson and Stephen Ashton

Chapter 10   Implications of Human Capital Issues   **258**

Derek Walker

Authors and Contributors

Authors

Chapters 1, 2, 3, 4, 5, 6, 7, 8, 9 and 10

Derek Walker is Professor of Project Management at the Faculty of Business, RMIT University. He worked in various project management roles in the UK, Canada, and Australia for 16 years before commencing his academic career in 1986. He obtained a Master of Science from the University of Aston (Birmingham) in 1978, a Graduate Diploma in Management Systems from Swinburne University (Melbourne) in 1985 and a PhD in 1995 from RMIT University (Melbourne). He has written over 70 peer reviewed papers and book chapters. His research interests centre on project management and project procurement systems. He is currently supervising five PhD and thirteen Professional Doctorate students in this area. He is also the Program Director of the Doctor of Project Management (DPM) program (details can be found on *http:// dhtw.tce.rmit.edu.au/pmgt/dpm.htm* or *http://www.rmit.edu.au/* then follow links to 'academic programs'). He recently contributed three chapters in the 1999 E & FN Spon publication *Procurement Systems: A Guide to Best Practice in Construction*. He was one of a team of four people (including Keith Hampson) who prepared the successful application for the Cooperative Research Centre (CRC) in Construction Innovation. He may be reached at derek.walker@rmit.edu.au and his web page is located at URL *http://dhtw.tce.rmit.edu.au/*

Chapters 1, 2, 3, 4, 6, 7, and 9

Keith Hampson led the national research team studying the National Museum of Australia project in Canberra. He is currently the CEO of the Australian CRC for Construction Innovation that has a budget of A\$64 million over seven years to investigate and advance innovation in the Australian construction industry. Keith has the responsibility of crafting a blend of commercial and public good outcomes on behalf of the twenty industry, government and research partners. He is a civil engineer and project manager with extensive industry experience having gained international construction management experience and knowledge from working and studying in Australia, England and the United States. Prior to his CEO appointment to the CRC CI Keith was Director of Research in Construction Management and Property at QUT and Team Leader of the QUT/CSIRO Construction

Research Alliance with senior responsibilities for teaching, research and business development in construction management. His interests include innovation in technology and procurement for more effective industry, international construction and strategic management. In addition to his Civil Engineering (Honours) qualification gained while serving a six-year industry cadetship, Keith completed an MBA, and a PhD in Construction Engineering and Management at Stanford University. He may be reached at k.hampson@qut.com or by reference to *http://www.construction-innovation.info*

Chapter 5

Martin Loosemore is Professor and Associate Dean (postgraduate Studies) at the faculty of the Built Environment, University of New South Wales, Sydney, Australia. He has published over 75 articles and three books on risk management and human resource management and is also director of a consultancy advising a wide range of companies within and outside the construction industry in this area. Martin gained his PhD in crisis management from the University of Reading in the UK and is a member of the Royal Institution of Chartered Surveyors, the Chartered Institute of Building and the American Society of Civil Engineers. Martin is also an Editor of the *Journal of Construction Procurement* which is an internationally refereed journal published from the UK. He can be contacted on *m.loosemore@unsw.edu.au*

Chapter 8

Michael Keniger is Professor of Architecture and Head of the School of Geography, Planning and Architecture at The University of Queensland. He is also the Queensland Government Architect, an advisory role. In 1998 the Queensland Board of Architects named him as 'Architect of the Year' in recognition of his outstanding contribution to architecture and to the community. He was the inaugural Chair of the Design Advisory Panel for Brisbane's South Bank, and a member of the Design Review Panel for the Sydney Olympics. He was an adviser to the construction alliance for the National Museum of Australia project in connection with the selection of the design, as a member of the Design Integrity Panel and as Chair of the Quality Review Panel. He has written extensively on contemporary architecture and urbanism in Australia. He is a Life Fellow and Past President of the Queensland Chapter of the Royal Australian Institute of Architects. He may be reached on *m.keniger@mailbox.uq.edu.au*

Chapter 6

Bruce Duyshart is the Asia Pacific eBusiness Development Manager with Lend Lease Corporation. He specialises in the design, development and implementation of IT solutions for the real estate industry. Since 1997 he has led an eBusiness team who have successfully implemented over 380 project collaboration websites, including the National Museum of Australia, collectively involving more than 12,000 project participants from 15 countries. He holds a Masters degree in Architecture from the University of Melbourne where he also established their undergraduate and industry continuing education CAD curriculum. Bruce is author of *The Digital Document* published by Butterworth-Heinemann. His other experience has included roles working in private architectural practice, IT consulting and developing websites for a number of design and construction industry publications. He may be reached at *bruce.duyshart@lendlease.com.au*

Chapter 6

Sherif Mohamed is a Senior Lecturer in Construction Engineering and Management at Griffith University's School of Engineering. He is a chartered civil engineer with sound experience in design and management. He holds an MSc, and a PhD in Civil Engineering from the University of Southampton, UK. His principal research interests lie in the broad area of re-engineering the construction process, with particular emphasis on the implementation of information technology (IT) in construction. He has authored and co-authored more than fifty refereed journal and conference publications, and is currently the Principal Supervisor of a research team covering a wide range of construction management topics. He can be reached at *s.mohamed@mailbox.gu.edu.au* or by reference to *http://www.gu.edu.au/school/eng/*

Chapter 9

Stephen Ashton is the Project Director of the architectural joint venture for the Acton Peninsula project and is the architectural representative on the Alliance Leadership Team. He is a registered Architect and a Director of Ashton, Raggatt, McDougal. He has extensive experience as Director in Charge on many institutional and public architectural projects, and on large-scale urban design and masterplanning schemes. Stephen was President of the Victorian Chapter of the Royal Australian Institute of Architects from 1990 to 1992, and lectures on practice and procurement topics. He may be contacted at *arm.melb@a-r-m.com.au* or by reference to *www.a-r-m.com.au*

Contributors

We are grateful to the following friends and colleagues who helped with production of various elements this book. We particularly thank Andrew Magub and Renaye Peters for their editorial comment, proofreading and general assistance with gaining permissions for figures and tables by other authors cited in this book.

Andrew Magub is a qualified Architect/Project Manager, a PhD candidate at the Queensland University of Technology, School of Construction Management and Property and is a Practice Director of the international Australian architecture firm Bligh Voller Nield. He also holds a Bachelor of Applied Science (Built Environment), Bachelor of Architecture (Hons) and a Master of Project Management. Andrew's area of research has previously included procurement of major stadium projects using alternative procurement methods. His current area of research is in the use of the internet for information sharing on construction projects, in particular the skill sets required to be effective with this new technology. He may be contacted at *Andrew_magub@bvn.com.au*

Renaye Peters was the Senior Researcher on site for the National Museum of Australia case study project. She has recently joined Leighton Contractors Pty Limited as a Design Manager. Her role with Leightons focuses on developing client relationships, identifying collaborative projects and nurturing clients through the delivery phases of complex projects. Renaye is a registered architect, and has many years' industry experience in hospitals, nursing homes, correctional facilities and public buildings. She tutors and lectures in architecture and construction management at the Queensland University of Technology. She has a degree in Architecture with Honours and is embarking on her PhD in Construction Management focusing on skill development in innovative procurement environments in construction. She may be reached at *Renaye.Peters@leicon.com.au*

Preface

This book is about relationships that lead to mutual benefit and the way alliances can be formed by groups who might otherwise be competitors fighting head-to-head in the construction industry where fierce competitiveness and fragmentation provide major obstacles to industry development.

The book draws upon two broad themes that help develop a relationship-based approach to project procurement. The systems and structures theme brings together a descriptive chapter on project procurement choices with chapters on the nature of enterprise networks, how the process of project alliancing partner selection takes place, how managing risk and crisis resolution takes place in a relationship-based context and how an IT infrastructure can enable improved business relationships. The second theme, attitudes and behaviours, helps us to understand the nature and reason for underlying practices that contribute to and sustain the relationship-based procurement approach described in the next four chapters. These include chapters on developing cross-team relationships, developing a quality culture through the relationship, developing an innovative culture and providing insight into how the implications of capitalising upon human capital can be achieved through a relationship-based procurement approach.

Readers of the book will not only gain insights into how collaboration and cooperation can add value to the procurement process but will also be able to refer to specific examples cited in this text. The book provides guidance, models and templates on how to put a relationship-based approach to procurement into practice. Several chapters provide specific prototypes that may be adapted. For example, the selection criteria and process for the National Museum of Australia project has a much broader stakeholder focus than has been the case for most construction projects procured to date. A template for developing a business case for supporting IT for projects is presented. We also present readers with a new and evolved approach to creating a quality culture. Perhaps one of the most useful contributions that will be appreciated is that this book presents evidence and lessons learned from a real and successful project. We believe that this book provides a credible demonstration of evidence that supports theory presented in this book through extensive presentation of results from the National Museum of Australia project research study.

This preface also provides an opportunity to highlight how relationships and alliances have worked in practice to produce this book. In research, each advance is a building exercise in which the achievements of one group are advanced through the efforts of others who take ideas forward, challenge these ideas and through that process discover value in viewing the ideas from other perspectives. Often connections and relationships from decades ago bear fruit in unexpected ways. Many unsung heroes contribute at different times and in different measure to advance theory and to test practice in order that each aspect of theory that does not work may be re-evaluated and improved. Writing a text such as this is a true collaborative and relationship-based process.

This book represents a step in a long journey on the road to developing an understanding of project procurement, particularly in the construction industry. Interest in construction procurement has grown since the establishment of the International Council for Building Research (CIB) Studies and Documentation Working Commission W92 (Procurement Systems) in 1990. Derek Walker was a contributor to an earlier text published in 1999, *Procurement Systems: A Guide to Best Practice in Construction* published by E&FN Spon, London, and edited by Steve Rowlinson and Peter McDermott. Derek co-authored three of the 12 chapters of that publication and wished to develop the breadth and depth of material on project procurement. Martin Loosemore, editor of the *Journal of Construction Procurement* (an internationally refereed journal published from the UK), continued his interest in publishing in this area by collaborating with Derek Walker on Chapter 5 of this book. Links between Derek and Martin can be traced to a common university in Wales, to a common heritage and sense of humour. This has facilitated them working together on a number of research publications. In writing, like many other examples of collaboration, tacit knowledge gained through shared interests, backgrounds and other relationships triggers creativity and helps authors draw together their unique perspectives to build a sum greater than its parts.

For quite some time there has been a keen interest developing in relationship-based procurement systems – more specifically partnering and project alliancing. In 1999, Keith Hampson led a national research team investigating the way that the National Museum of Australia project was being delivered. The research project team was itself an alliance led by Queensland University of Technology (QUT) with Derek Walker from RMIT University in Melbourne and several researchers from Australia's Commonwealth Science and Industrial Research Organisation (CSIRO) including Selwyn Tucker and Sherif Mohamed.

This demonstrates the power of relationships. The research team for the project evolved from an alliance between QUT, CSIRO and RMIT University. The National Museum of Australia Alliance Leadership Team (ALT) lent their wholehearted support and the Australian Department of Industry, Science and Resources (DISR) funded the research project and helped to disseminate the research outcomes.

The RMIT University link came about through a 20-year collaboration between Derek Walker and Tony Sidwell, (Head of the School of Construction Management and Property at QUT) into research relating to construction management and project procurement. This research project provided a stepping-stone to advance research into procurement practice and to disseminate research results.

At this point we must acknowledge the profound commitment by the National Museum of Australia project ALT including DISR to improve procurement systems and facilitate communication of how this may be undertaken. Their support has made the dissemination of many of the research findings contained in this book possible.

The National Museum of Australia research project linked Sherif Mohamed, with Keith Hampson. They collaborated on many occasions when Sherif was a researcher with CSIRO and when he later joined Griffith University on the Gold Coast near Brisbane, Australia. The National Museum of Australia research project also linked Bruce Duyshart from Bovis Lend Lease, who is a key driving force behind the development of ProjectWeb, with Keith Hampson and Derek Walker (see Chapter 6). The research project also linked Michael Keniger to Keith Hampson, Derek Walker and Renaye Peters through Michael's role in the Design Integrity Panel and as Chair of the Quality Review Panel for the National Museum of Australia project. This led to the collaboration between Michael Keniger and Derek Walker in writing Chapter 8 of this book. Stephen Ashton, a member of the National Museum of Australia project ALT and driving force behind the complex design of that project and innovations such as the development of 3D modelling and application of CAD/CAM also became linked to Keith Hampson and Derek Walker through the project. This resulted in their collaboration on Chapter 9 of this book.

This web of relationships and connections illustrates how relationship-based procurement can apply to research and practice publication as well as projects such as the National Museum of Australia. Two of the contributors to this book, Andrew Magub and Renaye Peters, are also linked through being graduates of QUT's Master of Project Management programme, formerly under the leadership of Keith Hampson, with both having worked on research projects under his direction. Renaye spent two years working half of her time on site as a researcher on the National Museum project observing and gathering valuable research data and being a key participant with others in writing the research project report so often cited in this book. Renaye also provided assistance in providing comments on several of the chapters. Andrew provided the majority of help with the onerous task of helping to provide editorial comment and opinion, proofreading, and assisting with obtaining permission for use of diagrams and tables produced by other researchers and cited in this book.

We have been able to extend the linkage of research and learning from the National Museum of Australia research project, through the development of links with Bruce Duyshart of Bovis Lend Lease to the Cooperative Research Centre (CRC) for Construction Innovation to this book.

We are indebted to numerous people and organisations in producing this book. Our universities have provided access to a wealth of literature through hard copy, and increasingly through electronic library access, allowing authors to work in hotel rooms, internet cafes, at home and in our workplace. These universities have also provided the links to students, colleagues and friends with whom we have been able to share and build ideas and test propositions. We also received wonderful support from the ALT and DISR in allowing us to independently research and report our findings of the National Museum of Australia project study. Andrew Magub and Renaye Peters contributed many sleep-starved nights and panic-filled days in helping us to produce this book.

Finally, all of us owe a great debt of gratitude to our families who have supported us throughout the production of this book, providing patience and forgiveness, allowing us to commit so much time and energy to this book.

We also would like to thank our publishers for their support and for allowing us the opportunity to demonstrate how alliances and a relationship-based approach can lead to what we believe is a worthwhile publication and text that will in turn be built upon by others in the future.

Derek Walker
Keith Hampson (Editors)
April 2002

Chapter 1
Introduction

Derek Walker and Keith Hampson

Recent attempts to define construction procurement reflect the changing and expanding nature of the scope of this important process in realising projects. The definition used in a recent guide to best practice in construction procurement states that it is 'the framework within which construction is brought about, acquired or obtained' (McDermott 1999, p. 4). CIB W92[1] was established in 1990 as a research group to investigate construction procurement and this global entity of international researchers have been researching and publishing work through symposia and workshops since that date. Thus, for over a decade a more comprehensive view of procurement has been emerging. A trend towards a more holistic and relationship-based systems view of procurement has become apparent in each of these symposia. This book continues that work by providing a clear focus on the nature of relationship-based procurement systems and the implications for the construction industry. It draws upon examples and concepts gathered from throughout the world and it is fortunate that a recently completed museum project in Canberra, Australia constructed under a project alliance provides a rich source of case study data. We are able to draw direct and clear inferences from this project because the authors of this chapter were part of a research alliance commissioned to track and study the project from the first on-site earthworks through to its completion in March 2001.

1.1 The structure of this book

We draw many examples and illustrations of aspects of procurement from our recent research report (Peters *et al.* 2001)[2] that we discuss in this book. We were fortunate to have had Australian government support[3] to undertake this work and full access to the work site, and access to the project management team with a senior researcher present on site for about 50 per cent of the time throughout the project duration. We have therefore been provided with a unique and rewarding opportunity to test theory against practice for this case study and to be able to

---

[1] International Council for Building Research Studies and Documentation Working Commission W92 (Procurement Systems).
[2] Keith Hampson led the national QUT/CSIRO Construction Research Alliance team studying and reporting on the National Museum of Australia project supported by Derek Walker from RMIT as Deputy Leader. Renaye Peters contributed as the on-site researcher in gathering data used in the project report cited above and in coordinating parts of this book.
[3] Department of Industry, Science and Resources, Commonwealth of Australia Government.

conduct in-depth conversations with the project alliance personnel engaged on the project over a prolonged period. This enabled us to gain valuable insights not only from our reflection on this evolving new sphere of study, but also from the reflections of senior staff engaged on that project.

We have structured this book to link ideas from several strands of management theory and practice. For example, we were able to link elements of systems and structures that support relationship-based procurement with the human elements of attitude, behaviours and organisational culture that facilitates relationship-based procurement systems.

This book comprises 10 chapters. Table 1.1 indicates how each chapter links with the concept of relationship-based procurement systems.

Table 1.1: Book structure

Systems and structures	Attitudes and behaviours
• Chapter 2 – Procurement Choices • Chapter 3 – Enterprise Networks, Partnering and Alliancing • Chapter 4 – Project Alliancing Member Organisation Selection • Chapter 5 – Managing Risk and Crisis Resolution – Business-as-usual Versus Relationship-based Procurement Approaches • Chapter 6 – Enabling Improved Business Relationships – How IT Makes a Difference	• Chapter 7 – Developing Cross-Team Relationships • Chapter 8 – Developing a Quality Culture – Project Alliancing Versus Business-as-Usual • Chapter 9 – Developing an Innovative Culture through Relationship-based Procurement Systems • Chapter 10 – Implications of Human Capital Issues

The classification provided in Table 1.1 can only, of course, be notional. It is impossible to write any chapter that does not offer examples or contain comments, analysis and discussion of both systems and structure, and attitudes and behaviours. However, this broad classification provides an indication of what the reader may expect. We start from the premise that there is a logical link between procurement choices, enabling systems and human behaviour. When harnessed collectively, this provides a facilitating environment that not only adds value to the procurers of projects but also to all stakeholders involved in project realisation. We argue that competitive energies can be collectively and collaboratively mobilised to provide best-for-project outcomes as well as providing reasonable returns for those investing time, energy, skills and knowledge to the endeavour. We do not believe that we are being naïve, simplistic or faddish in our conclusions and we draw supporting arguments of many authoritative writers from the literature to support our case.

One critical comment that can justifiably be made about the philosophy of a relationship-based procurement approach is that it is highly dependent upon sophisticated participants who believe that collaboration rather than competition best delivers value and benefits for them. This relies upon an approach of building solutions and leveraging upon the ideas of a diverse group of project stakeholders. It requires participants who welcome a different paradigm from the business-as-usual rationalist mindset. Collaboration is more about how to generate and build upon good ideas rather than extracting material gain at each and every stage of project realisation.

Large-scale construction projects engage many hundreds if not thousands of people and so not all participants will be uniform in their motivation or preferred ways of working together. Therefore, it is important to develop systems and cultures that reflect the objectives, aspirations and planned project outcomes. Organisations recognise that people are susceptible to 'back-sliding' and have historically put in place management systems to regulate this tendency. In a highly competitive system, regulation protects against cartels being formed. Organisational resource budget management systems attempt at the project level to control slackness and weakness of resolve to realise expected profit margins. Relationship-based procurement approaches also provide mechanisms and systems for organisations to facilitate the desired participant behaviours and to structure an environment where building solutions to problems is encouraged and facilitated.

The structure of this book is presented in two parts. We investigate in Part 1 the *what* issue of relationship-based procurement systems to clarify what is required and what choices are available. Part 2 concentrates on people issues, more specifically those relating to attitude and behaviour and *how* and *why* this approach delivers effective relationship-based procurement systems.

Part 1 is structured as follows:

- Chapter 2 presents the range of procurement choices that project initiators may make in the construction industry. These range from traditional fixed-price contracting to provide a product through to fully integrated design, build and operate systems that provide a service.
- Chapter 3 broadly introduces and discusses the development and range of enterprise networks that exist to procure projects. We concentrate upon partnering and alliancing, drawing many case studies and examples from the literature.
- Chapter 4 delves into the motivation of forming alliances. Examples are drawn from e-commerce, the manufacturing industry, the air travel industry, and the engineering and construction industry. The selection process for the National Museum of Australia project is discussed in depth, including the selection criteria used.
- Chapter 5 briefly explores risk and uncertainty with a detailed discussion on crisis management with a particular focus on the process of preparing for and managing crises. The body of literature on construction management indicates that while risk management is an important recognised part of project management, crisis management is poorly addressed. Research on crisis management indicates that it has a better chance of being addressed in a relationship-based procurement environment than in a business-as-usual one. We draw upon the National Museum of Australia project to illustrate how this may occur.
- Chapter 6 discusses how information and communication technologies (ICT) provide support and facilitation in linking project teams. While this is applicable to both relationship-based procurement systems and more traditional approaches it again has a better chance of being effectively addressed in a

relationship-based procurement environment. The National Museum of Australia project provides a useful case study to illustrate how this may occur.

The literature presented in Part 2 of this book clearly indicates that relationship-based procurement systems rely upon high levels of trust and commitment of individuals working as coherent teams to achieve common objectives. We investigate *how* and *why* this applies to relationship-based procurement systems. We do this as follows:

- Chapter 7 presents considerable detail about cross-team relationships. Much of what is discussed can also be applied to more traditional approaches to project procurement, however as trust and commitment is a fundamental driver of best practice in teamwork, it has particular relevance to relationship-based procurement systems.
- Chapter 8 concentrates upon how a quality culture can be developed and sustained for a relationship-based procurement approach. A key focus of this book is the issue of how to achieve a best-for-project mentality. This can be structured through developing a management system that delivers a quality culture and in this sense we contemplated this chapter being best placed in the first part of this book. The chapter, however, stresses that a quality culture is best developed through a focus on customer and stakeholder needs, including the vital quest of how to get the right people working on such projects. This 'people-issue' focus led us to decide to locate this chapter in Part 2. We use the National Museum of Australia project as a case study to explain how a quality culture was developed using that relationship-based procurement approach.
- Chapter 9 presents a detailed discussion on how innovation is important in delivering increasing value to project initiators and end-users. The relationship-based procurement approach provides participants ample opportunity for innovation to be encouraged and facilitated because it provides for shared risk-taking to be more systematically encouraged and rewarded. The literature indicates that more traditional approaches to project procurement reward risk-aversion and inhibit innovation. This is an important chapter on how the human side of project procurement can contribute to innovation and both product and process improvement.
- Finally, in Chapter 10 we consider the implication of human capital issues upon adopting a relationship-based procurement approach. The mindset of business-as-usual has been criticised widely in numerous reports as counter-productive for example see (Latham 1994; DETR 1998; ACA 1999; DISR 1999). Moving to a relationship-based procurement approach requires a change in the way that organisations act and facilitate trust and commitment. The National Museum of Australia project provided a useful case study to illustrate a model of systematically facilitating the qualities required of organisations and team members to leverage advantages that relationship-based procurement systems offer.

Before we launch into the next chapter, it is worth reflecting upon the justification for a change in mindset away from business-as-usual towards a relationship-based procurement approach. We need to explain why the start of the twenty-first century requires this change in mentality.

1.2 Implications of the current global work environment

The final edition of the Australian magazine *Business Review Weekly* (BRW) for the twentieth century reflected upon several major topical issues facing business and society. Generally, these revolved around how companies face competition and the nature of this competition. It becomes obvious from issues raised in this magazine and others published at the close of the second millennium that our understanding of competition has become more refined. Solutions that face up to competitive challenges have become more widely understood.

For example in one of the BRW articles it was noted, 'As companies try to grapple with globalisation, trade liberalisation and the explosion in information technology, they are forming complex networks of alliances. These alliances are blurring the lines between cooperation and collaboration, and challenging the whole notion of competition' (Ferguson 1999). Moreover in this same magazine issue discussion of the virtual organisation and the impact of outsourcing reinforced the notion that intangible assets (such as intellectual capital, talent and creative spirit) provide a clue to why it is necessary for companies to form alliances (Tabakoff 1999). Articles appearing in the USA based journal, *Harvard Business Review* (HBR), during 2000 also featured similar themes, for example (Dyer 2000; Womack and Jones 2000) where supply chain issues and organisational cooperation and collaboration were promoted citing large USA business interests.

Attracting and retaining talented people was given substantial attention in the BRW magazine by Hannen 1999; Schmidt 1999 and Way 1999. This also resonates with messages contained in HBR articles recently published, for example on the necessity of attracting and retaining talented people (Leonard and Rayport 1997; Leonard and Straus 1997; Amabile 1998; Hargadon and Sutton 2000; Bernick 2001) as well as recent books that focus on this pressing issue, one in particular describes this issue more as a war than a quest (Michaels *et al.* 2001).

The third major theme raised in the BRW magazine was the influence of e-commerce, IT and globalisation on business and society. In summary, globalisation has opened up previously protected and sheltered markets. IT and more particularly the internet, with e-commerce as its most recent manifestation, has facilitated inexpensive and effective access to a global customer base. Customer expectation has risen in response to the promise of the brave new internet-enabled world. The HBR has abounded with e-commerce articles that highlight the immense contribution to business performance that this offers, for example (Kim and Mauborgne 1997; Magretta 1998; Malone and Laubacher 1998; Hammer 2001; Haspeslagh *et al.* 2001). These articles illustrate how new businesses are being developed in the service sector and how the 'old' economy of manufacturing is being reinvented and transformed in the USA through clever, innovative and useful internet-based systems that link and connect organisations to customers and workers together in virtual organisations.

An interesting situation has arisen whereby customers perceive themselves to be like royalty by demanding numerous and varied solutions to their needs. Low product/service cost is one dimension of competition that has not yet disappeared but it has been eclipsed by the more sophisticated concept of value for money. Different customers perceive value in diverse terms. For some customers, value or service means saving time that would otherwise be wasted. For others, waste minimisation extends to business systems integration so that desires can be seamlessly fulfilled through integrated procurement delivery and payment systems – taking the pain and hassle out of contract administration. In such cases, there appears to be a work ethic of conservation of energy being applied in order that value is created for the customer in the simplest and most direct manner. Also, each member of a supply chain views each other member in that chain as an upstream or downstream customer. Thus, there is a tendency for people to form alliances to produce a complementary value chain.

The concept of strategic advantage through creating value was first more widely discussed in the late 1980s (Porter 1990). This was a novel approach as the conventional wisdom at that time was that business's purpose was to add value to the company's shareholdings through increased profits and company share value. The idea of a company's purpose as providing value to the customer has been widened during the later years of the twentieth century to encompass other stakeholders, most notably 'the community' (Elkington 1997; Way 1999). Increasingly, environmental issues, waste minimisation and ethical business practices all form part of the 'value' that customers seek and indeed demand. The whole issue of measuring success has been undergoing a transformation with growing interest in a balanced scorecard approach to measuring and analysing business performance (Eccles 1991; Kennerley and Neeley 2000; Kristensen *et al.* 2000; Murray and Richardson 2000; Tonchia 2000) and linking this to company strategy and reframing the organisation's customer focus (Kaplan and Norton 1992, 1998a, 1998b; Meliones 2000).

So, how is the traditional business enterprise to respond? Many still flounder in an attempt to both understand customer demands and to respond positively and satisfactorily to these. Some firms choose to acquire existing companies to gain access to customers or to attempt to match or shadow their culture, ethos, or other important characteristics. This approach is common in creating a regional or local presence but it can be expensive, time consuming and highly risky due to operating in unfamiliar environments. A similar and potentially risky response is forming joint ventures with local firms. Forming partnership arrangements and developing virtual organisations from separate organisations have been other responses. Partnered organisations seek to differentiate themselves though maximising value offered to customers by providing diverse resources and responding with diverse management approaches. In each approach, the aim is to provide customer value through offering something additional or more diverse than can be provided by the firm without resorting to permanent structural changes to the company.

What is required of these companies wishing to offer greater customer focus? The answer lies in adopting an intelligent approach through understanding customer diversity, its implications upon market demand and the supply of the firm's required resources. However, most companies have a relatively fixed workplace culture and organisational style at any point in time. Coping in an agile way to respond to diversity is difficult. It is much easier and more effective for firms to acquire an organisation that matches a particular customer's culture, style and characteristics. The problem with this approach is that establishing a new organisation or absorbing others through mergers or acquisitions demands a lot of management energy. It is far better to find an appropriate organisation with which to form a partnership/alliance relationship that already has complementary customer characteristics or can more easily acquire them. Relationship-based procurement systems and more specifically alliances provide a tool for creating agility in response to the diversity in skills, work culture and business practices that characterises customers. Thus an alliance's raison d'être is a market driven response to change, globalisation, business process alignment and aims to achieve customer focus.

Alliances can be more focused upon the achievement of commonly agreed goals than can looser arrangements such as one-off joint ventures. There are many similarities between alliancing and partnering. Partnering can be relatively distinct in terms of team co-commitment so that cooperation is the resulting relationship. Partnering can also be more organisationally integrated with teams collaborating closely operating under separate risk and rewards profiles to achieve aligned objectives. However, with partnering, teams or team members can still be left unaided to sink or swim if circumstances turn against them. With alliancing, there is a coalescence of organisational team members with shared risk and reward schemes that are dependent upon the level of project, not individual team success attained in meeting customer needs. The Australian National Museum Project is a prime example of this approach in the building industry (Walker *et al.* 2000). Other Australian examples, such as the Wandoo B Offshore Oil Platform and the Andrew Drilling Platform project in the process engineering industry, have been widely reported upon (KPMG 1998; ACA 1999). Alliancing arrangements mean that if agreed key performance measures are not met then both overhead and profits for the project will be at risk for all members of the alliance. Similarly, for successful outcomes, rewards are shared through agreed formulae to alliance members. Thus, in alliancing, there is greater likelihood of shared commitment resulting in a coalescence of teams than would be the case in other forms of cooperation and/or collaboration.

We attempt in this book to offer a path for future procurement decision makers to achieve a satisfactory result that meets the needs of a wider range of project stakeholders than was considered valid for much of the twentieth century. Examples drawn from the final BRW magazine of that century indicate a *sea change* in the way projects will be delivered in the twenty-first century. The manner in which coalitions of interests will be brought together to deliver projects, and the perceived extent of project success, will be judged by project stakeholders. These stakeholders will reflect a more diverse and complex group than the **paying customer** or **client** as they were generally understood in the twentieth century (Walker 2000).

We can nostalgically gaze backwards to a period when a handshake or pledge was a bond and commitment to honour an agreement. However, we must recognise that new times redefine the context in which we make and honour agreements. We have learned much from the lunacy of the litigious and win–lose mentality that was so heavily and consistently criticised in the twentieth century construction industry (NBCC 1989; Latham 1994; Office of Building and Development 1997; DETR 1998). Alliancing offers much that is attractive to both the customer and service/product provider.

In preparing this book we have summarised the literature and identified six challenges facing the delivery of projects. Table 1.2 illustrates this framework in which we offer six generic solutions to problems posed by these challenges.

Table 1.2: Six 'real' challenges facing facility procurement decision makers

Challenges	Problem solutions
1. The 'cheapest' initial capital price is seldom the most economic long-term solution.	1. Procure projects on the basis of 'best value' not 'cheapest initial price'.
2. Negative conflict-ridden approaches result in a litigious atmosphere in which win–lose mentality prevails locking out many creative solutions and win–win possibilities.	2. Use an agreed problem-solving approach and dispute resolution mechanism that recognises the validity of diversity of opinion and approaches providing a greater pool of solution possibilities.
3. Stakeholder-value generating possibilities are seldom revealed through a short-term profit gain or capital cost-reduction focus. This approach constrains solutions to a win–lose outcome and is not conducive to encouraging win–win outcomes.	3. Focus on satisfying the real needs of stakeholders. A focus on developing and maintaining long-term relationships often releases creative energies and synergies that reduces wasted energy and increases wider knowledge and experience for all project parties involved.
4. Project participants and their supporting communities often experience detrimental quality-of-life impact through an unhealthy focus on profit maximisation or initial cost reduction. Often supporting communities pay a high indirect cost for projects.	4. Business needs to recognise that its raison d'être is to increase community value rather than simply maximise its own wealth. Businesses survive and are sustained by supporting communities, which generates market demand and creates prosperity.
5. The environment is often degraded when the cheapest initial cost and bottom-line profits are relentlessly pursued. The consequences of waste generation are often borne by the community rather than those who have generated it.	5. A QM culture should be fully adopted. Each project must not degrade the external environmental or the supporting social system for which the project is intended to serve. Innovation and better use of knowledge must be factored into a project procurement system to provide both commercial organisations and project stakeholders with sustainable solutions.
6. Project stakeholders include a diverse group of individuals including project team participants and others who will ultimately be affected by the project.	6. Recognise that performance criteria extend beyond low initial cost or profit margins. A more balanced scorecard of project success should prevail that satisfies a wider scope of project stakeholder. The vital input of social capital should be better recognised and managed.

Individual organisations should recognise that while they may operate inside a formal alliance or other form of relationship-based procurement approaches as discussed in this book, they are subject to judgement of their actions by a wide community of interests. Project initiators need to appreciate the broad range of community values and address them to truly achieve long-term success by creating sustainable projects. The judgement of corporate or project success is increasingly being seen as a function of financial, environmental and social performance. This triple bottom line is an emerging issue highlighted as one of the twenty-first century's pressing challenges (Elkington 1997).

1.3 References

ACA (1999) *Relationship Contracting – Optimising Project Outcomes*, Sydney, Australian Constructors Association.

Amabile, T.M. (1998) 'How to Kill Creativity.' *Harvard Business Review.* **76** (5): 76–87.

Bernick, C.L. (2001) 'When Your Culture Needs a Makeover.' *Harvard Business Review.* **79** (6): 53–61.

DETR (1998) *Rethinking Construction*, Report. London, Department of the Environment, Transport and the Regions.

DISR (1999) *Building for Growth – An Analysis of the Australian Building and Construction Industries*, Policy Report. Canberra, Department of Industry, Science and Resources, Commonwealth Government of Australia: 88.

Dyer, J.H. (2000) 'How Chrysler Created an American Keiretsu.' *Harvard Business Review on Managing the Value Chain*. Boston, MA, Harvard Business School Press: 61–90.

Eccles, R.G. (1991) 'The Performance Measurement Manifesto.' *Harvard Business Review*, **69** (1): 131–137.

Elkington, J. (1997) *Cannibals with Forks*. London, Capstone Publishing.

Ferguson, A. (1999) 'Take your Partners.' *Business Review Weekly.* **21**: 53–54.

Hammer, M. (2001) 'The Superefficient Company.' *Harvard Business Review.* **79** (8): 82–91.

Hannen, M. (1999) 'How Business Chiefs See a New Century.' *Business Review Weekly.* **21**: 110–113.

Hargadon, A. and Sutton, R.I. (2000) 'Building an Innovation Factory.' *Harvard Business Review.* **78** (5): 157–166.

Haspeslagh, P., Noda, T. and Boulos, F. (2001) 'Managing for Value – It's Not Just About the Numbers.' *Harvard Business Review.* **79** (7): 65–73.

Kaplan, R.S. and Norton, D.P. (1992) 'The Balanced Scorecard – Measures that Drive Performance.' *Harvard Business Review.* **80** (1): 171–179.

Kaplan, R.S. and Norton, D.P. (1998a) Putting the Balanced Scorecard to Work. *Harvard Business Review on Measuring Corporate Performance*. Boston, MA, Harvard Business School Publishing: 147–181.

Kaplan, R.S. and Norton, D.P. (1998b) Using the Balanced Scorecard as a Strategic Management System. *Harvard Business Review on Measuring Corporate Performance*. Boston, MA, Harvard Business School Publishing: 183–211.

Kennerley, M. and Neeley, A. (2000) Performance Measurement Frameworks – A Review. *Performance Measurement – Past, Present and Future*. Neeley A. ed., Cranfield, UK, Centre for Business Performance, Cranfield University: 291–306.

Kim, W.C. and Mauborgne, R. (1997) 'Value Innovation: The Strategic Logic of High Growth.' *Harvard Business Review.* **75** (1): 102–112.

KPMG (1998) *Project Alliances in the Construction Industry*, Literature Review. Sydney, NSW Department of Public Works & Services.

Kristensen, K., Martensen, A. and Gronholdt, L. (2000) 'Measuring Customer Satisfaction: A Key Dimension of Business Performance.' *International Journal of Business Performance Management.* **2** (1/2/3): 157–170.

Latham, M. (1994) *Constructing the Team*, Final Report of the Government/Industry Review of Procurement and Contractual Arrangements in the UK Construction Industry. London, HMSO.

Leonard, D. and Rayport, J.F. (1997) 'Spark Innovation Through Empathic Design.' *Harvard Business Review.* **75** (6): 102–113.

Leonard, D. and Straus, S. (1997) 'Putting Your Company's Whole Brain to Work.' *Harvard Business Review.* **75** (4): 110–121.

Magretta, J. (1998) 'Fast, Global, and Entrepreneural: Supply Chain Management, Hong Kong Style.' *Harvard Business Review.* **76** (5): 103–114.

Malone, T.W. and Laubacher, R.J. (1998) 'The Dawn of the E-Lance Economy.' *Harvard Business Review*. **76** (5): 145–152.

McDermott, P. (1999) Strategic Issues in Construction Procurement. *Procurement Systems A Guide to Best Practice in Construction*. Rowlinson, S. and McDermott, P. eds, London, E&FN Spon: 3–26.

Meliones, J. (2000) 'Saving Money, Saving Lives.' *Harvard Business Review*. **78** (6): 57–65.

Michaels, E., Handfield-Jones, H. and Axelrod, B. (2001) *The War for Talent*. Boston, MA, Harvard Business School Press.

Murray, E.J. and Richardson, P.R. (2000) 'Shared Understanding of the Critical Few: a Parameter of Strategic Planning Effectiveness.' *International Journal of Performance Measurement*. **2** (1/2/3): 5–14.

NBCC (1989) *Strategies for the Reduction of Claims and Disputes in the Construction Industry – No Dispute*. Canberra, National Building and Construction Council.

Office of Building and Development (1997) *Partnering and the Victorian Public Sector*. Melbourne, Australia, Office of Building and Development, Department of Infrastructure, Victorian Government.

Peters, R.J., Walker, D.H.T., Tucker, S., Mohamed, S., Ambrose, M., Johnston, D. and Hampson, K.D. (2001) *Case Study of the Acton Peninsula Development*, Government Research Report. Canberra, Department of Industry, Science and Resources, Commonwealth of Australia Government: 515.

Porter, M.E. (1990) *The Competitive Advantage of Nations*. New York, Free Press.

Schmidt, L. (1999) 'A Time When "E" is Also for Ethics.' *Business Review Weekly*. **21**: 100–103.

Tabakoff, N. (1999) 'Virtual Companies: Evolution's Next Stage.' *Business Review Weekly*. **21**: 86–88.

Tonchia, S. (2000) 'Linking Performance Measurement System to Strategic and Organisational Choices.' *International Journal of Performance Measurement*. **2** (1/2/3): 15–29.

Walker, D.H.T. (2000) 'Client/Customer or Stakeholder Focus? ISO14000 EMS as a Construction Industry Case Study.' *TQM*, **12** (1): 18–25.

Walker, D.H.T., Hampson, K.D. and Peters, R.J. (2000) *Project Alliancing and Project Partnering – What's the Difference? – Partner Selection on The Australian National Museum Project – a Case Study*. CIB W92 Procurement System Symposium On Information And Communication In Construction Procurement, Santiago, Chile, Pontifica Universidad Catolica de Chile.

Way, N. (1999) Workplaces of the World, Fragment. *Business Review Weekly*. **21**: 92–97.

Womack, J.P. and Jones, D.T. (2000) From Lean Production to Lean Enterprise. *Harvard Business Review on Managing the Value Chain*. Boston, MA, Harvard Business School Press: 221–250.

Part 1
Procurement Systems and Structures

Chapter 2
Procurement Choices

Derek Walker and Keith Hampson

We generally use construction projects in this book to illustrate procurement examples, however, other business sectors embrace similar procurement forms. Manufacturing, car, ship and aerospace assemblies often use a form of design and construction (Graham 1999; Womack and Jones 2000). This second chapter explores procurement options using a cost risk perspective.

Figure 2.1 illustrates the continuum along which the project delivery systems can be categorised. At one extreme lies the traditional 'fixed cost' (and usually fixed time) project. At the other is the fully cost reimbursable project. Procurement options for clients can be viewed through a cost risk/relationship risk perspective. The initial tender cost can be fixed with all risk being absorbed by the contractor. Alternatively, the client can absorb a cost risk by letting variable sum contracts and adopting an open book philosophy in which incurred costs are verified or a formulated schedule of agreed rates for various aspects of the work is agreed upon.

2.1 Traditional procurement options – fixed price contracting

The traditional, or conventional, approach to procuring projects involves discrete design development, tender, contract award and construction delivery phases.

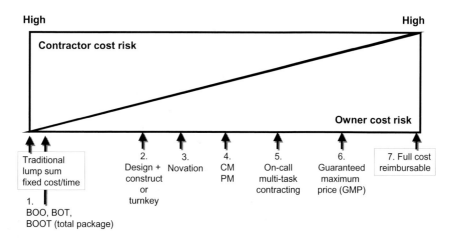

Figure 2.1: A construction cost continuum for project delivery

Each phase is, in theory, separate and distinct. The process begins with a client approaching the principal design consultant. This is generally the architect for building projects or a design engineer for engineering projects. The design is developed to as close to completion as possible before tenders are invited. In practice there are many design issues left incomplete and unresolved so there is often refinement and amendment of design details during the project delivery phase. Tenders are invited on one of two bases. Open tendering allows anyone to tender for the project. Closed or pre-qualified tendering restricts those invited who have met pre-tender qualification criteria such as financial soundness and relevant project experience.

Open or pre-qualified tender competition using the traditional approach almost always results in acceptance of the 'cheapest' fixed price for the specified work. Tenders are called after design completion and the construction cost is then assumed to be 'fixed'. However, in practice the design is almost never completely finalised and in most cases this results in many opportunities to claim for 'extras'. The final end cost of a project, however, also includes the costs of design changes approved during construction, charges based on errors or omissions in design and contract documents used during the tender period and other claims made for consequential delays arising from cost claims noted above. Thus, the 'traditional' procurement method often ends up delivering the tendered lowest price for a project and subsequent claims for additional works mean that many clients feel at the mercy of contractors seeking opportunities to create profit and additional revenue. This situation is not new and has been the subject of much discussion over decades both in Australia and overseas (NBCC 1989; Latham 1994; Heldt *et al.* 1997; Office of Building and Development 1997; DETR 1998).

The main criticism of the traditional lump sum approach has been that it invites a confrontational approach over disputes arising out of contract variations and what might be a fair price for these. The 'No Dispute' report was particularly scathing about this propensity (NBCC 1989). An entire *claims industry* has developed over past decades to advise contractors on how to claim for extra work, and client representatives on how to counter such claims. The traditional approach also casts roles in stone for all parties making it difficult to negotiate outside the risks of the contract. The system is widely held in disrepute, however, unsophis-ticated or inexperienced clients are generally unaware of the advantages of alternative procurement paths and/or such options appear more complicated than the 'traditional' approach. The level of familiarity with this approach for most contractors and clients can also be appealing, none of these parties particularly like 'the game' as played but each at least knows the rules. The apparent attraction of the 'traditional method' as delivering the 'market' or cheapest price, at least initially, is widely attributed to its continued popularity (Latham 1994).

One of the disadvantages that this system presents is that it removes the con-tractor from the design development phase and thus much management and constructability information is lost (Francis and Sidwell 1996; McGeorge and Palmer 1997). This has serious consequences in terms of both cost and relationship risk. The contractor has much to offer in terms of advice on how best to meet design specification in a cost and time effective manner because the contractor is closest to the workface with intimate knowledge of the production process (Francis and Sidwell 1996). Research on contract claims from the perspective of contractors and designers and employers has highlighted the impact of poor design coordination and subsequent design changes to make design details workable (Choy and Sidwell 1991; Yogeswaran and Kumaraswamy 1997). Similarly, relationship aspects are impaired under the traditional procurement system because the contractor is answerable to the principal design consultant with no formal direct access to the client to suggest improved design for constructability.

2.2   Total package options – BOO, BOT, BOOT

Another option that occupies the fixed-cost end of the project procurement risk spectrum is the total package option. In this procurement option a client's project need is met by an entity that contracts to design, build, operate, own for some period of time and transfer the facility back to the owner. In the *BOO*, *BOT*, *BOOT* the 'B' represents the word 'build', the first 'O' as 'operate' and the second 'O' as 'own' and the 'T' as 'transfer'. This option has been in operation for decades. There are many examples of these projects in most countries of the world. Two useful texts on this subject describe the use of BOT in Hong Kong and elsewhere (Walker and Smith 1995; Smith 1999). BOT requires that 'contracted parties must accept the conventional wisdom that risk should be assumed by the party within whose control the risk most lies. A major function of the BOT arrangement is, therefore, to recognise and provide a mechanism for the assignment and management of those risks' (Walker and Smith 1995).

With the BOT 'family' of procurement options an alliance or joint venture group forms to provide a facility for a client for which the client makes a concession agreement to fund the facility until that facility's ownership is transferred to the client. This arrangement is more common for infrastructure projects than buildings because the concession allows for tolls or other payments to be made by end-users to cover the cost of both procuring the facility and its operation. Extensions of this idea have been cited in which buildings have been renovated and leased back on this basis and others where the facility is required to be removed and the site returned to an acceptable environmental standard. One advantage of this arrangement is that it extends the ideas of constructability further to embrace lifecycle cost effectiveness. If the entity proposing the design solution is responsible for maintaining and operating the facility then they will have the incentive to reduce long-term costs and thus develop a highly cost effective product over the product lifecycle (Smith 1999).

The BOT entity undertakes financing, design and construction as well as operation and so the client is taking no direct cost risk other than the possibility that the facility does not meet its needs or that the concession agreement is unsatisfactory (Smith *et al.* 1994). The cost of establishing the arrangements can be considerable, as there are legal and financing costs to be met. These should be compared with the client's legal and finance costs in undertaking the project in other forms or options. Clearly, the option is unlikely to be viable for projects of small scope, however, governments are increasingly using this option for hospitals, prisons and other projects previously undertaken through other procurement scenarios outlined in this section (Grimsey and Graham 1997).

As a cautionary note, experience with BOO/BOT/BOOT schemes reveals some notable failures. Generally these have been based on failures of trust and/or communication. One of the most prominent failures has been the Bangkok Second Stage Expressway. The competitive tender phase took two years to find a successful bidder. Part of the complexity derived from the land procurement process to allow the project to proceed. At the date when work was supposed to start only 1 per cent of the land required was available. That problem was overcome but as the completion of the construction phase neared, other fundamental problems regarding the operation emerged. Toll rates were decreased instead of being increased to fund the project. This led to loans being frozen in response to financiers' concerns about the financial sustainability and then arguments broke out over the right to operate the expressway. The project environment subsequently dissolved into acrimony amongst accusations of corruption and widespread political interference. Thus, both relationship and cost risks were severely affected by the Thai government changing its mind (Ogunlana 1997; Smith 1999). Clients and stakeholders involved in such projects can at times be severely affected by the integrated nature of finance, design and construction.

2.3   Design and construct, novation and turnkey

A **design and construct** (D+C) procurement approach provides for an organisation to be contracted by a client to manage the design and construction processes with a single point of contact. There may have been preliminary sketch plans developed to indicate a generalised design solution or the design brief may be left fairly open for the D+C contractor to offer proposals. In **combined project management and construction management** (PM/CM) procurement options, an organisation undertakes to represent the client in leading the design team, and undertakes the management of the construction process including providing construction advice during design development. The PM/CM entity may act as advisor (in which case the managerial links are persuasive rather than directive) or may undertake the work under a contractual arrangement in which it carries financial risk. In a D+C arrangement, this team will hire both the design team members and the construction management team members either within the design and construction company entity or as subcontractors. The design team may be sourced from inhouse staff or, as is more frequently the case, sourced from the general pool of design consultants undertaking a variety of procurement forms (Walker 1993). Thus in many cases, the design and construction contractor subcontracts or forms a joint venture with design firms. It can be appreciated that PM/CM and D+C blur quite substantially as procurement process options so that a PM/CM entity that carries financial risk is really a D+C contractor.

In tendering for a D+C project, design consultants are generally contracted to develop concept (or concept advanced in terms of design developed) project solutions to enable design and construction tenderers to bid for the project. The client has the opportunity to work with the design team to develop a brief to a stage in which it can test the market for proposals that will develop a project solution based upon the concept design. A selection procedure is then put in place to evaluate the D+C proposals, however, designers from D+C groups submitting ideas and developed plans have the opportunity to provide innovative solutions and take the concept and provide their own footprint on the result. This has several potential advantages:

- it provides for innovative solutions to be tested by the market;
- it often results in a cost effective solution;
- it combines the expertise of both design and construction professionals; and
- it allows the client to have a single point of contact that manages the project if it wishes, or in the case of PM/CM, can maintain contractual control and essentially use this group as its internal design and construction group.

The relationship experience for project teams also moves towards treating design and construction supervision entities as being contributing partners in an enterprise to deliver a project solution that combines the skills of both design and construction groups. Dulami and Dalziel undertook a study of 37 managers of the construction team of whom 22 were involved on traditionally procured projects and 15 on D+C projects. Their analysis supports the hypothesis that design and construction improves project team integration (Dulaimi and Dalziel 1994). One advantage that design and construction variants offer was found to be better cooperation and communication between the design and construction teams. As design and construction contractors are primarily builders they have constructability input that can be of significant value, in much the same way that CM consultants can offer this advantage.

Research in Northern Ireland (where design and construction accounts for approximately 30 per cent of construction work) involving 30 client organisations and 30 major contractors revealed some interesting cautionary insights into design and construction.

- Competitive tendering is often based on minimal design work or site investigation, but with estimates inflated to cover consequent risks.
- Lifecycle costing receives limited consideration by contractors in pricing, particularly with poor client briefs.
- The quality of design and construction is related to the adequacy of the client brief and the professionalism of the project team. This professionalism can be strongly challenged by commercial pressures and poor communication.
- Clients are often ill-informed about the implications of the design and build approach, and about the importance of a comprehensive brief for the designer and contractor (Gunning and McDermott 1997, p. 221).

Novation has been developed from the design and construction concept. In design and construction, D+C responsibility for both design and construction is undertaken by one legal entity responsible for project execution. In tendering for this option design consultants are generally contracted to develop a design concept or advanced (in terms of design developed) concept solutions to enable design and construction tenderers to bid for the project. These design solutions, though substantially complete (perhaps up to 25 per cent) at the point of novation, are usually modified to suit constructability issues or for commercial reasons while maintaining design intent and integrity. This can pose serious problems for clients who have a more advanced design concept in mind than might be the case for the design and construction option. In these cases the client may commission a design solution to be outlined and partially developed which will form the basis for the project design solution but be open in terms of systems or more specific detailed design information. Thus, the cost of design development at the conceptual and preliminary design stage is already prepared and will not form part of any further fee structure. The means to meet the spirit and essential content of the design solution can be addressed by the entity that successfully takes over a D+C role to **novate** the design solution and take responsibility for the CM role in realising the project. This can represent a useful option, which enables clients to influence the project design to a greater extent than is possible in the standard D+C option.

The client's risk can be reduced because the contractor takes over design development after novation while accepting pre-novation design assumptions. This arrangement allows the contractor substantially to fine-tune the design to take account of its competitive advantage including constructability issues. For example in the A$200 million+ Australian Melbourne Cricket Ground (MCG) Great Southern Stand project in Australia, the structural design was changed from cast-in-place to precast concrete in response to an overheated cast-in-place concrete market and advantages inherent in an off-site fabrication structural design solution. Additionally, novation allows fast tracking and a fixed price, however the client forgoes some flexibility of making design changes without incurring potential cost penalties.

The contractor, by embracing the pre-novation design, cannot claim against design omissions or errors thus accepting contractual risk for the design. The design team, who frequently risk being simply 'passed' onto the contractor, may see their 'baby' amended to a level felt unacceptable by them. For some design professionals this may pose important issues of design integrity and ownership that may be difficult to accept. The system has had numerous successes, the Adelaide Entertainment Centre was another Australian project that successfully used a novated design approach (Chan and Tam 1994). In discussing the likely reason for this success the following is offered (Chan 1996):

- the brief being comprehensive;
- construction proceeding prior to completion of the design documentation;
- builders invited to bid having experience in this type of work;
- the building industry in Adelaide at that time being relatively busy and unlikely to 'buy' projects; and
- the project having some unusual or special features where building delivery innovation would provide a cost benefit.

The following comments about using novation are also provided (Chan 1996):

- for a limited marketplace with insufficient companies who do not have a proven record of both designing and constructing – perceived risk of taking over a design deters many would-be tenderers;
- by accepting a novated design companies accept errors and omissions and other potential problems including a design that may potentially prove unworkable;
- the client's right to nominate subcontractors or suppliers is removed under

novation, thus the company taking over both design and construction is free to make its own contractual arrangements as it sees fit;

- the architect will no longer supervise quality control or exercise sanction once novation occurs. This is difficult for many designers, as their reputation is closely associated with their work, which may be modified in a way that could upset them;
- the client severs direct communication links with the design team once novation occurs; and
- once novation occurs, the contractor pays the design team. This may pose a financial risk to the design team if they believe that the contractor is not financially sound.

In a survey of 54 Hong Kong architecture firms and 30 building contractors opinions were sought on attitudes towards variants of design and construction. Results of the study indicated three categories of D+C existing that provide a useful typology of how the concept of novation has evolved (Mo and Ng 1997).

(1) *Traditional* D+C is where the contractor is responsible for complete design and construction.
(2) *Enhanced* D+C is where the contractor is responsible for design development, working details and construction.
(3) *Novated* D+C is where the contractor is responsible for design development, working details and construction with the assignment of the design consultants from the client.

Generally, novation is seen as a D+C variant. The client commissions a design team to undertake a partially complete design instead of each design and construction company separately developing a design to a proposal stage. Upon successful negotiation of the construction contract, the design team and its design is passed to the successful bidder. The difference between novation and D+C is principally one of client control and influence. The client has more influence and control to shape the desired outcome in novation than with traditional D+C.

Turnkey procurement systems provide for the company supplying D+C services to also finance the project. The client makes staged or periodic payments towards the value of work completed for PM, CM, and D+C work. In turnkey projects, however, the company is generally paid upon completion of the commissioning and testing. Literally, the client pays for the project and gets the key to gain access to the project. While this shares some similarities with 'BOT family' projects the contractor does not undertake to operate the constructed facility. The turnkey approach suits many clients who wish, for tax or other financial purposes, to only make a payment upon delivery of an acceptable product.

2.4 Construction management (CM) and project management (PM) and PM/CM

Non-traditional procurement methods allow for early contractor involvement in the design development process. This has the benefit of allowing contractor expertise to be made readily available to the design team. This buildability or constructability advice is crucial to the development of design solutions that maintain value in terms of the quality of product as well as providing elegant solutions to production problems (Francis and Sidwell 1996; McGeorge and Palmer 1997). One non-traditional procurement method proved popular over the past two decades is **construction management** (CM). The term 'management contracting' used in the UK is synonymous with the term CM used in Australia or the USA (Sidwell and Ireland 1989). Under the CM method the contractor acts as consultant builder providing significant advice on the practicality of the design and expected construction methods to be employed. The CM will also provide services such as construction planning, cost control, coordination and supervision of those who have direct contracts with the owner to carry out operational work. CM teams often undertake the management of the construction process for a fee. Trade or work package specialists physically undertaking the work under separate contracts with the client which are coordinated, supervised and managed by the CM team. An alternative within this option allows for the CM team to take responsibility for the construction works as head contractor. Either way, the CM team provides design development advice and supervises and manages the construction process.

There are two forms of CM that can be adopted, **agency** and **direct**. Agency CM applies where the CM undertakes the work as a consultant for a fee providing constructability advice and coordination of the construction works. A number of large USA construction management firms operate on this basis. A second form is **direct CM** where the CM undertakes work for a guaranteed maximum fee or negotiated price, usually when the design is sufficiently advanced to address issues of risk adequately. Sidwell and Ireland suggest this is generally at a 50–90 per cent design completion stage (Sidwell and Ireland 1989).

Advantages of using the CM approach are:

- reduced confrontation between the design teams and the team responsible for supervising construction;
- early involvement of construction management expertise;
- overlap of design and construction;
- increased competition for construction work on large projects due to work packaging and splitting the construction activities into more digestible 'chunks';
- more even development of documentation;
- fewer contract variations;
- no need for nominated trade contractors; and
- public accountability.

This approach lies between the fixed/tendered price and variable price ends of the project cost risk spectrum. This form also introduces the possibility for addressing relationship issues stemming from unequal influence to promote improvements, particularly in design. Under the traditional procurement arrangement the work is designed, then tendered, then constructed, thus the design team has an established relationship with the client and dominates the design development phase. In the CM, PM or PM/CM arrangement, the design team and CM team share design development input (often with the client or client representative). This results in a more balanced power and influences the relationship developing between these parties as they jointly solve problems, recognising each other's capacity and willingness to contribute ideas.

During the late 1960s and early 1970s, an emerging form of **project management** (PM) structure was identified that arose out of the move towards multidisciplinary design practices and inhouse project teams undertaking government projects. The traditional client representative, generally an architect, was the lead design team member. The client representative's major role was to coordinate design activities and oversee construction operations as a non-executive project manager (Walker 1993). This position has become more complex with variations in responsibility accepted and remuneration varying from a fee for non-executive PM services to a guaranteed fixed price with contractual relationships between PM and contractor teams (Barnett 1998/9). Responsibility varies from being an 'advisor' to 'executor'.

The PM team generally takes responsibility for design coordination and supervision of those responsible for undertaking the work packages. If these responsibilities are not supported by a contractual arrangement, or where the PM team advises a client willing to take a more active management role, the PM will use a power base of persuasion and expert knowledge to influence others. In such cases major decisions are directly presented to the client, if an executive, or addressed to a project board of directors for formal approval if the group is charged to represent a client's interests. The PM team has a contractual arm's-length relationship with both the design teams and the construction team in 'pure' PM arrangements. If the 'advisor' role is adopted, then the PM managerial influence will take the form of persuasion rather than authoritative direction.

2.5   Sequential negotiated work packages – on-call contracting

The idea of agreeing general principles, terms and conditions of reward systems for contracted services and then refining agreements for specific work tasks is not new. In a sense this is very much like the arrangements that prevail with much of maintenance contracting where skilled workers are contracted for on-call services. Similarly, medical practitioners have been reported to contract their services on an on-call basis (Jensen and Hall 1995).

On-call contracting is a procurement strategy where the owner initially signs a master contract with one consultant for a project then divides the project work into task orders (TOs) that are released to the consultant in phases (Shing-Tao and Ibbs 1998). It is based on the premise that the owner or client representative knows the nature of the work better than the consultant does at the start of a project but the consultant knows more at the execution stage and should assume more cost risk at that time. It is different from the cost plus (reimbursable) or guaranteed maximum price (GMP) concept. The advantages are said to include the capacity within an uncertain project environment to freeze design of work packages into discrete TOs. Table 2.1 illustrates some on-call operating characteristics.

This approach enables the TO work to be sufficiently defined, planned for and budgeted to control the work packages as if they were mini-contracts. However, because much of the base negotiations regarding work rates and other conditions have been agreed in the master contract the TOs are easily accommodated. If the

Table 2.1: Typical on-call contracting characteristics of design projects		
Characteristic	**Master contract**	**Task order (TO)**
Planning mode	Pre-project planning	TO planning
Requirements covered	General requirements	Specific requirements
Contract analogy	Single prime contract	Multiple contracts
Payment method	Reimbursable	Fixed Price (recommended)
Cost risk	Risk mainly with the owner	Risk mainly with the consultant

Source: Shing-Tao and Ibbs (1998, p. 36).

contractor is unavailable or cannot meet the schedule required, a second contractor can be given the opportunity to negotiate that TO.

This option has attractive features where the client is the best entity to control project risk and control project scope but the contractor's detailed operational knowledge is available and can be used to fix costs at an agreed sum. In many ways this is similar to fast-tracking, though the number of TOs would be generally greater in number than work packages in a fast-tracking approach.

A fast-track approach entails overlap of detailed design development from the more generalised design development to allow separate work packages to be tendered while construction proceeds. In theory this offers advantages of project delivery speed because construction work can proceed on some work packages (foundation work for example) while design development continues on later stages of the project (services or finishes for example) (Revay 1988; Sidwell and Ireland 1989; Miles 1995). This approach has been criticised for leading to both time and cost over-runs. For example, in a study of 150 projects in Alberta, Canada, it was found that the start of construction on many fast-track projects could have been delayed by up to six months and would still have finished ahead of the date actually completed (Revay 1988). Others have identified potential problems with fast-tracking including greater propensity for errors and omissions, making communication and coordination difficult that results in potential for much re-work and time delays (Miles 1995). The issue of poor communication, confusion and re-work, however, has cast fast-tracking in a poor light. While a fast-tracking approach can easily lead to this situation, good design management practice overcomes these problems releasing the underlying advantages of overlapping design and construction.

On-call contracting with the detailed planning and design involved in the establishment of TOs, can enforce, or at least relies upon, good design management and implementation planning. On-call contracting differs from a fast-track approach in detail. The overall work is pre-planned in general and complex work packages planned in detail once the specific requirements are known for both approaches. However, with on-call contracting these work packages are smaller in scope, more numerous and planned more fully and in detail. This obviates some of the criticism of fast-tracking. Fast-tracking requires much of the detailed planning to be left to contractors to deal with after each contract package is awarded, however detailed planning is a part of the negotiation process with on-call contracting (Shing-Tao and Ibbs 1998).

One drawback of on-call contracting is that it is highly planning intensive with added administrative expenses, however, greater detailed planning effectiveness has been shown to contribute significantly to construction time performance (Walker 1994). Bonus incentive schemes can be successfully used to reward satisfactory performance on milestones (Abu-Hijleh and Ibbs 1988). This has been suggested to be effective when linked to TOs as a reward for the planning contribution made in developing the detail of each TO. Interesting observations about on-call contracting include:

- a hybrid of single and multiple contracts will retain advantages and avoid or minimise disadvantages of traditional procurement strategies;
- TO planning supplements incomplete pre-project planning by continuing the scope definition during the design;
- on-call contracting can use the reimbursable method for the master contract and fixed price for the task orders, which breaks the rule of thumb of adopting reimbursable contracts for complex projects; and
- the TO approach is expensive in terms of administration and planning effort but this generally reaps compensating rewards in facilitating improved results from better project coordination, communication and planning (Shing-Tao and Ibbs 1998).

2.6   Guaranteed maximum price (GMP)

In the *GMP* arrangement, the client accepts a share of the cost risk. The client will agree to reimburse the consultant or contractor only up to a negotiated guaranteed maximum amount. After that, the consultant or contractor bears the risk (Gordon 1994). This method provides the knowledge and expertise of the client to influence the budget-making process to provide a reimbursable amount for the work but potential mismanagement on the part of the contractor is guarded against through a guaranteed maximum limit. The contractor either accepts any expenditure over that amount or another suitable arrangement is negotiated.

The GMP arrangement seems to be gaining popularity in the USA when used in conjunction with a D+C approach. The arrangement is described as follows (King 1996, p. 420):

On a GMP project, the contractor bases its bid on partially-completed documents and, extrapolating from them, warrants to the Owner that the price will not exceed a certain sum. The work is then paid for at the contractor's actual cost plus a fee, until the GMP is reached. After that, the contractor absorbs additional costs. If the actual cost is less than the GMP the Owner keeps the savings (or sometimes a portion of them with the contractor as an incentive).

This procurement approach shares similarities with D+C and novation in that the design is partially developed by the client's design team. Unlike novation, the contractor may or may not decide to take over the design team and its initial design concept. In this regard it shares similarities with D+C. The major difference with the GMP approach, as opposed to other procurement systems, is that in GMP the design is fixed, as is the negotiated price limit, but with open-book reimbursement from the owner/client. The client takes most of the risk by guaranteeing the cost reimbursement, however, the contractor also bears the risk of a cost over-run and is, therefore, deterred from acting in a way that is extravagant or wasteful.

2.7 Full cost reimbursable procurement

There are occasions when a client wants or needs to maintain total control over the design result and the construction process. This may be for security reasons, because the design solution is highly complex (thus contractor risk is prohibitive), or because of the client's rapidly changing requirements. In such cases, one option available to the client is to provide an inhouse D+C facility, however this may not always be feasible. It takes more than capital funding to establish a contracting and/or design organisation, it requires high levels of expertise and organisational knowledge.

Thus, a client can in effect 'rent' such skills and capacity through contracting to pay a firm all the costs of production plus an agreed fee for providing the expertise to advise on production techniques and coordinate implementation – on a **full cost recovery basis**. This option provides for a contractor to be chosen to undertake the work on a cost reimbursable basis with an agreed allowance for profit and overhead (Ireland 1987). The project cost, scope and other performance aspects can be shaped during the design and development phase. This option is suited to highly uncertain or risky projects where design details are unknown at the time of tender, other aspects of the external environment are subject to great change, or the client prefers to maintain the right to be able to discharge the contractor during the construction phase. This is suitable for experimental projects where novel problems may be encountered. The client is essentially paying a fee for the contractor's expertise and knowledge and using the contractor's organisation as if it were its own by paying for work through the open-book method.

The disadvantage of this approach is that the client takes all risk for the final cost and must therefore have very high confidence in both the contractor and the design team, and also have appropriate auditing systems to ensure that the facility being constructed represents value for money. An open-book approach potentially invites exploitation. This approach requires high levels of compliance, supervision and independent monitoring. These costs have to be considered when choosing this procurement option. This alternative requires the highest level of trust and confidence in the contractor of any procurement option discussed thus far.

2.8 Discussion – procurement choice as design influence

Some researchers argue that procurement form is largely irrelevant and that the real issue is how the procurement option enhances or inhibits team members to maximise their constructive input to achieve project goals (Walker 1996, 1997, 1998). While the traditional approach will be argued to inhibit positive interaction generally, there are many factors that impact upon performance of non-traditional procurement choices.

Figure 2.2 illustrates this concept of examining the procurement continuum through the lens of client influence over the design process. At one extreme the

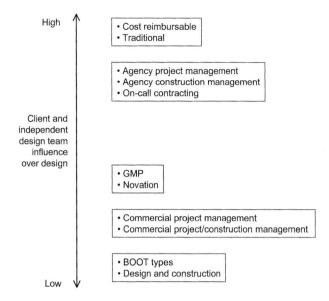

High

- Cost reimbursable
- Traditional

- Agency project management
- Agency construction management
- On-call contracting

Client and
independent
design team
influence
over design

- GMP
- Novation

- Commercial project management
- Commercial project/construction management

- BOOT types
- Design and construction

Low

Figure 2.2:   Client influence over design and procurement type

client may trust the design influence being exercised by the contractor more than that exercised by an independent design team under the direction of the client. At the other extreme the client may wish to retain influence and control over the project's aesthetics and functionality.

If clients have high levels of confidence in their own expertise to exert this influence effectively they will tend to exercise direct authority over an independent design team. If they have low levels of confidence in their own expertise they may devolve this authority to a client representative or allow the independent design team high levels of autonomy. At the high end of the influence spectrum, a traditional procurement method choice rests on trust that the 'lowest tendered price' will deliver the least expensive functionally acceptable alternative design. This choice represents trust in market efficiency/effectiveness and design team efficiency/competence with high client or client representative influence. At the low end of this spectrum, the client or client representative choice for D+C procurement or BOOT type forms indicates that the contractor should exert greater influence on the design solutions. In accepting BOOT type projects, the client trusts that the market provides a better total project delivery and operational solution than the client could develop.

Figure 2.3 summarises issues of power and influence as well as design flexibility associated with procurement forms. Competition on cost, contractor selection method and who supervises the construction influence the contractor/client risk acceptance arrangements. Client flexibility to influence design is presented in terms of design overlap, team member responsibility for design and whether buildability advice is available.

Clearly, procurement choice is a complex issue involving not only technical and legal aspects but also considerations of power, influence, risk acceptance and desired design flexibility. Partnering and alliancing may be chosen independent of

	POWER + INFLUENCE			DESIGN FLEXIBILITY			INDICATIVE RISK TO CONTRACTOR (remainder to client) Low — High
	COMPETITION ON COST	SELECTION OF CONTRACTOR	SUPERVISION OF CONSTRUCTION	OVERLAP OF DESIGN AND CONSTRUCTION	CONTRACTOR RESPONSIBLE FOR DESIGN	BUILDABILITY ANALYSIS DURING DESIGN	
LUMP SUM (TRADITIONAL)	Competition	Open or select	Architect	No	No	No	2 1
DESIGN + CONSTRUCT/ GMP/ NOVATION	Competition	Usually negotiation	Contractor	Yes	Yes	Yes	2 1
TURNKEY + BOOT/BOT etc	Competition	Usually negotiation	Contractor	Yes	Yes	Yes	2 1
DIRECT CM	Competition at trade level	Usually Negotiation or competition on fees	CM	Yes	No	Yes	3 4
AGENCY CM	Competition at trade level	Usually negotiation or competition on fees	CM	Yes	No	Yes	3 4
PROJECT MANAGEMENT	Competition at trade level	Usually Negotiation or competition on fees	PM	Yes	Yes	Yes	3 4
ON-CALL CONTRACTING	Competition at TO level	Usually negotiation	PM	Yes	Yes or no	Yes	5 6
COST REIMBURSEMENT	No competition	No competition	Architect or contractor	Partial	Yes or no	Yes	3

Risk notes: 1 no inflation adjustment   2 with inflation adjustment   3 no GMP provided   4 GMP provided
5 TO consultant/contract risk   6 Client's representative/PM

Figure 2.3: Summary of differences between procurement options
Source: Adapted from Ireland (1987, p. 188).

any chosen procurement path. However, partnering and alliancing is extremely difficult when applied to a traditional procurement approach. This is because it generally establishes the groundwork for an adversarial environment and restricted influence between design and construction teams.

2.9   Chapter summary

Before clients commit themselves to deciding upon entering partnerships or alliances they will need first to review their capacity to become involved as project partners in realising their project. We saw in Figure 2.1 that there is a continuum of risk acceptance between the contractor and clients that favours a particular procurement form. Aside from the extreme where the contractor or the owner takes all the risk, there is much flexibility of choice in the procurement approach for the owner/client.

We also saw in Figures 2.2 and 2.3 how the client or owner may actively choose to participate in the procurement process by exercising influence over other teams. Naturally their capacity to do so is limited by their capacity to contribute positively to the decision-making processes that are required throughout the project realisation phases. Interaction and participation in various phases of the project delivery process by the client, design and construction teams as a cohesive group have a direct influence on the quality of their relationship.

The choice of procurement form reflects the client's capacity, desire, or confidence in establishing a framework where decision making and initiative are distributed amongst these three identified groups. The choice of procurement outcome is a demonstration of trust in gaining sufficient design flexibility to achieve delivery of the desired project outcome for the client. More specifically:

- a trust in the market to deliver the 'best procurement solution';
- trust in the design team's capabilities for design; and
- the client's own trust assessment of the level of its internal expertise to add value to the design process.

We will see in the next chapter how trust, commitment, organisational design, and management style affects relationships in which teams are required to work closely together to pursue shared goals. This will establish a sound understanding of how partnering and alliancing are grounded in relationship building and maintenance between individuals and teams.

2.10   References

Abu-Hijleh, S.F. and Ibbs, C.W. (1988) 'Schedule-based Construction Incentives.' *Journal of Construction Engineering and Management*, **115** (3): 430–443.

Barnett, A.M. (1998/9) 'The Many Guises of a Project Manager.' *The Australian Institute of Building Papers*. **3** (1): 119–134.

Chan, A.C. (1996) *Novation Contract – Client's Roles Across the Project Life Cycle*. CIB W92 – Procurement Systems North meets South – Developing ideas, Durban, RSA,

Department of Property Development and Construction Economics, University of Natal, Durban.

Chan, A.P.C. and Tam, C.M. (1994) *Design and Build through Novation*. East meets West – CIB W92 Symposium, Hong Kong, University of Hong Kong.

Choy, W.K. and Sidwell, A.C. (1991) 'Sources of Variations in Australian Construction Contracts.' *The Building Economist*. December: 24–29.

DETR (1998) *Rethinking Construction*, Report. London, Department of the Environment, Transport and the Regions.

Dulaimi, M.F. and Dalziel, R.C. (1994) *The Effects of the Procurement Method on the Level of Management Synergy in Construction Projects*. East meets West – CIB W92 Symposium, Hong Kong, University of Hong Kong.

Francis, V.E. and Sidwell, A.C. (1996) *The Development of Constructability Principles for the Australian Construction Industry*. Adelaide, Construction Industry Institute Australia.

Gordon, C.M. (1994) 'Choosing an Appropriate Construction Contracting Method.' *Journal of Construction Engineering and Management, ASCE*. **120** (1): 196–209.

Graham, R. (1999) 'Managing the Project Management Process in Aerospace and Construction: A Comparative Approach.' *International Journal of Project Management*. **17** (1): 39–46.

Grimsey, D. and Graham, R. (1997) 'PFI in the NHS.' *Engineering Construction and Architectural Management*. **4** (3): 215–231.

Gunning, J.G. and McDermott, M.A. (1997) *Developments in Design & Build Contract Practice in Northern Ireland*. CIB W92 Procurement Systems Symposium 1997, Procurement – A Key To Innovation. The University of Montreal, CIB.

Heldt, T., Hampson, K.D., Murphy, S., Wood, P., Deck, S. and Tucker, S. (1997) 'Innovative Project Procurement In The Queensland Government: The Woodford Correctional Centre.' *Journal of Project and Construction Management*. **3** (2): 57–69.

Ireland, V. (1987) *The Choice of Contractual Arrangement*. The Emerging Profession – Project Managers Forum. Adelaide, Project Managers Forum.

Jensen, P. and Hall, C. (1995) New Arrangements for Radiology at Sydney Hospital. *The Contracting Casebook – Competitive tendering in action*, eds Domberger S. and Hall, C.. Canberra, Australian Government Publishing Service: 87–97.

King, V. (1996) *Constructing the Team: A US Perspective*. Glasgow, Scotland, E&FN Spon.

Latham, M. (1994) *Constructing the Team*, Final Report of the Government/Industry Review of Procurement and Contractual Arrangements in the UK Construction Industry. London, HMSO.

McGeorge, W.D. and Palmer, A. (1997) *Construction Management New Directions*. London, Blackwell Science.

Miles, R.S. (1995) 'Twenty-first Century Partnering and the Role of ADR.' *Journal of Management in Engineering, ASCE*. **12** (3): 45–55.

Mo, J.K.W. and Ng, L.Y. (1997) *Design and Build Procurement in Hong Kong – an Overview*. CIB W92 Procurement Systems Symposium 1997, Procurement – A Key To Innovation, The University of Montreal, CIB.

NBCC (1989) *Strategies for the Reduction of Claims and Disputes in the Construction Industry – No Dispute*. Canberra, National Building and Construction Council.

Office of Building and Development (1997) *Partnering and the Victorian Public Sector*. Melbourne, Australia, Office of Building and Development, Department of Infrastructure, Victorian Government.

Ogunlana, S.O. (1997) *Build Operate Transfer Procurement Traps: Examples from Transportation Projects in Thailand*. CIB W92 Procurement Systems Symposium 1997, Procurement – A Key To Innovation, The University of Montreal, CIB.

Revay, S.G. (1988) *Managing Productivity*. 5th Canadian Building and Construction Congress, Montreal, Canada, National Research Council of Canada.

Shing-Tao, A. and Ibbs, C.W. (1998) 'On-call Contracting Strategy and Management.' *Journal of Management in Engineering*, **14** (4): 35–44.

Sidwell, A.C. and Ireland, V. (1989) 'An International Comparison of Construction Management.' *The Australian Institute of Building Papers.* **2** (1): 3–12.

Smith, A.J. (1999) *Privatized Infrastructure – The Role of Government*. London, Thomas Telford.

Smith, N.J., Merna, A. and Grimsey, D. (1994) *The Management of Risk in BOT Projects*. Internet 94, 12th World Congress on Project Management, Oslo, Norway, International Project Management Association.

Walker, A. (1993) *Project Management in Construction*. London, Blackwell Science.

Walker, C. and Smith, A.J. (1995) *Privatised Infrastructure – The BOT approach*. London, Thomas Telford.

Walker, D.H.T. (1994) An Investigation Into Factors that Determine Building Construction Time Performance. PhD, Department of Building and Construction Economics. Melbourne, RMIT University.

Walker, D.H.T. (1996) 'The Contribution Of The Construction Management Team To Good Construction Time Performance – An Australian Experience.' *Journal of Construction Procurement.* **2** (2): 4–18.

Walker, D.H.T. (1997) 'Construction Time Performance and Traditional Versus Non-traditional Procurement Systems.' *Journal of Construction Procurement.* **3** (1): 42–55.

Walker, D.H.T. (1998) 'The Contribution of the Client's Representative to the Creation and Maintenance of Good Project Inter-team Relationships.' *Engineering and Architectural Management*, **5** (1): 51–57.

Womack, J.P. and Jones, D.T. (2000) From Lean Production to Lean Enterprise. *Harvard Business Review on Managing the Value Chain*. Boston, MA, Harvard Business School Press: 221–250.

Yogeswaran, K. and Kumaraswamy, M.M. (1997) 'Perceived Sources and Causes of Construction Claims.' *Journal of Construction Procurement.* **3** (3): 3–26.

Chapter 3
Enterprise Networks, Partnering and Alliancing

Derek Walker and Keith Hampson

Choices of procurement method were presented in the second chapter. It was noted that other than when using the traditional procurement method, the adopted choice should not affect the capacity of a partnering or alliancing strategy to achieve the project goals through teamwork. In Chapter 7 we present three essential features of partnering or alliancing – mutual objectives, continuous improvement and problem resolution arrangements. We also discuss characteristics of the underlying human relationship ethos that must be engendered for it to be successfully applied. In this chapter we explore the features and characteristics of enterprise networks and explain how these may prepare project teams more fully to realise their potential in the twenty-first century.

Table 3.1 illustrates an interesting taxonomy of management evolution culminating in a four-**management blueprint** model for success in dealing with the turbulence and uncertainty of today's competitive climate (Limerick *et al.* 1998, p.

Table 3.1: The four management blueprints				
	First blueprint	**Second blueprint**	**Third blueprint**	**Fourth blueprint**
	Classical	Human	Systems	Collaborative organisations
Organisational forms	Functional Mechanistic Organic	Inter-locking Matrix	Contingency Divisional	Loosely coupled networks and alliances
Management principles	Hierarchy	Supportive relationships	Differentiation	Empowerment and collaborative individualism
Managerial processes/forms	Management functions	Democratic leadership	Open systems analysis	Management of meaning
Managerial skills	Person-to-person control	Goal setting Facilitation	Rational/diagnostic	Empathetic Proactive
Managerial values	Efficiency Productivity	Self-actualisation Social support	Self-regulation	Social sustainability Ecological balance

Source: Limerick, D., Cunninton, B. and Crowther, F. (copyright © 1998) *Managing the New Organisation: Collaboration and Sustainability in the Postcorporate World*. Warriewood, NSW, Business & Professional Publishing.

30). The fourth management blueprint is the recommended direction for the immediate future as we have now entered the twenty-first century.

Many companies have moved to the third blueprint, but have difficulty in moving forward. This may be due to fear of higher levels of management losing control over their management authority prerogative and fear over loss of competitive advantage through networking and outsourcing. The fourth blueprint also relies upon considerable bases of mutual trust and respect requiring readiness or 'maturity' from management and partner organisations stemming from loosely coupled organisations. This requires a greater capacity for real rather than espoused empowerment than third blueprint managers can cope with. Indeed, Limerick *et al.* (1998) describe an uncomfortable staging post between third and fourth blueprint organisations where the worst of cases prevail. 'Neocorporate bureaucracy' is a new form of corporatism, still embedded in the major paradigm of the hierarchical corporate organisation, but with an attempt to apply some precepts of the fourth blueprint. The effect is like grafting the legs of a gazelle onto an elephant with managers becoming more risk-averse. Furthermore, by holding onto hierarchy and in attempting to reduce costs through delayering organisational structure while maintaining mechanistic control, they merely push more of the cost and effort of formal accountability to lower levels of management. This results in greater stress for those remaining stemming from lowering of flexibility and disempowering through bureaucratic control (Limerick *et al.* 1998, p. 84).

Such approaches snatch defeat from the jaws of potential victory. Instead of releasing energy from devolution of authority, the result is further strategic control that stifles initiative and results in cynicism – as this process is seen as **pseudo-devolution**. These approaches also alienate those valuable members of teams who can constructively critique performance and enact organisational learning. Such people view the hierarchy as 'corporate nazis' and withdraw their valuable input from the decision-making process. Organisational cultural diversity is lost in the neocorporate bureaucracy.

Limerick *et al.* 1998, p. 91) provide sound advice on how to establish a fourth blueprint organisation. This is centred on forming strategic, loosely coupled alliances with a management emphasis based upon:

(1) Liberating managers.
(2) Developing boundary roles.
(3) Developing communication systems (particularly effective IT systems that allow such groups to share meaning).
(4) Getting the mindset right to benefit from alliancing.
(5) Establishing the alliance carefully (choose partners carefully with compatible organisational structures and synergy of contribution).
(6) Defining the focus.
(7) Managing the soft issues (trust, commitment, etc.).
(8) Managing the hard edge processes too (get commitment on a mutual set of expectations and understanding of acceptable behaviours of each partner)
(9) Managing the network control systems (provide sufficient resources for effective IT and effective human contact).

Another intrinsic element of the fourth blueprint is organisational learning and team learning. This is achieved through sharing the diversity of available views within groups characterised by independent collaborative individuals with high levels of communication and people skills. Companies that get the most out of alliances are those that learn from each other (Hamel *et al.* 1989). Limerick *et al.* (1998, p. 179) argue that companies have to become action-learning organisations, that are self-reflective and can transcend and critique their own identity, values, assumptions and missions that are initiated and controlled by line managers themselves. Such organisations do this through not only supporting critical appraisal but also, and more importantly, by providing feedback for lessons learned, to be transformed into subsequent action. This requires organisations to welcome both challenge and experimentation through the establishment of the organisation as a learning community. This would be composed of both inside-organisation people and informed external participants who are free of the internal assumptions and mindsets of organisational members. The approach exemplified by the fourth blueprint is strongly supported by management theorists and commentators. For example, in the Karpin Report many examples are cited of a gradual global shift taking place towards this new paradigm (Karpin 1995). Characteristics of the fourth blueprint model are offered as current world best practice (Karpin 1995; Lendrum 1998, Section II).

The best way forward for gaining advantages for this business arrangement has been identified as the formation of relationship-based enterprises. The actual form of these varies, but all have in common a basis of partnership with varying degrees of autonomy, flexibility of action and physical proximity. At the most isolated and fragmented extreme of this continuum lies single firms who compete and take on projects under a variety of contractual forms. At the other extreme of this continuum lie firms who bring together their resources to create alliances and/or joint ventures which act as a single organisation but maintaining their separate identities. In the case of joint ventures a new legal entity will be created seconding resources from host firms. Both joint ventures and alliances often have the customer as a partner in this arrangement – either on a once-off basis for a specific project or to develop a continuing relationship. The on-call contracting method described in Chapter 2 can be considered as both a procurement method and a relationship type. Relationship entities undertaking projects may choose to follow any of the procurement paths described in Chapter 2 apart from the traditional process.

3.1   The development of enterprise networks

The concept of networks coming together to complete projects is not new. Indeed some of the most early historically significant buildings in Europe and elsewhere were constructed using workers from craft guilds. These business entities did not constitute legal entities as recognised today (such as businesses and firms). They did, however, act as firms and operated as clusters of workers drawn together on a significant enterprise often sharing resources, certainly sharing knowledge, mutually adjusting to take advantage of opportunities and relating on a sound basis of trust. Working in close proximity has been recognised as a competitive advantage for firms. Michael Porter has studied the competitive advantage of cluster organisations serving customers. He cites relevant examples – including the wine industry in California, the entertainment industry in Hollywood, shoes and fashion apparel in Italy, and cork and wood products in Portugal. In understanding the advantage of such clusters, he offers insights into why successful alliances of firms provide competitive advantage, and both survival and growth for their participants. He argues that it is because their location is based on a competitive advantage of infrastructure (Porter 1998).

The ability to exchange ideas rapidly is one principal basis of communication infrastructure. Surface communication such as road, rail and water-based transport and efficient air transport facilities are very important but it is their ability to get people together solving problems, building relationships and sharing ideas on improvement that is the underlying factor contributing to communication infrastructure. Over the second half of the twentieth century, the term communications has shifted from surface, sea and air to telecommunications. The ability to send messages by telex and later fax has revolutionised businesses. Telephony, particular cellular phones, has had a major impact on both business and private interaction. The rise in general use of the internet during the close of the twentieth century has had a profound impact upon the exchange of ideas and perceptions. With electronic data interchange (EDI), transfer of graphics, video images, text and data files adding to the repertoire of communication media, the provision of infrastructure to support this kind of knowledge exchange is crucial to successful networks. Clusters of individuals and firms are increasingly linked together through electronic infrastructure to exchange information, data and ideas rapidly. Trust, commitment, and aspirational dimensions of relationships are built and maintained effectively through a combination of factors but face-to-face contact is nevertheless the preferred option.

Many of the clusters that Porter cites are formed through local engagement. Members of these clusters are in fairly close proximity where road, rail, plane and other forms of transport infrastructure are important. Communications infrastructure and social links are essential for clusters working collectively to upgrade, improve and add value (Porter 1998). Combining communication infrastructure, local expertise and cluster development can lead to significant gains in productivity and competitiveness. This has been demonstrated by the manufacturing sector where clusters develop and spawn facilities. In this arrangement, offshore factories enhance their value adding activities through fourth blueprint principles, intelligent use of ideas-sharing and local network cluster development. These successful factories progress from being a low-cost manufacturing offshore plant, to one in which new products are designed, developed and exported. They migrate from being a **server** or **outpost factory** to being a **lead factory** and centre of excellence (Ferdows 1997).

If this model is applied to the construction industry then a cluster relationship approach may lead to adding value to the supplier and subcontractor part of the supply chain. Skill improvement for the design disciplines can also be developed through more open relationships with the delivery and assembly part of the supply chain for construction clients. Alliancing and joint venturing is a logical development from design and construction to operation and maintenance.

The case for bringing together parts of supply chain clusters has been persuasively argued where design and construction teams share a physical working space. Luck and Newcombe (1996) cite benefits of improved integration and coordination as well communication using authoritative references from the management literature. In their concluding remarks they make an important point about the role of informal mutual adjustment where personal contact and positive interaction leads to problems being addressed as a matter of course rather than through formalised and ritualised mechanisms. This is an important point in understanding how trust is developed in clusters of workers, where not only information and ideas are exchanged but bonds are developed through personal contact and where immediacy of access is made possible through a communication infrastructure.

We have seen how **enterprise networks** are based on a foundation of effective interaction using good quality communication infrastructure of IT delivery, enabling people to interact freely to share ideas. This as we argue in Chapter 7, is the fundamental feedstock of building trust in relationships and is illustrated in Figure 3.1.

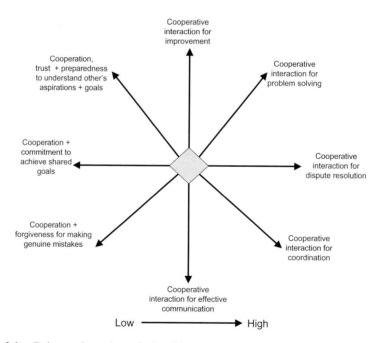

Figure 3.1: Drivers of trusting relationships

Factors that drive the quality of a relationship include sharing information, joint problem solving and sharing ideas to arrive at common goals that meet project objectives. Figure 3.1 serves to remind us how any working relationship is built and what elements of the interactions need to be heeded to maintain the relationship.

3.2   Types of joint enterprise arrangements

Just as it was important to document the types of procurement systems in Chapter 2 for us to gain an understanding of the continuum of options available, it is also important to review the alliance form options.

This can be viewed from a continuum based upon a high risk, high cost maximum use of human resources to a low risk, low cost minimum use of human resources. The continuum presented below is based on the typology described by Segil (1996, p. 16) and is taken from a marketing and manufacturing perspective, although this can be compared to construction project alliance forms.

(1) Takeover/merger – where one company undertakes a complete acquisition of another so that the two entities become a merged single entity.

(2) Joint venture/equity – where two companies cooperate on the creation of a new separate company.

(3) Research and development technology transfer – where two companies join in a research and development project to promote a new technology/product or service.

(4) Original equipment manufacturer (OEM) licensing arrangement – where one company develops a product or service to be marketed and sold by another, perhaps with better access to distribution channels.

(5) Joint marketing/distribution – where one company joins with another to market, sell, and distribute a product.

At the apex (1) lies the vertical integration of the supply chain where one company may voluntarily **merge** with another firm to maintain an indefinite and continued business alliance which is internalised to the extent that the joined companies become one. For example, a developer may merge or buy out a construction arm and perhaps several key subcontractor suppliers. This arrangement remains as a permanent situation until various 'arms' are sold off or otherwise disposed of. Joint ventures are common in the construction industry particularly in forming consortia to undertake a BOT/BOO/BOOT project. The above research and development venture may be considered analogous to a BOT/BOO/BOOT partner that uses their specialised skill as an equity stake to establish with others the project partnership network and subsequently transfer their stake in the project to others. Such an entrepreneurial stake may include identifying the customer need, developing the project concept, locating the financing entities, developing the legal structure for a project ownership group and negotiating the operating concession. This type of entrepreneur would sell their stake to take an interest in another project. The **original equipment manufacturer** (OEM) equivalent is hard to find in the construction industry unless a franchised consultancy fits this category but such examples are rare. The traditional partnering or alliancing arrangement appears to be the closest equivalent in the construction industry to **joint marketing/distribution**. Mergers and/or acquisitions in the construction industry seem to be centred on funding issues, technological transfer or access to markets. The changing market in Australia with large German contractors merging or acquiring Australian companies is an example. These companies may be better placed to diffuse global best practice, better able to commit equity on BOT/BOO/BOOT projects, better able to withstand a cyclical industry notorious for its booms and busts and better able globally to centralise research and development activities.

While the above describes different sorts of joint relationships that facilitate a project, however, it lacks any description of mental models that participants may consciously carry about the relationship they will experience in delivering a product or project. Figure 3.2 provides a useful description of these mental models (Lendrum 1998, p. 12).

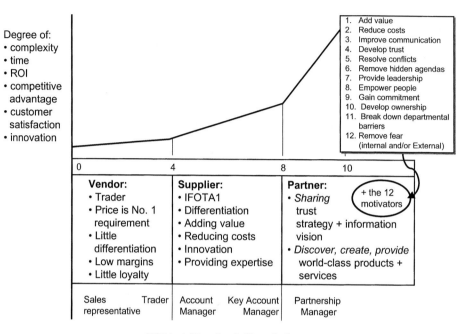

IFOTA1 = In full, on time, to A1 specifications

Figure 3.2:   The alliancing and partnering mindset
Source: Lendrum, T. (copyright © 1998) *The Strategic Partnering Handbook*, McGraw Hill, Sydney.

The **vendor** relationship mindset is based upon the single transaction mentality where both sides attempt to gain maximum financial advantage. The customer wants the goods at the cheapest cost, the supplier seeks to maximise the profit through possible clawbacks of the quality, taking advantage of making claims for anything not explicitly agreed to at the time of sale. The 'fine print' prevails in complicated contracts. The success of winning a contract is based on an open tender system and loyalty on either side is virtually absent.

The **supplier** relationship relies on continued sales to the customer and attempts to offer extra value into products, service through innovation, support and quality management. Many subcontractors have reached this level as 'favoured' to qualify for pre-selection to their contractors though contracts may be tendered and a multi-criteria-based decision may determine the success of selected tenders. The person responsible for negotiation and marketing would be anxious to maintain credibility to deserve continuing this supplier relationship.

The **partnership** relationship requires all the attributes of the supplier – but more.

In addition, there needs to be the mutual trust, loyalty and commitment so often absent in full in the other two categories. This is achieved because the customer believes the partner to be a world-class performer who can provide innovative products that stretch beyond incremental improvement gains. The 12 motivator identifiers illustrated in Figure 3.2 facilitate this trust and commitment. Figure 3.3 illustrates how the alliance mindset delivers benefit (Lendrum 1998, p. 20).

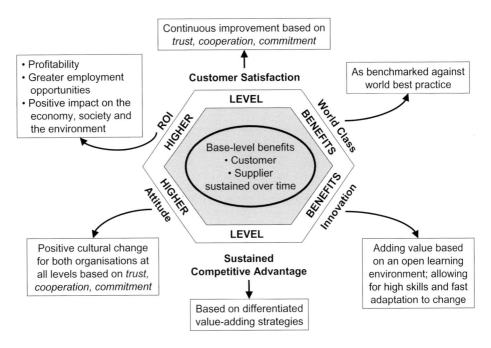

Figure 3.3: How alliances deliver value
Source: Lendrum, T. (copyright © 1998) *The Strategic Partnering Handbook*, McGraw-Hill, Sydney.

Table 3.2 presents base level benefits for customers and suppliers (Lendrum 1998, p. 19). These are typical of sound management practice and one would hope to see many of these in any event. Benefits from benchmarking, innovation, improved attitude and customer satisfaction requires hard work and commitment from both sides, so the business maturity levels must be very high to facilitate this process.

3.3 Australian partnering research study results

The Construction Industry Institute Australia (CIIA) stress that there is no partnering contract as such, rather an agreed partnering charter forms the basis of a working agreement that is intended to shape a non-adversarial culture to promote win–win working relationships between partners. This is achieved through the aim to … 'foster cooperative and mutually beneficial relationships among project stakeholders by developing an explicit strategy of commitment and

Table 3.2: Customer and supplier base level benefits

Customer Base level benefits	Supplier Base level benefits
• Improved quality, fewer rejects, less waste • Lower operational costs • Reduced inspection times • Customer non-conformance complaints drastically reduced • Lower prices in real terms (total costs) • Superior performance or effect at lower, equivalent (or even higher) prices, i.e. greater value for money • Improved reliability, flexibility and dependability of supply • Improved cash flow and reduced working capital costs • Lower inventory and cycle times • Reduced product/service development time • Fewer hassles and less frustration • More time and resources available for downstream customers • Increased margins (i.e. increased total value) • Improved communication and people relationships • Increased market share • Aggregate purchasing • Supplier-managed inventories • Early supplier/extended range of products and services • Elimination of waste associated with tenders, annual auctions and multiple suppliers • Elimination of litigation and adversarial confrontation	• Larger volumes of products and services (domestic and/or export) • Longer-term stability of supply • Greater stability of forecasts • Improved production efficiencies/cycle times • Higher quality at lower operational costs • Lower costs in real terms • Fewer hassles less frustration • Improved skills from joint training • Increased margins • Fewer customer complaints/less waste • Improved communication and people relationships (internal and external) • Price premium over the competition (i.e. greater value for money) • Achievement of preferred supplier/preferred relationship status • Increased market share and access to new markets • The partnership becomes a benchmark for other customer/supplier relationships • Greater responsiveness and flexibility in fulfilling customer expectations and resolving customer complaints • Improved rate of product/service development • Improved logistics and delivery systems • Greater integration of activities between divisions/departments' etc. • Fewer process steps and less complexity • Early involvement in product or service development • Scrapping of the dreaded tender system • Elimination of litigation and adversarial confrontations

Source: Lendrum, T. (copyright © 1998) *The Strategic Partnering Handbook*, McGraw Hill, Sydney.

communication. These goals are documented in a charter that stands alongside legally-binding contractual arrangements' (Lenard *et al.* 1996, p. 11). The CIIA report is based upon a comprehensive study of 32 Australian projects with some 131 questions asked of senior members of partnering teams. This study is significant as a thorough research output in which industry contributors and senior academics worked together on a research task force. The conclusions of the report list three clusters of useful findings (Lenard *et al.* 1996, p. 33–35):

(1) Good communication and high levels of trust between partners obviated much of the conflict and divisiveness that leads to a litigious outcome. An alternative dispute resolution to the *business as usual* legal wrangles included an agreed dispute escalation mechanism.

(2) The results indicated that 'partnering had reduced claims, disputes, delays and the need for reworking, whilst improving safety and profit margin.'

(3)   Technology transfer was seen to promote innovation diffusion when partnerships were established early on in the project lifecycle.

Interesting insights were provided from the 1996 CIIA report. Public sector clients accounted for 91 per cent of projects investigated. Responses to the question 'partnering in this project has been a great success' yielded the following results:[1]

- Strongly agree   38.7%
- Agree   35.5%
- Neither agree nor disagree   3.2%
- Disagree   6.5%
- Strongly disagree   16.1%

Almost all (85 per cent) of respondents said that they would undertake another partnering project. In 42 per cent of the cases partnering was adopted pre-award and in 88 per cent of these a pre-tender meeting was held with potential contractors. In 81 per cent of the case studies the end-user had been part of the partnering arrangement. In cases where partnering had been deemed a success, 87 per cent of respondents agreed that the partnering arrangement had led to lower administrative costs because of the elimination of defensive case building (Lenard 1996, p. 31).

The above results are supported by work undertaken on strategic alliance experiences gained from a survey of 51 companies involved in alliances with the Queensland public sector as client. In that study the 3 highest ranked benefits out of 13 measured were cooperation, resolution of problems and coordination (Kwok 1998, p. 137). It is interesting though that 'inter-organisational managerial skills' and 'access to technologies' ranked twelfth and last out of the 13 identified benefits tested[2] (Kwok 1998, p. 158). Tables 3.3 to 3.8 summarise results from the 1996 CIIA report on partnering in Australia.

Essential components of partnering were investigated and the CIIA 1996 study results are presented in Table 3.3 (Lenard and Bowen-James 1996, p. 20). An important aspect of the components presented is that the partnering arrangement requires action plans. The need for workshops and an external facilitator is an interesting issue as it implies that survey participants see the need for guidance from a mutually trusted person to gain commitment and confidence in the process. This accords with other recent literature (KPMG 1998; Lendrum 1998). Table 3.4 illustrates the CIIA study results regarding content covered in workshops (Lenard and Bowen-James 1996, p. 21).

It appears that the importance of action plans is reinforced by the gap in need for them when comparing successful projects versus unsuccessful ones. In essence this complies with the adage 'failing to plan, planning to fail'. Where there was also a perceived need for planning and action that was reported to have widely differing

---

[1] Scores indicated that the mean perception of success was 3.83 on a 5-point scale.
[2] Kwok notes that 8 out of 12 of the respondents that had been engaged in an alliance in the past had abandoned it, mainly on grounds of poor performance subcontractors (Kwok 1998).

Table 3.3: Essential partnering plan

Partnering plan components	Survey agreement	Comments
1. Independent facilitator	84%	• There is a definite need for an action rather than rhetoric approach
2. Commitment of senior management	100%	• Commitment, particularly from senior management is essential
3. Charter	85%	• Problems are expected but so is a rational and reasonable way of dealing with them
4. High level workshop participation	94%	
5. Dispute resolution plan	93%	
6. Implementation plan	84%	
7. Continuous partnering evaluation	97%	
8. Finalising workshop	84%	

Source: Adapted from CIIA, (Lenard and Bowen-James 1996, p. 20). With permission of the Construction Industry Institute of Australia.

perceptions, there was a corresponding result in perceived success or otherwise with the project (items 4 and 6 in Table 3.4). Following on the theme of a need to plan for success, the CIIA study reveals interesting insights into their perceived types of plans needed for successful projects.

Of the successful projects in the CIIA study, Table 3.5 illustrates the respondent's perceptions of what elements of the partnering plan were necessary. While the interesting element is the low level of need for mechanisms for sharing risk and benefits (24 per cent) and a process for orienting new partner members (43 per cent), the other elements seek a formal structure with all being able to easily and transparently understand their roles and interaction.

Table 3.4: Partnering workshop content

Content covered in workshops	Projects perceived a success	Projects perceived a failure	Comments
1. Self-perception exercises	56%	43%	Significant differences between projects (perceived as a success or a failure)
2. Training in team skills	39%	43%	
3. Development of goals and objectives	96%	86%	• Dealing with problems as they inevitably arise
4. Dispute resolution plan	89%	**43%**	• Commitment to training and development appears poorly cultivated
5. Anticipated problems	78%	71%	
6. Action plan to address problems	78%	**57%**	
7. Development of a charter	100%	100%	
8. Celebration	89%	**29%**	

Source: Adapted from CIIA, (Lenard and Bowen-James 1996, p. 21). With permission of the Construction Industry Institute of Australia.

Table 3.5: Partnering plan study elements

Elements of the partnering plan	Survey agreement	Comments
Roles and responsibilities of each organisation as it relates to them	71%	The elements that attract strong agreement are general issues of good management
Measurable objectives relating to each partnering goal	67%	• The disagreement on mechanisms to share risks and benefits appears to indicate an underdevelopment of trust and commitment
Written description of the dispute resolution process	76%	
Mechanism for sharing risks and benefits	**24%**	• Orientation should aid both project understanding and commitment
A process to orient new team members	**43%**	

Source: Adapted from CIIA, (Lenard and Bowen-James 1996, p. 22). With permission of the Construction Industry Institute of Australia.

Where partnering was considered unsuccessful, low tenders correlated with partnering failure. This reinforces the ideal of value for money rather than cheapest price. Continuous evaluation for the successfully partnered project was evident for 90 per cent of that group of respondents. Table 3.6 illustrates the types of evaluation and the extent.

The CIIA study Lenard and Bowen-James (1996) presents benefits of partnering, illustrated in Table 3.7 below. Clustering and summarisation of these benefits correlates positively with factors contributing to overall project success. These again fall into two major groups. First, it presents a group of general good management practices such as time, cost and quality. Second, the perceived benefits of partnering are solidly placed in reduced energy being expended on battling entrenched positions. Disputes are more effectively dealt with through open communication and dispute resolution procedures minimise legal and administration transaction costs. While these are clearly the espoused values, aspects of

Table 3.6: Study partnering evaluation on successful projects

Type of partnering evaluation	Survey agreement	Comments
Regular formal partnering meetings	91%	• Again many areas of agreement relate to general good management practice
Minutes kept of partnering meetings	91%	
Regular meetings with subcontractors	67%	• The relative reticence in holding workshops to focus on resolution of problems indicates some evidence of denial about problems in partnerships and how these may be brought out into the open and resolved
Daily information meetings	**45%**	
Periodic monitoring against mutually agreed goals	100%	
Evaluation sheets filled out with results discussed at meetings	77%	
Periodic workshops held to focus on unresolved issues and problems	59%	

Source: Adapted from CIIA, (Lenard and Bowen-James 1996, p. 23). With permission of the Construction Industry Institute of Australia.

Table 3.7: CIIA study partnering benefits

Partnering benefits	Survey agreement	Comments
Reduced exposure to litigation through open communications and issue resolution strategies	91%	• Again many areas of agreement relate to general good management practice
Lower risk of cost overruns because of better cost control	61%	• The strong focus on dispute resolution and conflict management indicates a defining difference between partnering and non-partnering
Lower risk of time delays because of better time control	65%	
Better quality product	78%	
Lower administration costs because of elimination of defensive case building	87%	
Increased opportunity for a financially successful project because of the non-adversarial attitudes	96%	

Source: CIIA, (Lenard and Bowen-James 1996, p. 25). With permission of the Construction Industry Institute of Australia

practice (illustrated as % in bold in Table 3.4, Table 3.5 and Table 3.6) indicate a gap in application of the rhetoric of planning and managing disputes.

Table 3.8 presents the CIIA results on partnering problems. Several of these potential problems were evidently not realised, for example, none of the participants saw the effort and cost of partnering as a real problem. There was low agreement on potential problems related to dispute and problems not being adequately dealt with. This may be explained by the strong agreement on the prevailing conditioning in the construction industry towards entrenched positions in readiness for a win–lose approach and a lack of trust and commitment being adequately realised. There was 100 per cent agreement on commercial pressure problems. This indicates a level of maturity and ethical integrity in using partnering by recognising the danger of compromising trust when monetary gain comes before principled action. Design problems and design communication issues still remain problematical to the partnering participants. This may be explained by a lack of including the design team fully in the partnering process. In 71 per cent of cases where partnering had failed, participants felt that the design consultants should have been more thoroughly included in the process, whereas in successful partnering projects this figure was only 17 per cent (Lenard and Bowen-James 1996, p. 26).

The CIIA survey (Lenard and Bowen-James 1996) provides useful and rigorous data that is current and relevant to the development of partnering arrangements in the Australian construction industry. Clearly the question of maturity in terms of readiness to embrace such cooperative arrangements is in a positive growth stage but as yet not fully realised.

Table 3.8: CIIA study problems encountered	
Problems encountered with partnering participants	**Survey agreement**
Being conditioned in a win–lose environment	71%
Fully understanding the partnering concept	**43%**
Seeing staff training as essential to partnering	**14%**
Project should have been more carefully selected for its suitability for partnering	57%
Key subcontractors should have been more thoroughly included in process	57%
Too many problems with plans and specifications from design consultants during the construction stage	86%
Design consultants and other consultants should have been more thoroughly included in the process	71%
Continuity of open and honest communication not achieved	86%
Technology problems (incompatible hardware/software)	**43%**
Dealing with a bureaucratic organisation impeding effectiveness	71%
Commercial pressures compromise the partnering attitude	100%
Issues and problems allowed to slide and escalate	**43%**
Partners not willing to communicate outside the contract discussions	**29%**
Partners not willing to compromise on craft team solutions	57%
Up front time required and cost for partnering process overwhelming	**0%**

Source: CIIA, (Lenard and Bowen-James 1996, p. 26). With permission of the Construction Industry Institute of Australia

3.4   USA partnering research study results

The evidence presented in the literature suggests that the trend towards partnering is advanced in the USA. One of the major champions of partnering has been the US Army Corps of Engineers. Larson, for example, reports on a study of 280 partnering projects of varying type and scope from heavy process engineering through to hospital extensions (Larson 1995). Over half of these projects were awarded to an open, competitive, low-bid process. Larson provides a taxonomy of relationships based on perceived attitudes of project partners and categorises the following distribution of projects as:

(1)   *Adversarial.* Participants perceive themselves as adversaries with each party pursuing their own concerns as their prime priority and objective. There were 78 (28 per cent) of these in the sample.

(2)   *Guarded adversarial.* Participants cooperate strictly within the bounds of the contract. Superiors using the formal interpretation of contractual obligations to resolve major disputes. There were 66 (24 per cent) of these in the sample.

(3)   *Informal partners.* Participants sustain the relationship beyond the boundaries of the contract. Disputes are resolved through mutual adjustment that at least partially satisfies both sides. There were 77 (28 per cent) of these in the sample.

(4) *Project partners.* Participants treat each other as equals, on the same team and working closely together to solve problems and improve processes. There were 59 (21 per cent) of these in the sample.

Figure 3.4 illustrates Larson's findings. These were analysis of variance (ANOVA) results for success criteria (1–5 scale, low to high) by owner–contractor relationship type. They consistently demonstrate a pattern of the higher level of collaboration and cooperation relationship having higher success ratings than lower ones. Interestingly, on the issue of low bid performance of the 280 projects, 142 were awarded on a lowest bid, competitive basis. Of the six success factors investigated, bid status was an insignificant factor at the 0.05 significance level. However, their analysis revealed significant interaction effects at the 0.05 significance levels on the overall results for owner–contractor relationships for avoiding litigation. For these two factors, non-lowest bid price as opposed to lowest-bid price, was consistently associated with effectiveness for adversarial, informal partners and partners, and less effective for guarded adversarial relations. Refer to the paper for a fuller description of definitions and terms (Larson 1995). This indicates that selecting by the lowest tender is less effective than taking a more comprehensive approach to the project's probable completion cost.

Success for owner-contractor relationship

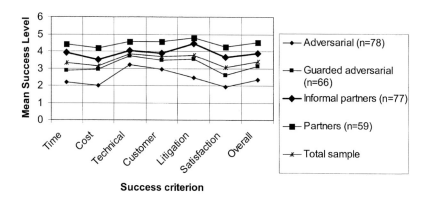

Where Time = meeting schedule; Cost = controlling cost; Technical = technical performance; Customer = meeting customer needs; litigation = avoiding litigation; Satisfaction = satisfaction of participants; Overall = overall results.

Figure 3.4:   Larson study – success factor results by relationship type
Source: Adapted from Larson (1995, p. 34) with permission of the *Journal of Management in Engineering* – American Society of Civil Engineers/Engineering Management Division.

In 1992, a study of US Army Corps of Engineers construction projects found that 31 out of 37 domestic districts used partnering (Weston and Gibson 1993). The survey highlighted 19 partnering projects and used data available for 16 of these (12 civil and 4 military engineering projects) which represented 85 per cent of projects undertaken using partnering at that time. Using criteria for success of cost change, change order cost, claims cost, value of engineering savings and duration change, a comparison of these were made with 28 non-partnering projects using 't' tests. The results indicate that the mean cost change for partnered projects was 2.72 per cent for partnered projects and 8.75 per cent for non-partnered projects – an improvement of 6.03 per cent for partnered projects (with a 0.01 'P' value on a two-tailed test indicating a very high level of statistical reliability). These results showed a similar difference for reduction in construction time change with a 6.46 per cent mean value but this was of low statistical reliability. The mean difference between partnering and non-partnering projects' contract change costs were 3.89 per cent and 7.74 per cent respectively – an improvement of 3.85 per cent for partnered projects (with a 0.07 'P' value on a two-tailed test indicating a very high level of statistical reliability).

When 't' tests were undertaken to test the result at the 90 per cent significance level both cost and change order (contract variations) criteria indicate significant improvements through use of partnering. Also none of the respondents interviewed who were involved in partnering projects were dissatisfied with partnering. Their subjective data analysis revealed that intangible benefits to partnering included:

(1) Reduced administrative paperwork.
(2) More enjoyable project work environment.
(3) Reduced communication barriers.
(4) Less adversarial relationships.

Both studies from data gathered on US Army Corps of Engineers projects indicate significant gains from the use of partnering. This indicates that higher quality of cooperation and collaboration pays dividends to the client and parties concerned. The potential benefit of partnering is illustrated in Figure 3.5 below in terms of the extent that project objects are aligned through partnering.

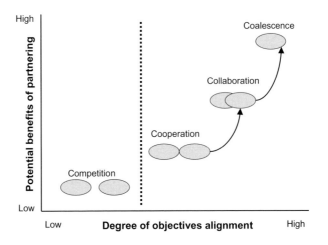

Figure 3.5:   The project continuum
Source: Thompson and Sanders (1998, p. 74) with permission of the *Journal of Management in Engineering* – American Society of Civil Engineers/Engineering Management Division.

Further evidence based upon the USA Construction Industry Institute (CII) study of partnering experience gathered from 21 partnering relationships involving more than 30 owners indicates an overwhelmingly positive experience with partnering and validated the choice to partner (CII 1996). Results are illustrated in Table 3.9. The taxonomy proposed closely follows Larson's three evolutionary stages of the alignment of project objectives illustrated in Figure 3.5.

Finally in better understanding partnership or alliance options it is worth returning to the 1996 CIIA study which provides a model of partnering options that can be categorised as experimental partnering, packaged partnering and committed partnering.

3.5   Types of project partnering – from cooperation to coalescence

It is clear from both the Australian and USA case studies and research results that the partnering systems and their variants provide worthwhile benefit. Table 3.10 and Table 3.11 demonstrate the need for the owner/client to be knowledgeable enough about procurement options to appreciate the importance of:

- defining project goals;
- identifying resources required to provide the partnering infrastructure;
- knowing how to evaluate potential project partners; and
- understanding relative benefits of different types of relationship arrangements.

The critical element of partnering and alliancing studies indicates that there are vital components of the relationship that differentiates alliancing from partnering (section 3.8 explains alliancing in detail) – or at least those more committed levels of partnering. These elements generally fall into the following categories:

- level of trust and commitment;
- degree to which the relationship is planned and nurtured rather than forced or required as a condition of contract;
- way in which the relationship is initiated, fostered and maintained as part of an integrated procurement process;
- the degree to which transparency/open-book philosophy is maintained; and
- the way in which risk and reward is treated.

Table 3.11 presents a four-level relationship typology that addresses the issue of dispute resolution methods for different types of partnering arrangements.

Much of the partnering rhetoric can be manipulative, indeed in many of the studies cited in this section there were cautionary comments about clients or contractors that use the guise of partnering as a means of taking unfair advantage of those required to form a partnering arrangement. In Australia both Kwok (1998) and Lenard *et al.* (1996) cite commercial disadvantage as a principal cause of these types of relationship failure. Clients who enforce a partnering arrangement after tender award may be courting problems. As relationships edge towards Levels 3 and 4, specific plans and procedures are put in place to develop and maintain an atmosphere of trust. This can be contrasted with the Levels 1 and 2

Table 3.9: CII USA study results

Relationship	Characteristics	Comments
Competition (low objectives alignment)	1. No common objectives; they may actually conflict 2. Success coming at the expense of others (win–lose mentality) 3. Short-term focus 4. No common project measures between organisations 5. Competitive relationship maintained by coercive environment 6. Little or no continuous improvement 7. Single points of contact between organisations 8. Little trust, with no shared risk; primarily a defensive position	• Business as usual
Cooperation (low/medium objectives alignment)	1. Common objectives that are project specific 2. Improved personal interpersonal relationships 3. Team members who are likely to be involved in projects outside the partnering relationship 4. Partnership measures that may or may not resemble organisational measures used on other projects 5. Multiple points of contact 6. Limited trust and shared risk: guarded information sharing	• Schedule reduction – 10.5% • Cost reduction – 16.3% • RFI turnaround time: 14 days versus 30–60 days
Collaborative (medium/high objectives alignment)	1. Long-term focus on accomplishing the strategic goals of involved parties 2. Multi-project agreement: long-term relationships without guaranteed workload 3. Common measurement system for the projects and the relationship 4. Improved processes and reduced duplication 5. Relationship-specific measures tied to team incentives 6. Shared authority 7. Openness, honesty, and increased risk sharing	• 40% reduction in man-hours needed per project completion (identical projects) • 17% reduction in staff man-hour/craft man-hour ratio • 21% reduction in staff payroll expense/craft payroll expense ratio • 10% improvement in worker utilisation rate • 10% reduction in overall project cost • 100% success in meeting budget/time • 50% reduction in engineering rework • 50% reduction in sales expense
Coalescing (high objectives alignment)	1. One common performance measurement system 2. Cooperative relationships supported by collaborative experiences and activities 3. Cultures integrated and directed to fit the application 4. Transparent interface 5. Implicit trust and shared risk	• 15% reduction in equipment and construction cost • 33% reduction in engineering rates • 100% acceptance of risk by the owner in exchange for a low fee charged by the engineer

Source: Adapted from Thompson and Sanders (1998) with permission of the *Journal of Management in Engineering – American Society of Civil Engineers/Engineering Management Division*

Table 3.10: Australian partnering forms

Partnering type	Partnering description	Partnering outcome
Experimental partnering	• Charter, workshop, small number of follow-up meetings • Usually first partnering experience • Minimally resourced • Often seen as a 'toe-in-the-water' exercise	• Often unsuccessful, generally because of lack of clear understanding, commitment and structure
Packaged partnering	• Offered as part of a contractor's tender or imposed upon the contractor after the tender is accepted • Often involves only the client and contractor • This model is used very successfully as a marketing tool	• Problems may arise from lack of commitment and understanding of each stakeholder's objective • A client–contractor relationship perceived to be cooperative at the outside of a project may not necessarily last for the duration of the contract
Committed partnering	• Often developed as a result of first, unsuccessful experience • Incorporates as many stakeholders as possible in a tight, well facilitated dispute resolution mechanism • Well resourced	• Problems may arise from lack of commitment and understanding of each stakeholder's objective. • A client–contractor relationship perceived to be cooperative at the outside of a project may not necessarily last for the duration of the contract

Source: CIIA (Lenard, Bowen-James *et al.* 1996, p. 17). With permission of the Construction Industry Institute of Australia

environment in which disputes can be resolved amicably without the usual levels of hostility, gamesmanship, posturing, and generally responding with ambit claims. With the Level 3 approach, there should be evidence of a paradigm shift from project participants seeing themselves as being in separate teams to being groups who are part of a single project team.

Table 3.12 illustrates how developing a partnership relationship has been categorised as a five-phase process at the Level 3 partnering relationship. It is important to stress that parties should be planning for expected success rather than potential or inevitable failure. Also, active support from senior level management is essential for the relationship success.

3.6   Requirements for Level 4 – synergistic strategic partnerships

Figure 3.6 provides an illustration of the requirements of such high-level partnerships and the business relationship maturity and sophistication needed. In this model, four clusters of factors and two sets of processes contribute to a high level of maturity and readiness to abandon the short-term view of a vendor and reach beyond the limitations of the supplier mentality (Lendrum 1998, p. 23).

One interesting feature of the model is the workplace reform element. Strategic partnering/alliancing requires a number of workplace infrastructure elements to be in place before it can be successfully developed. These are underpinned by a quality management focus. Workplace reform measures and total quality man-

Table 3.11: Levels of Partnering and ADR

Level 1 Adversarial Arm's length Contractual	Level 2 Collaborative Team-oriented	Level 3 Value-added Integrated Team	Level 4 Synergistic Strategic Partnership
• Competition • Each side has clearly established responsibilities • Client monitors and inspects contractor • Little or no trust	• Cooperation • Each side knows and commits to the goals of the project and to each other's goals – requires degree of trust	• Collaboration • One integrated team consisting of both client and contractor personnel is created – requires high trust • This team has one set of goals for a successful project • Team often creates a separate organisational entity for the life of the project	• Coalescence • Elements of shared risk also defined • Joint sharing of liabilities for project failure • Joint sharing of gains from project success • Both sides share their goals and cost – requires extremely high trust
• Often adversarial • Often creates disputes, sometimes litigation	• Significant energy in communications and 'win–win' conflict resolution • Disputes typically resolved in some degree of compromise and harmony	• Accountability is collective among the integrated team • Both client and contractor provide senior level 'sponsors' to remove barriers and support the project	• Curve on benefits is logarithmic – based on meeting and then exceeding project goals • The essence of the relationship is to increase the mutual profitability of both parties • Neither at the expense of the other • Both at creating new synergistic solutions
• Both sides plagued by schedule slips and cost over-runs	• Established for early positive intervention • Projects often accomplished on schedule and within budget	• Typically includes some incentive for exceeding project goals	• Requires extensive communication, collaboration and organisational commitment and sponsorship • Creates the opportunities for major breakthrough

Source: Adapted from Ellison and Miller (1995, p. 46) with permission of the *Journal of Management in Engineering –* American Society of Civil Engineers/Engineering Management Division

agement (TQM) philosophy for excellence facilitate the liberation of creative energy of the workforce and technicians undertaking hands-on project work. So quality management needs to support continuous improvement, innovation and breakthrough invention. This requires an attitudes and values cultural infrastructure to facilitate high-level performance including nurturing by a management environment and leadership characteristic that allows creative energies to flow. The theoretical basis for this requirement is extensively discussed in Chapter 7. This concentrates on workplace and management flexibility. It aspires to more than maintaining multi-skilling to achieve lower costs and strives beyond this

Table 3.12: Developing a partnership relationship	
Phase	**Actions**
1. Needs analysis	• Describing the current status of the project • Defining the roles of key participants • Defining potential opportunities and liabilities • Developing a framework including guidelines or criteria for the work
2. Partnership structure and scope	• Identifying the core structure • Naming the principal contacts • Establishing a charter with mission goals, and objectives: roles, responsibilities, and formal authority; and incentives to meet and exceed goals
3. Relationship with other stakeholders	• Defining the roles of major subcontractors, outside agencies, community organisations, decision makers • Identifying the means of minimise disputes and to build compromise solutions
4. Sharing risk/ rewards	• Identifying contractual issues and defining the relationship among the various stakeholders • Establishing the tools for both measurement and sharing of liabilities • Defining the incentives for measurement and sharing gains/ liabilities
5. Continuous improvement	• Joint assessment of progress • Evaluation of changing needs and expectations • Analysis and application of lessons learned • Prescribing actions to respond to changes, correct course, and seize opportunities

Source: Ellison and Miller (1995, p. 46–47) with permission of the *Journal of Management in Engineering* – American Society of Civil Engineers/Engineering Management Division

empowering workers to develop their skills so that they can offer advice on developing and implementing innovative processes and product improvement that contributes to breakthrough innovations/inventions. This model is based upon a paradigm shift from an us-and-them approach to firm internal partnership where gains are shared and increases in productivity rewarded.

An infrastructure of external suppliers and a client that understands support infrastructure requirements for this level of partnering is also needed. There needs to be a focus on external customers in the supply chain with trust and commitment and world-class practice standards prevailing. Supply chain management for value and the intertwining of customers, suppliers and subcontractors is well advanced in the automotive, aerospace and other manufacturing sectors (Hamel *et al.* 1989; Womack *et al.* 1990; Hamel and Prahalad 1994; Doz and Hamel 1998). Thus, there is a need for sophisticated clients, suppliers and subcontractors as well as sophisticated project managers that can optimise the supply chain as well as managing the alliance relationship. This requires high-level technical/process and human relations skills.

There also needs to be an underpinning technology infrastructure to enhance communication, planning and decision making, and to support control of the project – using the advantage of flexibility of action through the support of the technology infrastructure. The **internet** is proving to be a significant enabler for electronic communications with electronic data interchange (EDI) to allow just-in-

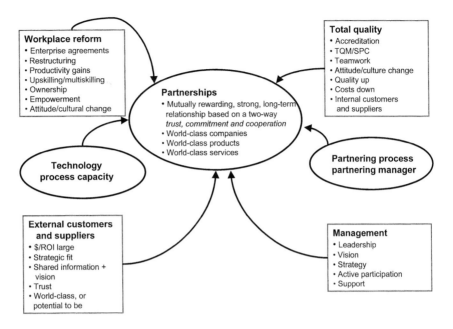

Figure 3.6:   Strategic partnering requirements
Source: Lendrum, T. (copyright © 1998) *The Strategic Partnering Handbook*, McGraw-Hill, Sydney.

time supply not only for physical products but also for information products. **Extranets**, for example using web-based systems, are useful for sharing design development information and more mundane but essential information such as requests for information (RFIs) (Roe and Phair 1999).

The management process requires a high level of partnering management to create and sustain partnering/alliance relationships. Earlier sections in this chapter detailed results of empirical studies supporting this view. Partnership management and leadership practices need to support partnerships or alliances. Figure 3.6 represents a considerable paradigm shift for most industries – the construction industry is no exception – where shifting away from selecting the cheapest tender to a criterion of best value. Calls for this shift are growing in many countries (CIDA 1994; Latham 1994; Hampson and Kwok 1997; DETR 1998; KPMG 1998).

3.7   Reasons for embarking on Level 4 – synergistic strategic partnerships

Drago, in evaluating strategic alliances in the USA IT industry states that '... The roles of each partner must be understood and actions taken through the alliance must be managed and monitored to the satisfaction of all parties. This, in turn, can require considerable time and effort from managers within the organisations of the alliance ... each partner must give up some flexibility over that part of its domain to reduce either external or internal uncertainty.' He also offers observations for the IT industry on why strategic alliances make sense (Drago 1997). This has particular relevance to construction industry suppliers and subcontractors because they tend to be at the end of the queue in terms of influencing designers and the client. Further, they have much to offer in innovative potential but often they fail to grasp opportunities to convince clients or the design teams of the merits of their propositions. Project partnering and alliancing provide the potential for such opportunities. Drago's observations follow.

(1) Small organisations are more likely to enjoy the benefits of strategic alliances than larger ones – due to their lack of resources to 'go it alone', less access to knowledge, and less influence over their competitive environment.

(2) Organisations competing in highly innovative industries or industrial segments gain greater benefit than others – through achieving greater access to customers and increased ability to create the 'rules of the game'.

(3) Firms producing component parts for larger technological systems can *both* reduce uncertainty for product specialisation *and* increase their influence with product design decision makers.

(4) Firms competing with 'pure' innovation strategies are more likely to enjoy benefits that those with 'limited' innovation – due to greater market uncertainty in developing and bringing innovations to market.

(5) Firms with goals of becoming 'technology sponsors' are likely to enjoy significant benefits through alliancing – through having the opportunity to develop industry default standards on their own terms thus reducing uncertainty rather than having an open system of giving away intellectual property to influence accepted standards.

(6) Firms entering new markets are more likely to enjoy benefits – by spreading risks for greater market and operational uncertainty.

(7) Firms that generally suffer from a lack of critical resources can partially overcome this constraint – by sharing resources with others to mutual advantage, for example R+D facilities, key staff, training facilities' etc.

Full alliancing (Level 4 strategic partnerships) with its selection process first based on service criteria, then on price consideration as a second order issue (in the assumption that best value will follow) has its sceptics. In a recent article in *Building Australia* reporting upon a seminar by KPMG undertaken on project alliancing, the NSW Auditor-General is quoted as stating that from the NSW Government's point of view, it has to justify projects on the basis of cost. 'If, as in project alliancing, you enter a project and don't know the cost or do not have the capacity to manage the risk of cost over-runs, it would be hard, as a government agency, to justify the decision to undertake the project on such a delivery method' (Pratley 1999, p. 37). Pre-qualification issues are nevertheless seen as vitally important for those with alliancing experience. Eric Kolatchew,[3] is quoted as saying that 'an intense review needs to be carried out of the pre-qualification organisations, taking into account the way they operate, previous jobs, the opinions of past and present clients and compatibility of the key personnel proposed' (Pratley 1999, p. 35).

Many of the alliancing projects recently reported upon have featured an innovative pricing and cost structure methodology which is a radical departure from

---

[3] Project Director for the BHP Port Hedland HBI project – judged an unsuccessful alliance project.

the business-as-usual case that has attracted so much criticism. It exhibits and demonstrates a radical paradigm shift in owners' attitudes. In the traditional procurement as well as alternative forms discussed earlier, the owner/client clearly sees project delivery in product terms. A price is determined for the end product – the project. Even in many forms of partnering, the output is seen as a finished product. The exciting change in perception of alliancing is that professionals centre the contractual focus on service delivery, which delivers both product and knowledge. The theory behind this paradigm shift is that creative knowledgeable professionals, when brought together in a synergistic environment will create innovative or breakthrough solutions that deliver better value for money than a lowest-cost design and constructed solution. Additionally, the creative process will not only enhance the project's delivery solution but also deliver new techniques, knowledge and practice to the industry. Thus, the client/ owner gains from this process – organisations exposed to this process gain both organisational and personal knowledge for those involved and the industry in general gains.

3.8   Developing a strategic alliance relationship

A continuum of partnering relationships was introduced in Table 3.10 and Table 3.11. It can be appreciated that this continuum is a function of the degree of *joint* rather than *shared* commitment of parties undertaking a project. In non-partnering or non-alliancing arrangements, parties may share a commitment to project goals. Indeed on most projects there is an incentive for teams to work together, namely of achieving project success. Under those arrangements, and indeed under partnering, one team may 'sink or swim' without necessarily affecting the business position of other teams. One team may make profits from a project while other partnered firms/teams may actually make a financial loss.

With alliancing, there is a *joint* rather than *shared* commitment. Parties agree their contribution levels and required profit beforehand and then place these *at risk*. If one party in the alliance under-performs then all other alliance partners are at risk of losing their rewards (profit and incentives) and could even share losses according to the agreed project painsharing/gainsharing model. Thus, alliance members form a quasi-joint venture because they operate at one level as a single entity, however, they do not merge their companies in any legal or official way. They remain truly independent companies but they must help each other satisfy **key performance indicators** (KPIs) to realise the rewards at risk. This provides a powerful incentive to achieve project goals – indeed to perform beyond expectations where incentive schemes encourage them to do so.

The important distinction between partnering and alliancing is that with partnering, aims and goals are agreed upon and dispute resolution and escalation plans are established, but *partners still retain independence* and may individually suffer or gain from the relationship. With alliancing the alliance parties form a cohesive entity, that *jointly shares risks and rewards to an agreed formula*. Thus if the project fails to meet agreed project KPIs then *all partners* jointly share the agreed penalty. Rewards are likewise bestowed for successfully exceeding expectations. Risk and reward issues are pivotal in providing immediate monetary or financial motivation to meet or exceed KPI's on alliancing projects. Several variations on a consistent model are evident from the literature (KPMG 1998; ACA 1999) and are illustrated in Figure 3.7.

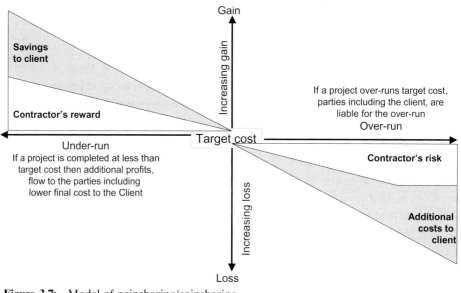

Figure 3.7:   Model of painsharing/gainsharing
Source: Australian Construction Association (ACA 1999, p. 19).

In one case of an alliancing project, the Australian National Museum, the project partners developed the project budget based on a cost plus formula of construction costs plus project preliminaries and a corporate profit margin. Budgeted profit was based upon profit levels over several years past plus an agreed bonus level. Past profit levels were independently audited. The preliminaries and profit margin were then placed *at risk* subject to acceptable project performance. The project budget process involved developing rigorous and challenging key KPIs. If these are met then the actual construction cost plus agreed overhead and profit levels will be paid to alliance partners. If KPIs are exceeded the alliance team shares a bonus based upon a predetermined agreed formula (Walker *et al.* 2000). This arrangement is similar to other alliancing projects, for example the Wandoo gas field project (KPMG 1998, Appendix 2; Pratley 1999, p. 34). This is an interesting approach as it encourages different alliance partners to help each other achieve the KPIs and when problems in achieving these become evident there is a group incentive to work collaboratively towards obviating problems to overcome the problem and achieve or exceed the KPI.

Strategic alliances are effective when partner selection is based upon a world-class product or service (Lendrum 1998). The alliance partners for the Australian National Museum project were selected for their expertise and ability to meet stringent performance criteria *before price issues were considered*. The successful alliance was required to demonstrate itself to be a trustworthy, committed and world-class group of professionally competent firms. They were then invited to join with the owner/client to develop the project. In doing so they formed an alliance of talented professionals, pooling resources to achieve the project goals. This alliance then fully developed a project price target through design development and agreed upon the risk and reward sharing arrangements. The client ensured value for money by independently verifying the project estimate. Expected cost savings were derived from improved value for money being obtained through leverage of skills and expertise of the alliance partners in developing the project concept through to delivery. Lifecycle cost considerations and value engineering exercises also featured as cost management techniques applied.

Strategic alliancing requires top level commitment and a management structure in place where a senior manager accepts responsibility and commitment not only to develop the necessary alliance relationships but also to maintain them (CII 1996; Lenard *et al.* 1996; KPMG 1998; Lendrum 1998; ACA 1999). The alliancing manager needs to act with integrity to engender trust and openness so that issues can be brought out into the open and investigated rationally without intimidation, and that parties feel that they are not likely to be exploited or manipulated. Alliances are more focused in ensuring that *all* allied companies perform as opposed to partnering arrangements. The typology of relationship maintenance in Table 3.13 provides useful guidance here.

Teams of people ensure that project goals are realised – that is why the alliance arrangement requires the paradigm shift illustrated in Figure 3.6. Therefore, there is a need for a programme of workplace reform that ensures removal of barriers that inhibit individuals from working cooperatively. Workplace agreements, for example, should enable and support flexibility of job roles. Thus, skills, training and wider task responsibility and innovation support processes can challenge more rigid industrial award provisions and expectations (Lendrum 1998, p. 22). Building solid relationships with all members of the supply chain, from supplier through to client, is important to strategic alliancing because there are many synergies that can be capitalised upon – such as sharing administration systems that ensure IT compatibility aids communication effectiveness. Sharing information more broadly about customer needs and feedback from parts of the supply chain together with creative ideas are a key feature of strategic alliancing (Doz and Hamel 1998).

Alliancing is highly strategic, connecting the activities of world-class operators with complementary skills not only to manage risk, but also to encourage and develop incremental improvement through both innovation and breakthrough inventiveness. The way in which networks of firms come together and ways they interact provide both added value and value for money to the client – this replaces cheapest capital cost as a prime objective.

An important aspect of the partnering or strategic alliance philosophy is innovation, capture of knowledge and lessons learned. Innovation has been identified as a critical need for today's construction industry (Hampson 1993; Lenard 1996; Hampson and Tatum 1997). Similarly, the need for procurement systems to target organisational learning and knowledge as a project output has also been proposed elsewhere (Walker and Betts 1997; Love *et al.* 1999; Walker and Lloyd-Walker 1999) and more generally for other industries (Argyris and Schön 1978; Hamel *et al.* 1989; Kanter 1989; Senge 1992; Field and Ford 1995; Argyris and Schön 1996; Pedler *et al.* 1996; Savage 1996; Sveiby 1997; Wind and Main 1998).

Lendrum (1998, p. 123) advances a mode, illustrated in Figure 3.8 of creating effective alliance partnerships. This model illustrates how creating an effective alliance can be instigated to achieve 12 motivators identified in Figure 3.2.

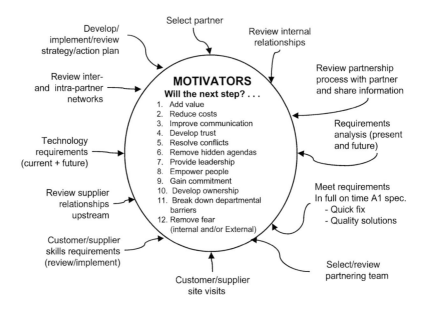

Figure 3.8: Creating effective alliance partnerships
Source: Lendrum, T. (copyright © 1998) *The Strategic Partnering Handbook*, McGraw-Hill, Sydney.

A matrix of key selection criteria for potential alliances should be prepared before selection takes place. The Australian National Museum project is a prime example of this process. The following 12 criteria were prepared to assist selection of the project alliance. Potential alliance partners were required to respond to these by demonstrating, through citing evidence from past projects, their attainment level for these criteria. The criteria listed below are explained more fully in Chapter 4.

(1) Demonstrated ability to complete the full scope of works including contributing to building, structural mechanical and landscaping design.
(2) Demonstrated ability to minimise project capital and operating costs without sacrificing quality.
(3) Demonstrated ability to achieve outstanding quality results.
(4) Demonstrated ability to provide the necessary resources for the project and meet the project programme.
(5) Demonstrated ability to add value and bring innovation to the project.
(6) Demonstrated ability to achieve outstanding safety performance.
(7) Demonstrated ability to achieve outstanding workplace relations.
(8) Successful public relations (PR) and industry recognition.

(9) Demonstrated practical experience and philosophical approach in the areas of developing ecological sustainability and environmental management.

(10) Demonstrated understanding and affinity for operating as a member of an alliance.

(11) Substantial acceptance of the draft alliance documented for the project including related codes of practice, proposals for support of local industry development, employment opportunities for Australian indigenous peoples.

(12) Demonstrated commitment to exceed the project objectives.

Several of the points clearly relate to traditional project management excellence criteria: quality, cost and time. It is interesting that the criteria for the Australian National Museum project include KPIs that relate to the 'triple bottom line', which has been offered as the future direction for responsible corporate governance. The triple bottom line includes reporting on KPIs related to environmental and community stakeholders as well as financial performance (Elkington 1997; Walker 2000). The above criteria move beyond the limited measures for traditional project management success to a broader model of excellence.

The above has shown that considerable attention is required in choosing an alliance partner. There is an urgent need to confront and address the critical 'soft' management issues such as alliance relationship quality as well as some of the more traditional performance measures noted within the 12 point criteria for the Australian National Museum project. The significant differences between project alliancing and partnering is that the determination of the successful team is based upon:

- demonstrated performance ability rather than price;
- the budget being only finalised after the alliance team is appointed;
- the risk/reward structure being similarly determined only after selection of the alliance team;
- the alliancing principles being part of the contract; and
- the operational phase of the project involves open-book accounting with a shared risk and reward formula being applied.

Most partnering and alliance literature stresses the importance of selection of suitable partners. It is usual to shortlist these from a group of potential alliance partners (based on a set of selection criteria such as the Australian National Museum project 12 point criteria) and only to interview one or two alliance groups. Figure 3.9 illustrates the alliancing selection process.

This is a radical departure from other forms of procurement – despite its similarity to cost plus or on-call multi-task contacting. One of the more tangible outputs from the initial stages of team establishment is the final alliance agreement. This is very similar to a partnering charter. The aim is to specify the goals and values that will govern the relationship between participants. Lenard *et al.* caution against a charter being drafted in a way that conflicts with contracts between parties. In the CIIA study 86 per cent of participants involved in failed partnering relationships cited a conflict between the partnering charter and work contracts (Lenard *et al.* 1996, p. 21). Partnering agreements and strategic alliance charters flow from an agreed set of principles arising out of the partnering workshop.

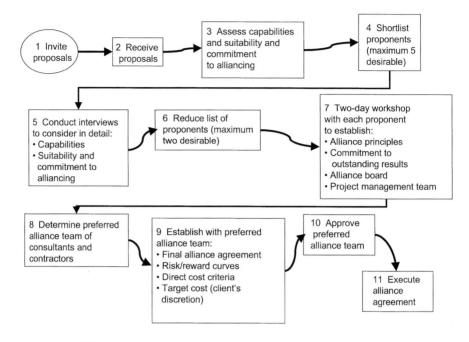

Figure 3.9: Alliancing selection process
Source: Adapted from KPMG (1998, p. 25) with permission of KPMG.

Figure 3.9 illustrates in step 7 of the flow diagram the two-day workshop in which an independent facilitator helps the client and other participants develop the set of principles that will provide the underlying work culture that will define the project's team working environment. The preferred alliance team will emerge and this team together with other consultants and/or contractors that are brought together in the alliance will establish an alliance board, project management team and develop measures to be used to gauge commitment and quantify required results. This method is very similar to that described for a project in the UK developed by a management contracting organisation with subcontractors and suppliers in which final pricing was agreed at the end of the process (Matthews *et al.* 1996). KPMG recommends that 'the workshop be used to provide the client with the opportunity to gauge the suitability of the contending teams to success-fully work together in an alliance environment to see whether:

- The client believes that the team can be trusted;
- The teams have people with the right leadership and culture;
- There are any weak links within the team;
- The individuals have a proper appreciation of the project; and
- There is enthusiasm for the concept of alliancing' (KPMG 1998, p. 27).

Once the alliance team is selected, it will agree target costs for the project and the gainshare/painshare formula with the client. The client will generally be advised by external consultants to ensure that probity and value for money are maintained. The difference between partnering and alliancing is evident at this stage because typically partnering agreements are established and negotiated after individual project partners have tendered or negotiated their separate contracts (Larson 1995). The purpose of the partnering agreement is to establish a non-adversarial and generally civilised way of working together to achieve mutually agreed project goals. Thus, the partnering charter determines an organisational culture and determines the relationship climate or framework within which the project team will work. Typically, escalation paths for dispute resolution are established to enable disputes that are unable to be resolved at one level of management to be escalated upwards through the firm's hierarchy so that it can be resolved without having to resort to legal action taking place (Weston and Gibson 1993; Larson 1995; Miles 1995; Lenard, *et al.* 1996; Thompson and Sanders 1998).

With strategic alliances, the alliance takes group responsibility for developing the project costs and maintaining that cost or even improving upon it. Disputes and disagreements are treated in a similar manner to partnering but the added incentive of alliance partners' sharing gain or pain, as a group is more effective – see Figure 3.7 for an example. It is in the interest of all alliance parties to help a struggling member perform at a high standard in order that all may share in any bonus. It is an all or nothing case that encourages genuine cooperation.

The Australian National Museum project alliance charter provides a good example of what might be agreed upon by alliance partners.

This Alliance will create an exceptional Australian cultural precinct on Acton Peninsula, Canberra. The Project is the flagship for the Century of Federation and will be a source of pride for all Australians. The way we go about it will lead the way for construction projects in the future. We are therefore committed to the following principles.

We are committed to:
- Continually strive for innovation and breakthroughs;
- Honest, open and ethical communications and actions;
- Timely and forthright resolution of all issues;
- Equitable risk and reward;
- Public accountability and good governance;
- A culture of responsibility;
- Listening with intensity and speaking with responsibility;
- Supporting all team members;
- Collective ownership of decisions;
- Trust, integrity and respect; and
- Achieving a Balanced Quality of Life.

It is interesting that the above agreement should encompass aspects of community values and in particular the balanced quality of life issue. These are part of the concepts expressed in an emerging movement for companies to aim beyond maximising the financial bottom line by creating value for shareholders. The triple bottom line concept aims for firms to also achieve sustainability and social responsibility measures of success (Elkington 1997). The alliance charter and the 12 selection criteria listed earlier indicate a conscious pursuit of a broader agenda for success. It is also interesting that the Australia Commonwealth Government, by instigating the Australian National Museum project using an alliancing delivery mechanism, also fully accepts the right for companies to make reasonable and normal profit from their involvement in the project. The alliance reward system allows for normal corporate profit to be paid on the open-book cost reporting of project costs and site management overheads for achievement of project cost and quality targets with added rewards for exceeding these targets.

3.9   Sustaining a strategic alliance relationship

As competing transforms into the coalescence of the relationship development process, a fully developed partnership maintenance program needs to be developed with the most senior level of management support.

The importance of each of the elements of a strategic alliance, its initial development as well as its maintenance has been briefly outlined above. It is important to understand that much of the alliancing concept revolves around alliance partners as a group collectively taking responsibility for the success or otherwise of a project. This is manifested through a shared risk and reward system predicated upon a 'no dispute' philosophy where firms use their energies by working together to solve problems rather than attributing blame.

This feature of parties agreeing not to sue each other as a means of settling differences of opinion has raised concerns (KPMG 1998; Uher 1999). There is significant doubt about the legitimacy of requiring one party to sign away their right to legal redress. Generally, in partnering/alliance arrangements such as the above, only in the event of 'wilful default' does a party have an express legal cause of action against another participant under the terms of the agreement. Typically wilful default is defined as (KPMG 1998, p. 33):

> A deliberate or intentional failure by a participant (or a director, employee, officer or subcontractor of a participant) to:
> (A)   perform a legal or contractual duty; or
> (B)   take proper action when such action is required,
> having regard to what is reasonable in all circumstances and the failure is persisted in with reckless disregard as to the likely consequences of such failure; but such wilful default does not include:
> • an honest mistake;
> • mere oversight, inadvertence or error of judgement; or
> • an accidental, involuntary or negligent act or omission,
> made in good faith by a participant (or a director, employee, officer or subcontractor of a participant)

Clearly, trust and transparency are key issues that need constant attention to ensure effective relationship maintenance.

Trust is enhanced by an open-book philosophy whereby partners and the client representatives have total access to inspect any partner's costs, time, occupational health and safety (OH+S), quality and other project information. This helps to clarify and support 'facts', to facilitate cooperation and open communication and joint problem solving. Alliances, and indeed all partnering-type relationships, need to understand the nature and quality of their internal communication systems to determine whether the organisation has what it takes to develop and maintain customer–supplier partnerships. Using a technique of mapping models similar to Figure 3.10 can assist in identifying potential relationship problems and trigger the development and operation of relationship maintenance plans.

Objective - Understanding the nature and quality of internal communication, to determine whether your organisation has what it takes to develop and maintain customer/supplier partnerships.						

Groups	A	B	C	D	E	F
A						
B	++					
C	$$	!!!				
D	$$	++	$$			
E	!!!	?	$$	++		
F	?	?	!!!	$$	!!!	

!!!	A disaster !
++	Needs improvement !
$$	Works well !
?	A bit of a gamble !

Process:
1 Draw up map (as above)
2 Involve a valid sample sized group to map relationships
3 Share information results widely to test and to action response
4 Work on understanding WHY the results are so - research!
5 Work on deep and rich research to understand fundamentals
   such as trust, loyalty, commitment, barriers and drivers etc.

Figure 3.10:   Relationship mapping
Source: Adapted from Lendrum (1998, Chapter 6).

Mapping not only the relationships between alliance partners but also documenting the quality of their relationships facilitates the establishment of targets. It also serves to assist monitoring relationship maintenance plans and improving them.

It is clear from the above map that there are groups with strained and fragile relationships. The next step that an alliance would take after an audit revealing this kind of result would be to further investigate the drivers and inhibitors of the recorded situation to develop plans and actions remedying the situation. Often this involves numerous follow-up workshops undertaken throughout the project delivery phase. Other team-building measures may also be considered. Managing alliance relationships to maintain creative energies released through the alliance formation is a crucial task of the alliance's team leaders.

3.10   Case studies in project alliancing

The concept of strategic alliances has been operating for several decades in the manufacturing industry. The automotive industry provides the first glimpses of how they might function. Generally, the major shift from a procurement system in which many hundreds if not thousands of suppliers and subcontractors are used to produce an end product for a particular project has been well reported (Womack *et al.* 1990). Other industries such as the aerospace industry, the air service providers and the electronics industries have provided well documented case studies for analysis (Hamel *et al.* 1989; Hamel and Prahalad 1994; Doz and Hamel 1998). Moreover, a wide range of providers of services and products have formed strategic alliances (Ferdows 1997; Porter 1998). Few global examples of these occur in the construction industry, which appears slow in grasping opportunities offered through strategic alliances. Strategic partnering has, however, been a feature of the construction industry long enough to be able to report on case studies and to comment on their implementation. It is useful to include both partnering and alliancing together in this exercise because a core element of both of these is trust and commitment as well as a well formulated and agreed method of resolving differences in opinions and direction and resolving disputes.

The CII in the USA has proved a useful vehicle for project partnering relationship investigations. Table 3.10 summarised and commented upon one set of findings (CII 1996; Thompson and Sanders 1998). Examples of similar studies of the US Army Corps of Engineers involved 19 partnering projects (Weston and Gibson 1993) with another study of partnering reporting on 280 construction projects (Larson 1995) as discussed earlier. Interesting insights gained from these have been commented upon earlier (see Figure 3.4 and Table 3.9). Australian case studies were also reported upon with detailed commentary on results presented from a CIIA study of 32 Australian projects (Lenard *et al.* 1996).

Each of the above confirmed and highlighted the benefits to be derived from improving the working relationship of project team members, including the client representatives. Each study highlighted the need for trust and commitment with agreement on project goals and civilised resolution mechanisms for disputes and disagreements. Inspection of the web sites of several companies widely reported to have strategic alliances or strategic partnering relationships with clients reveals at least one example from the USA

http://www.bechtel.com/aboutbech/partners.html

and the UK

http://www.bovis.com/Services/srvAlliances.asp?ServiceID=Alliances

Studies of alliancing projects were also reported upon. The ADR continuum was presented as levels of partnering and again this was based upon case study work of the US Corps of Engineers experience (Ellison and Miller 1995). The Australian study conducted of alliancing with the Queensland Government was also reported upon (Hampson and Kwok 1997; Kwok 1998). The KPMG report to the Government of New South Wales (KPMG 1998) also provides useful insights in the concept and practice of alliancing in Australia. The experience of BOOT projects is also of value as BOOT partners undertake a relationship-based approach in forming the BOOT consortium that uses many of the features of partnering and alliancing in forming the project delivery entity. Shepherd draws interesting parallels between BOOT projects (referring to them as public/private partnership projects) and alliances. In citing the Melbourne City Link projects he suggests that alliancing principles may be applied in the selection of those delivering BOOT projects (Shepherd 1999).

The literature cited in this book suggests that partnering has been successfully used on civil and process engineering projects as well as general building projects. The evidence on project alliances indicates that while an increasing number of engineering projects have been completed, the approach has not been attempted for general construction other than the Australian National Museum project. A study of contracting relationships and competitive advantage in public sector projects in Queensland (Kwok 1998) appears from the data presented to resemble strategic partnering more closely than strategic alliancing as it is now more generally being understood (Bennett and Jayes 1995). The distinction between advanced level or synergistic strategic partnering (Ellison and Miller 1995), coalescence as opposed to collaboration (Thompson and Sanders 1998) and strategic partnerships as described by Lendrum (1998) provides a blurring between partnering and alliancing that is often confusing.

The report provided by KPMG (1998) provides a more useful distinction between partnering and alliancing. The key distinction appears to be the way in which the client undertakes the alliance selection process. In project alliancing, the client seeks a contract for services between the client and the provider on the basis of selection by proven performance against stringent performance criteria followed by development of committed cost limits and gainsharing/painsharing formula to ensure value for money (KPMG 1998). Using this distinction, the Australian National Museum project would appear to be the first project alliancing building construction project to be undertaken in Australia, possibly the world during the twentieth century. There are no building construction case studies that can be used to compare project performance with the Australian National Museum project. There are, however, several engineering projects that have been undertaken using alliancing principles and practices. These will be briefly described and commented upon to help illustrate the nature and experiences of alliancing that can help us better understand how alliancing can function.

3.10.1   Wandoo B Offshore Oil Platform – Western Australia (*Sources:* KPMG and Australian Construction Association (KPMG 1998; ACA 1999))

The oil and gas industry has a history of recognising the advantages of strategic alliances (Doz and Hamel 1998). Ampolex, Brown & Root, Keppel Fels, Leighton Contractors, and Ove Arup & Partners formed an alliance to develop a West Australian offshore oil field located in 55 metres of water. It was a complex project with engineering, construction and development uncertainties to overcome. There appears to have been a need for very rapid deployment of a project team to develop the field. There was also a group of companies, including Ampolex, evidently prepared to take on the risk of undertaking the project on an alliance basis.

Each participant had valuable knowledge and expertise to contribute in finding an intelligent win–win solution to the problem of rapid development of the oil field. Sanction and stretch targets plus a gainshare plan were developed. Stretch targets are intentionally difficult and demanding targets were set to challenge team participants to develop innovative ways in which to achieve objectives. The purpose is not simply to make participants work harder or longer hours but to develop innovate inventive ways of achieving these targets.

The final gainshare of A$16 million proportions were agreed as Ampolex (50% per cent), Brown & Root (20 per cent), Leighton Contractors (16 per cent), Keppel Fels (12 per cent), and Ove Arup & Partners (2 per cent). The project was designed, constructed and commissioned in 26.5 months (against an industry norm of 34 months) with a sanction target cost of A$377 million, stretch target of A$305 million and actual cost of A$364 million. Safety was not compromised and it is interesting that the sanction target of no Class 1 injuries was met.

The main rationale for alliancing was that a small team could effectively deliver the project at world-class standards of performance, very quickly and with enthusiasm. The client and client representative were sophisticated and also able to demand and verify world-class standards of project delivery. The project was able to deliver an environment where innovation and excellence prospered. All parties were expected to benefit. In the event, the alliancing arrangement delivered above expectation, benefits were shared accordingly. This project proved to deliver a project under budget to all quality and performance expectations. It achieved this in a faster than expected duration in a win–win atmosphere of collegiate representation on a Project Alliance Board. This board was able to sanction decisions within a framework of trust and commitment without exploitation by one party over another.

3.10.2   The Andrew Drilling Platform – North Sea UK – (*Source:* KPMG 1998)

This project provided a model for Wandoo B to follow. The project began its existence in 1990 with a focus on ensuring project viability in a deteriorating business environment for oil extraction in the UK North Sea oil and gas fields. The challenge was to deliver a cost effective oil platform. Over the course of 1991 the conceptual platform design was defined and developed in cooperation with a leading engineering contractor at an estimated cost of £450 million – too high to be commercially viable for the oil field.

The challenge was to find a way of delivering the project to British Petroleum (BP) at a cost that could be sanctioned and still be profitable for those concerned in delivering the project. The approach was to develop a set of 10 minimum conditions of satisfaction (MCOS) against which to judge prospective companies offering proposals for the design and delivery of platform facilities and subsea hardware and pipelines. These extended the usual high levels of technical competence to include relationship factors now more readily recognisable as alliance relationship requirements. The tender document was brief (50 pages) requesting performance rather than prescriptive delivery of BP-stated engineering solutions. Brown and Root identified potential capital savings of at least 20 per cent and was able to stipulate its own satisfaction requirements for negotiation. At the end of 1992, Brown and Root's proposals were successful (based upon the 10 MCOS alone) and did not include a financial evaluation for the Andrew's topsides, jacket and subsea design, plus procurement and project management support. This was to define a distinct difference in approach to that adopted in strategic partnering and other cooperative forms of project delivery.

As the project development progressed additional contractors joined the alliance each bringing their own creative and critical thinking abilities to contribute to the project design and delivery solution. The pre-sanction estimate – the target costs that determine a project's ability to justify itself as a business case – is an important and critical milestone in delivering engineering projects. The alliancing approach also included an agreement to challenge and interrogate suggestions as they emerged so that the most intelligent solutions survived scrutiny. This demanded a highly professional approach by all concerned and a lack of being 'precious' about standard or 'business as usual' approaches. The result was that the design encapsulated detailed planning and feasibility analysis as a part of the decision-making process. It appears to have been effective in transferring some of the cost of estimating to forming the alliance. This approach was evidently worthwhile to the participating contractors as other transaction costs were minimised through using a collaborative approach. For example, the project's commitment to minimising field inspections, eliminating expediting, pursuing functional specification and reducing vendor documentation was achieved leading to a radical new strategy for interfacing with suppliers.

A 30 per cent reduction across the project total acquisition cost for equipment and materials was aimed for. This allowed Andrew to be properly planned and a more focused and reliable estimate to be produced. Also, the presence of only 13 BP personnel integrated into the team reinforced the fact that Andrew was definitely not a business-as-usual approach for BP. The more enlightened management approach of collaborating with BP as a partner bore fruit because contractors took responsibility for their actions and proposals as well as determining self-imposed targets and standards. Trust and commitment flourished and the results were demonstrable.

The Andrew platform and pipeline facilities estimate was £334 million with a £39 million contingency sum as opposed to the previous BP estimate of £450 million. BP's cost and risk experts subjected the alliance estimate to an independent check and their assessment was that the project had a 38 per cent probability of being completed for that price (based on their historical data for the North Sea). BP would normally expect a 50 per cent probability as an acceptable sanction figure and it would traditionally bear additional cost over-runs. The alliance partners had proposed that under the risk and reward approach they would share an over-run should it occur, up to a cost of £50 million. At this additional cost the probability of achieving the estimate rose to 80 per cent but at this level, the contractor's profits would be outweighed by their expenditure. Their commitment won the alliance their proposal. Relative shares of responsibility are illustrated in Table 3.13 below.

Each of the alliance contractors entered into an individual commercial contract with BP, against well-defined scopes of work, totalling some £217 million. Payment methods followed one of two main types: either as manhours, reimbursed at cost with fixed overhead and profit, or as fixed lump sum contracts with milestone dates (similar to the concept of on-call contracting (Shing-Tao and Ibbs 1998) described in Chapter 2). In addition, the costs of as yet unawarded contracts were included along with BP's own management costs, to reach the work estimate. In a radical departure from conventional project practice, the alliance contractors were asked to take responsibility for a percentage of any cost savings or over-runs that Andrew might potentially produce. BP claimed that the project was completed 6

Table 3.13: Andrew risk/reward sharing model

Participant	Responsibility area	Responsibility share
BP	Operator	46%
Brown & Root	Project management and detailed engineering	22%
BARMAC	Jacket, template and piling fabrication	6%
TJB	Deck and topside fabrication	12%
Saipem	Platform transportation and installation	6%
Allseas	Pipeline installation	4%
Santa Fe	Platform drilling facilities and well construction	3%
Emtunga	Accommodation	1%

Source: Adapted from KPMG (1998, p. 11), with permission

months ahead of schedule and came under the target cost by approximately £80 million. The risk/reward sharing arrangement, which clearly demonstrated potential advantages of the alliance concept, was illustrated in Table 3.13.

3.10.3   East Spar Development – Western Australia (*Source:* Australian Construction Association (ACA 1999))

This project involved development of a gas condensate field 40 kilometres offshore, west of Barrow Island in Western Australia. Design and construction took place between February 1995 and November 1996 at a cost of A$270 million. A number of consortia submitted an expression of interest to the client Western Mining Corporation (WMC). Three contracting parties Clough/Kvaerner the joint venture proposers, and the owner/operators, formed an alliance with the owner. An alliance board was then established with two representatives from all three parties to the alliance. Their role included selection of the Project Manager and Section Managers. A rigorous reporting system was established. The design concept and budgets were then fully developed by the alliance with full open disclosure and airing of all issues. This included the development of a risk and reward model to share profits and losses as illustrated in Figure 3.11 below.

In establishing a target cost there was some difficulty involved in the owner and contractors agreeing the target cost. The parties reached an acceptable compromise with two targets approximately 10 per cent apart. The agreement allowed for a sharing ratio of 1:6 between the two-target estimate ranges and 1:2 outside those ranges. The alliance arrangement allowed flexibility of design detailing within an integrated owner/contractor team. Even with a number of major design changes the costs fell close to the client's target point.

3.10.4   Fluor Daniel SECV – Victoria (*Source:* Lendrum 1998, pp, 321–328)

Fluor Daniel pioneered partnering in engineering, construction and maintenance in the USA. In October 1992 the then State Electricity Commission of Victoria (SECV) reviewed its competitiveness and concluded that considerable cost saving could be made through reform of the way it operated and procured services. It identified a 35 per cent cost reduction being required to maintain competitiveness with New South Wales and Queensland electricity generating and supply markets.

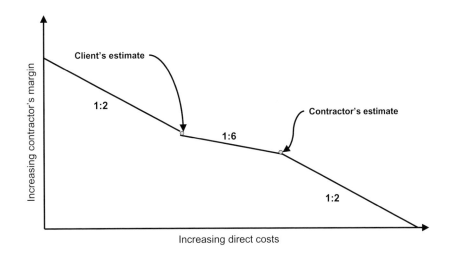

Figure 3.11:   East Spar alliance risk/reward model
Source: Australian Constructors Association (ACA 1999, p. 31).

With the prospect of a national electricity grid emerging, the challenge was real and vital. The Victorian Government decided upon a policy of outsourcing non-core business and tender reform. Maintenance was targeted as a prime candidate for reform. Several key managers of the SECV visited Fluor Daniel's USA operations and customers. In July 1993 Fluor Daniel was awarded the maintenance of Loy Yang A Power Station in Victoria followed by two other power stations comprising 71 per cent of the La Trobe Valley generating capacity. The scope of activities included maintenance services, major and minor unit overhauls, small and large capital initiatives, management of preferred contractors and suppliers, and procurement and stores management. This was accomplished while re-deploying a skilled SECV workforce committed to the values and objectives of the partnership. This required a special flexibility because much of the unscheduled and emergency maintenance work required committed staff to be deployed at short notice for weeks and sometimes months at a time. Thus, the strategic partnership arrangement required excellent industrial relations skills and the negotiation of an enterprise agreement to facilitate the levels of flexibility required.

This case study is different from previous ones discussed in a number of respects. The strategic partnering arrangement spans numerous sub-projects and so it is more of an operational alliance than a construction project specific alliance. It is also bound up with the outsourcing and general reform of the procurement process issues. There are pertinent similarities although:

- Costs are reimbursable with unneeded contingencies eliminated;
- Fluor Daniel's profits are based on a structured incentive program based upon innovation and technological change, both large and small sub-project outage performance, and specific client business targets;
- Open book access and shared goals; and
- Shared vision and common objectives.

3.10.5   Other alliance projects undertaken in Australia

Alliancing is an increasingly popular procurement project delivery choice in the USA, UK and Australia. Experience thus far is limited but encouraging. The KPMG report cites the NSW Rail Access Corporation's Maintenance Contracts, Sydney Water Corporation's Northside Storage Tunnel Contract and W.A. Water Corporation's Waste Water Treatment Plant Contract as 1997 and 1998 examples (KPMG 1998). In a recent article on project alliancing, several alliance projects are mentioned including cautious comment about cost over-runs on the Northside Tunnel Project and BHP's Port Hedland HBI Project (Pratley 1999). In that same article there is also a brief description of the ElectraNet SA alliance with Kilpatrick Green and Burns and Roe Worley in South Australia. Clearly, despite some perceived shortcomings, alliancing is recognised as a valid and useful option which continues to attract considered support and attention.

3.11   Chapter summary

Alliancing has emerged as a procurement choice in response to several stimuli. Companies form alliances as a competitive tool to further their advantage. Their motivation may be threefold in response to the globalisation of business and competition (Doz and Hamel 1998).

First, firms use alliances to build critical mass of expertise and service/product offerings. Critical mass in the construction industry allows a firm to provide very high levels of expertise across a wide range of specialisation. Critical mass also refers to access to specialised equipment management systems and balance-sheet financial resources. For large-scale projects the cost of tendering alone may be daunting – so the risk response to sharing risk with compatible and complementary partners makes sense. This may be achieved through co-opting potential competitors to strengthen a perceived weakness in any single firm to be able to compete with the strongest. In non-strategic alliancing conditions this is often achieved through a joint venture (JV) arrangement. A JV would be a more common response for undertaking large-scale BOOT infrastructure projects. A critical mass of specialisation skills may be lacking for highly complex or highly specialised projects so in those cases an alliance makes sense. If a firm builds a hub of alliance partnerships with other firms, it can place itself in an equally strong position to larger more highly resourced firms with a wider range of skills, expertise and organisational knowledge (Drago 1997).

Second, firms may wish to reach new markets through forming alliances with local firms and developing the specialised local knowledge and networked links of local companies. This is similar to the idea of using local factories in the manufacturing sector as a strategy for building new strengths to compete against 'foreign' companies. This approach also leverages skills and diversity of 'local' firms to allow development of new products/services that can be exported or returned to the host firm's location (Ferdows 1997; Porter 1998; El Higzi 2000). Thus, new opportunities and new products can be created (Doz and Hamel 1998).

Third, skill gaps are strengthened. This is different from forming alliances to provide a skill gap. This is related to the internalisation of valuable knowledge from a partner through alliancing. When firms form alliances they work together cooperatively sharing knowledge, ideas and feedback on systems and processes. Thus, by being exposed to diverse sources of critique and to new skills, firms can capture that knowledge resource, digest it (internalise it) and use this to competitive advantage. New competencies are built in this way (Doz and Hamel 1998).

Clients of construction projects may use alliancing for the same reasons as stated above for other firms. Government departments, large multinational enterprises and other significant developers have their own limited inhouse expertise to manage projects. They can effectively use alliancing to gain and internalise knowledge (to be better at their involvement in overseeing future projects as more sophisticated clients) and/or to build an inhouse capacity.

Case studies and much of the alliancing and partnering literature indicate that alliancing and strategic partnering are appropriate for highly complex projects, where there is simply insufficient time independently to develop a brief and design and proceed through the traditional tendering process (Hamel and Prahalad 1994; Bennett and Jayes 1995; Doz and Hamel 1998; Lendrum 1998). Moreover, strategic alliancing has also emerged as a response to the need to use the specialised knowledge of contractors and designers to minimise wasted time and redundant design in achieving buildable solutions that are leveraged from the contractor's practical knowledge. The value of constructability/buildability and value analysis has been well understood for many years now (Walker 1993; Morris 1994; Francis and Sidwell 1996; McGeorge et al. 1996; McGeorge and Palmer 1997). Thus, clients appear to favour alliancing as a response to project complexity under circumstances of tight timeframes where cost budgets can be better formulated and controlled in an atmosphere of trust and commitment. This approach also seems to deliver aligned joint project objectives.

This chapter began with a discussion on the changing framework for organisational design to facilitate project goals. It was seen that with higher levels of rapid change, complexity and need for rapid responsiveness, traditional organisational structures are not adequately coping with current demands. The relevance of post-corporate organisational structures to managing today's projects was discussed and four management blueprints were outlined. The Fourth Blueprint was offered as an appropriate model by implementing strategic alliances to cope with the new dimensions of business complexity facing clients in the twenty-first century.

The linking discussion of enterprise networks and team spirit was important because this linkage forms the foundations of strategic partnering and strategic alliances. Qualitative aspects of relationships are pivotal to successful alliancing, so it is appropriate that the drivers of trusting relationships should be explored and explained. The logical successor to this discussion should be a discussion on types of partnerships and alliances. The work of Lendrum (Lendrum 1998) is useful in this context. A detailed discussion of forms of partnering and alliancing followed. This brought forward the opportunity to discuss in detail the results of the CIIA study of partnering (Lenard et al. 1996). This is expected to be demonstrated as a seminal study that will claim its place in the literature of relationship contracting in Australia. The study brings with it interesting and valuable insights into the way that partnering operates in the current climate. At the same time, other partnering studies from overseas were discussed. This explains the important differences between partnering and alliancing – a concept that has been poorly understood.

It was also important to explore the journey when undertaking alliancing. The idea of strategic relationship development and maintenance was discussed. It is of little use to develop a relationship, an organic entity dependent upon sustenance and nurturing, without considering how such a relationship can be maintained. To do otherwise would be to take a shallow and unrealistic approach. The discussion moved on towards a model of creating effective alliance partnerships. The selection criteria used on the Australian National Museum project was discussed as it provides an interesting, novel and advanced concept congruent with the notions expressed by triple bottom line theory (Elkington 1997). Finally some 'how to' aspects were discussed as they apply to the literature on alliancing. A selection process was offered (KPMG 1998) and case studies were provided to elaborate on issues of practice as well as theory.

In documenting history, particularly the history of a rapidly evolving story such as alliance procurement systems, there are bound to be errors of omission. Reviews of the literature are constrained by an attempt to maintain a brief yet a pertinent coverage of the topic. The role of continued study of a topic is to continually question what is observed and experienced and to attempt to seek relevance to current issues, and to place experience into historical context. It is easy and perhaps trite to label various developments as 'fads'. Partnering has been shown to be a genuine attempt to solve a deteriorating problem in procurement whereby working relationships between contracted partners dissolve into a morass of legalese and confrontationalist approaches. This book attempts to discern between strategic partnering and alliancing. This may be judged by future researchers to be an irrelevant concept or a defining concept. Time will tell.

The evidence provided for indicating where a relationship-based procurement approach is appropriate is both strong and well supported. Projects such as the Australian National Museum will provide deeper insights and it is fortunate that the Australian Government has funded research to identify, track and critically examine the performance of alliancing and its intricacies as experienced on that project. Any review of the literature is snap-frozen in time, influenced by the availability of information and the culture of the time. It is hoped that this publication will provide a useful basis for study and re-evaluation of theory and practice based upon reflections of actual experiences and the expanding body of knowledge of this topic.

3.12   References

ACA (1999) *Relationship Contracting – Optimising Project Outcomes*. Sydney, Australian Constructors Association.

Argyris, C. and Schön, D. (1978) *Organizational Learning: A theory in action perspective*. Reading, MA, Addison-Wesley.

Argyris, C. and Schön, D. (1996) *Organizational Learning II: Theory, method, and practice*. Reading, MA, Addison-Wesley.

Bennett, J. and Jayes, S. (1995) *Trusting the Team*. Reading, UK, Centre for Strategic Studies in Construction, The University of Reading.

CIDA, Construction Industry Development Agency (1994) *Two Steps Forward, One Step Back – Management Practices in the Australian Construction Industry*. Sydney, Construction Industry Development Agency.

CII (1996) *The Partnering Process – Its Benefits, Implementation, and Measurement.* Austin, Texas, CII, Bureau of Engineering Resources, University of Texas at Austin.

DETR (1998) *Rethinking Construction,* Report. London, Department of the Environment, Transport and the Regions.

Doz, Y.L. and Hamel, G. (1998) *Alliance Advantage – The Art of Creating Value Through Partnering.* Boston, Harvard Business School Press.

Drago, W.A. (1997) 'When Strategic Alliances Make Sense.' *Industrial Management & Data Systems,* (97/2): 53–57.

El Higzi, F. (2000) *The Influence of Environmental Factors on the International Market Entry Strategies of Australian Construction Industry Companies.* Masters by Research, Brisbane, Queenland University of Technology, School of Construction Management and Property.

Elkington, J. (1997) *Cannibals with Forks.* London, Capstone Publishing.

Ellison, S.D. and Miller, D.W. (1995) 'Beyond ADR: Working Towards Synergistic Strategic Partnership.' *Journal of Management in Engineering,* **11** (6): 44–54.

Ferdows, K. (1997) 'Making the Most of Foreign Factories.' *Harvard Business Review.* **75** (2): 73–88.

Field, L. and Ford, W. (1995) *Managing Organisational Learning – From Rhetoric to Reality.* Melbourne, Longman Australia.

Francis, V.E. and Sidwell, A.C. (1996) *The Development of Constructability Principles for the Australian Construction Industry.* Adelaide, Construction Industry Institute Australia.

Hamel, G., Doz, Y.L. and Prahlad, C.K. (1989) 'Collaborate With Your Competitors – And Win.' *Harvard Business Review,* **67** (1), 133–139.

Hamel, G. and Prahalad, C.K. (1994) *Competing for the Future.* Boston, MA, Harvard Business School Press.

Hampson, K. and Kwok, T. (1997) 'Strategic Alliances in Building Construction: A Tender Evaluation Tool for the Public Sector.' *Journal of Construction Procurement.* **3** (1): 28–41.

Hampson, K.D. (1993) Technology Strategy and Competitive Performance: A Study of Bridge Construction. PhD, Civil Engineering. Stanford, Calif., Stanford University.

Hampson, K.D. and Tatum, C.B. (1997) 'Technology Strategy and Competitive Performance in Bridge Construction.' *Journal of Construction Engineering and Management.* **123** (2): 153–161.

Kanter, R.M. (1989) *When Giants Learn to Dance: Mastering the Challanges of Strategy, Management and Careers in the 1990s.* London, Simon & Schuster.

Karpin, D. (1995) *Enterprising Nation – Renewing Australia's Managers to Meet the Challenge of the Asia-Pacific Century, Taskforce Report.* Canberra, Industry Taskforce on Leadership and Management Skills – Ministry for Employment, Education and Training, Australian Federal Government.

KPMG (1998) *Project Alliances in the Construction Industry,* Literature Review. Sydney, NSW Department of Public Works & Services.

Kwok, T. (1998) *Strategic Alliances in Construction: A Study of Contracting Relationships and Competitive Advantage in Public Sector Building Works.* PhD, Faculty of the Built Environment. Brisbane, Queensland University of Technology.

Larson, E. (1995) 'Project Partnering: Results of Study of 280 Construction Projects.' *Journal of Management in Engineering,* **11** (2): 30–35.

Latham, M. (1994) *Constructing the Team.* Final Report of the Government/Industry Review of Procurement and Contractual Arrangements in the UK Construction Industry. London, HMSO.

Lenard, D.J. (1996) *Innovation and Industrial Culture in the Australian Construction Industry: A Comparative Benchmarking Analysis of the Critical Cultural Indices Underpinning Innovation.* PhD, Faculty of Architecture. Newcastle, University of Newcastle.

Lenard, D.J. and Bowen-James, A. (1996) *Innovation: The Key to Competitive Advantage.* Adelaide, Australia, Construction Industry Institute Australia, University of South Australia.

Lenard, D.J., Bowen-James, A., Thompson, M. and Anderson, L. (1996) *Partnering – Models for Success.* Adelaide, Australia, Construction Industry Institute Australia.

Lendrum, T. (1998) *The Strategic Partnering Handbook.* Sydney, McGraw Hill.

Limerick, D., Cunninton, B. and Crowther, F. (1998) *Managing the New Organisation: Collaboration and Sustainability in the Postcorporate World.* Warriewood, NSW, Business & Professional Publishing.

Love, P.E.D., Li, H. and Hampson, K.D. (1999) *Cooperative Strategic Learning Alliances in Construction.* Proceedings of the CIB W55 and W65 Symposium 1999, Cape Town, South Africa, CIB.

Luck, R.A.C. and Newcombe, R. (1996) *The Case for Integration of the Project Participants' Activities within a Construction Project Environment.* Glasgow, Scotland, E&FN Spon.

Matthews, J., Tyler, A. and Thorpe, A. (1996). 'Pre-construction Project Partnering: Developing the Process.' *Engineering, Construction and Architectural Management,* **3** (1–2): 117–131.

McGeorge, W.D., Chen, S.E., Barlow, K., Sidwell, A.C. and Francis, V. (1996) 'Current Management Concepts in the Construction Industry.' *Australian Institute of Building Papers.* **7**: 3–12.

McGeorge, W.D. and Palmer, A. (1997) *Construction Management – New Directions.* London, Blackwell Science.

Miles, R.S. (1995) 'Twenty-first Century Partnering and the Role of ADR.' *Journal of Management in Engineering,* **12** (3): 45–55.

Morris, P.W.G. (1994) *The Management of Projects: A New Model.* London, Thomas Telford.

Pedler, M., Burgoyne, J. and Boydell, T. (1996) *The Learning Company: A Strategy for Sustainable Development.* London, McGraw Hill.

Porter, M.E. (1998) 'Clusters and the New Economics of Competition.' *Harvard Business Review.* **76** (6): 77–90.

Pratley, J. (1999) 'Project Alliancing: Does it work?' *Building Australia.* July: 33–37.

Roe, A.G. and Phair, M. (1999) 'Connection Crescendo.' *Engineering News Review.* May: 22–26.

Savage, C.M. (1996) 'Knowledge Management: – A new era beckons.' *HR Monthly – The Australian Human Resource Magazine,* November: 12–15.

Segil, L. (1996) *Intelligent Business Alliances.* London, Random House.

Senge, P.M. (1992) *The Fifth Discipline – The Art & Practice of the Learning Organization.* Sydney, Australia, Random House.

Shepherd, A. (1999) 'Project Alliancing on a Public/Private Partnership Project.' *Building Australia*: 14–17.

Shing-Tao, A. and Ibbs, C.W. (1998) 'On-call Contracting Strategy and Management.' *Journal of Management in Engineering,* **14** (4): 35–44.

Sveiby, K.E. (1997) *The New Organizational Wealth: Managing and Measuring Knowledge-based Assets.* San Francisco, Berrett-Koehler Publishers, Inc.

Thompson, P.J. and Sanders, S.R. (1998) 'Partnering Continuum.' *Journal of Management in Engineering,* **14** (5): 73–78.

Uher, T. (1999) 'Partnering Performance in Australia.' *Journal of Construction Procurement.* **5** (2): 163–176.

Walker, D.H.T. (1993) 'Facilitating complex projects in Victoria, Australia and the rôle of the Major Projects Unit – A case history.' *Journal of Real Estate and Construction.* **3**: 51–63.

Walker, D.H.T. (2000) 'Client/Customer or Stakeholder Focus? ISO14000 EMS as a construction industry case study.' *TQM,* **12** (1): 18–25.

Walker, D.H.T. and Betts, M. (1997) *Information Technology Foresight: The Future Application of the World Wide Web in Construction*. CIB W78 Workshop, Information Technology Support for Construction Process Re-Engineering IT-CPR-97, James Cook University, Cairns, Queensland, CIB.

Walker, D.H.T., Hampson, K.D. and Peters, R.J. (2000) *Project Alliancing and Project Partnering – What's the Difference? – Partner Selection on The Australian National Museum Project – a Case Study*. CIB W92 Procurement System Symposium On Information And Communication In Construction Procurement, Santiago, Chile, Pontifica Universidad Catolica de Chile.

Walker, D.H.T. and Lloyd-Walker, B.M. (1999) Organisational Learning as a Vehicle for Improved Building Procurement. *Procurement Systems: A guide to best practice in construction*, eds Rowlinson, S. and McDermott, P. London, E&FN Spon, 1: 119–137.

Weston, D.C. and Gibson, G.E. (1993) 'Partnering – Project Performance in US Army Corps of Engineers.' *Journal of Management in Engineering*, **9** (4): 410–425.

Wind, J.Y. and Main, J. (1998) *Driving Change – How the Best Companies are Preparing for the 21st Century*. London, Kogan Page.

Womack, J.P., Jones, D.T. and Roos, D. (1990) *The Machine that Changed the World – The Story of Lean Production*. New York, Harper Collins.

Chapter 4
Project Alliancing Member Organisation Selection

Derek Walker and Keith Hampson

In Chapter 3 the nature of enterprise networks and how they deliver value was discussed with particular emphasis on partnering and alliancing. The nature of different types of relationship-based procurement strategies was explained and latter sections of that chapter concentrated upon alliancing with several case studies, mainly from the engineering industry, being provided to focus discussion. However, little of the motivation of alliance members was discussed. The selection process for National Museum of Australia was introduced (see Figure 3.10) and the 12-point criteria for that project were outlined, general underlying principles that impel alliances to be formed as a response to external and internal drivers were left for further discussion. Such drivers must inevitably shape and determine the preferred method of selecting alliance member organisations and the way that these organisations may be selected and evaluated.

Chapter 3 provided considerable detail in what alliancing may be and why that option may be chosen. In this chapter the general motives and drivers that lead to alliances being formed, selection procedures and practices are discussed with examples of real projects being provided. Alliances are discussed from a number of industry sectors. The National Museum of Australia project is also used as a detailed case study to illustrate and highlight the criteria used and motivational rationale for choosing these particular criteria. The selection process is described in detail to explain how and why it proceeded in the manner designed. Again this provides a link between motivation of the client and alliance members. Implications of this process upon the construction industry and other industries choosing this approach are also discussed. The National Museum of Australia project provides a useful case study to focus upon. While a case study approach does not provide statistically sound results that can be generalised it does provide a means to analyse lessons learned that may be applied to other cases (Yin 1994).

4.1 Introduction – Why form an alliance? – General motivational issues

We have seen earlier in this book that enterprise networks such as joint ventures, strategic alliances, or partnering arrangements, provide a mechanism for sharing risk and optimising scarce resources that may be difficult (if not impossible) for a single entity to obtain or focus upon one project. In this sense the critical motivation is to cope better with turbulence in the external environment – it is often a response to uncertainty, often to market forces or client demands, and a need to jointly provide necessary skills, resources and competencies to satisfy the needs of a customer thus providing improved business opportunities.

Primary purposes of alliances are said to include **co-option**, **co-specialisation**, and **learning** and **internalisation** (Doz and Hamel 1998). Co-option helps to neutralise potential rivals or bring onboard organisations with complementary goods or services that fill a difficult-to-fill gap. Co-specialisation creates value through synergising unique and different resources that become more valuable when bundled than kept separate. Rather than just supplying a missing link, co-specialised products offered as part of an alliance offer the product in a unique and special way in which knowledge, expertise and technical know-how are often implicitly included in the 'deal'. Learning and internalisation provides an avenue for organisations to learn, internalise and embed new products, features or aspects. This may be motivated by a future potential to compete against that alliance member, to understand that member better so as to improve current and future working relationships or to gain access to new business opportunities otherwise unavailable (Doz and Hamel 1998, p. 4–5). They argue that alliance motivation for these three purposes may be used as a result of strategic motivation (Doz and Hamel 1998, p. 107). Organisations that they cite provide goods to customers rather than deliver projects for a customer.

The purpose or motivation for co-option, co-specialisation and learning and internalisation provides a useful introduction to alliance motivation. It helps explain alliance members' motivation when working on single projects for a customer/client entity as opposed to how they might be motivated for general-public consumers. Co-option, co-specialisation and learning and internalisation concepts have been extended in an interesting way with relation to networking systems, primarily IT applications. Mohanbir Sawhney and Deval Parikh argue that in the networked world of the twenty-first century, value can be effectively created and enhanced at each end of the various interfaces between customer and suppliers as well as through re-thinking how the infrastructure that connects them is configured (Sawhney and Parikh 2001). They see new value opportunities where new customer interfaces and intermediary services can play a role and 'back-end' intelligence being reconfigured onto powerful, efficient, reliable distributed servers. They recognise that new concepts in shared infrastructure are emerging – internet service providers storing people's e-mail and digital information, e.g. images (photographs) and other documents for example.

There is a rise in the number of alliances being formed between providers of network services such as order processing, tracking orders and other business functions. The ability to coordinate these has also created new alliances in project management and logistics. Companies like Cisco and Hewlett-Packard are offered as examples of organisations that are re-shaping themselves into hubs with arrays of alliance partners specialising and providing co-specialised services that are difficult to develop rapidly (Sawhney and Parikh 2001). Thus in this kind of organisation, the main motivation is reacting to customer demands with a new configuration of services characterised by modular connected niche specialisations orchestrated through a hub organisation in an apparently seamless manner. In other cases proactive organisations form alliances again each with critical specialisations to meet a perceived customer need through design. Many dot-coms fall into this category as creative new ideas for a simple application have developed into an integrated application. A re-configuring of supply chains through alliancing is an emerging trend that seeks to unearth value. This has developed from concepts of value-chain management first widely publicised in the late 1980s and early 1990s (Porter 1990; Porter 1998).

The concept of value and its impact upon why alliances are useful has been more widely appreciated with growth in electronic business-to-business (B2B) activity – the rise of e-commerce. In an article on the emerging B2B landscape at the start of the twenty-first century, Richard Wise and David Morrison offer interesting insights into how e-commerce will develop from trends that already emerged (Wise and Morrison 2000, p. 93). They base their projections on the history of the development of the financial services sector and they have developed a model of how buyers and sellers will be linked. This is explained as follows:

(1) **Mega-exchanges** will dominate the B2B landscape acting as central hubs for buyer–seller communication. They may have started out to facilitate auctions, grown out of retailers or operated as a buyer–seller entrepreneurial intermediary. Their required capabilities will include a capacity for large-scale transaction processing, ability to establish industry standards and be perceived as being neutral brokers. Their sources of profits may have evaporated due to intense competition so they may derive benefits from other sources such as becoming not-for-profit organisations. Their originators would probably have re-invented themselves as spin-off enterprises, as e-speculators for example. They will be linked directly between buyers and sellers with specialist originators, e-speculators, and solution providers offering input to the products and services available to an array of customers and supply-chain entities.

(2) **Specialist originators** will standardise and automate the buyer decision-making process for more complex products and then send the transactions to the exchanges for execution. They will produce complex or expensive products that require deep knowledge of the product. These may be electronic, advanced manufacturing components from automotive, aerospace or other engineering sectors or perhaps from insurance or other specialised service deliverers. Their source of competitive advantage includes deep product knowledge, effective use of decision-support software, access to qualified suppliers and/or ability to bundle transaction volume.

(3) **E-speculators** will participate in or run exchanges, gaining real-time information in order to take direct or derivative positions. Like the stock traders of today, they will be dealers in buying and selling options, hedging, and other sophisticated transactions based on their real-time knowledge of the dynamics of the markets they transact in. They will function as facilitators and risk managers. There is already an expanding market in buying and selling telephony, power and other services today. This group will probably occupy a similar B2B market niche as these traders of today.

(4) **Solution providers** will operate separately from open exchanges by embedding the product sale in a suite of unique valuable services. They will require strong technical and problem-solving skills. Their sources of competitiveness

will be in brand/reputation strength, their rich stream of innovative offerings and by locking in customers through relationships and alliances.

(5) **Sell-side exchanges** will arise out of sellers having high fixed-cost assets needing to derive maximum usage. These will require strong supplier relationships, ability to offer additional relevant services and be perceived as being neutral. They will gain efficiency by swapping and reselling orders amongst a closed set of suppliers. For example, a construction or plant-hire company may be involved in short-term emergency supply of critical supplies or trade one service/good of one supplier with another much as barter economies operate today. Another example is transport tracking of 'empty space' for back-loading or maximising capacity for deliveries.

The above is actually happening in an unconnected and disparate manner today. IT will be the great enabler to draw together alliances in a manner that transcends bartering, arbitrage, hedging/trading and other non-over-the-counter forms of buyer–seller relationship. Alliances and stronger alliance behaviour are pivotal in the development of the model proposed above.

4.1.1   Manufacturing and distribution industries (primarily product focused)

The literature indicates that alliances have been particularly prevalent in the automotive manufacturing industry (Womack *et al.* 1990; Womack and Jones 2000) and in aerospace (Doz and Hamel 1998, pp. 62–63). The formation of alliancing motives has been classified into four clusters: financial, technological, managerial and strategic (Whipple and Gentry 2000).

Financial motives
Financial motives include economic and financial performance reasons. Profit increase, stabilisation, maintenance or protection can accrue from supply or demand factors. General efficiencies and cost savings can be achieved from reduction in duplication and waste. Price reductions may follow or be achieved through greater purchasing power through alliance membership. Some alliances are purely formed on these grounds and internet purchasing has accelerated this. The formation of an alliance to establish a web portal for bulk purchasing of construction inputs (Construction Industry Trading Exchange CITE) of which Bovis Lend Lease, the global developer/constructor, is a member provides one example. Leverage of capital can also be achieved from joint investment to gain greater scope, depth or breadth of commitment to investments such as equipment, hardware or software systems or skills training for example. Alliances can also achieve financial advantage from inventory/stock reductions.

Technological motives
Technological motives include access to new technologies that would be too expensive to develop as a stand-alone operation, access to a specialised expertise and skills base that is difficult to obtain and retain, information sources, and knowledge. Sometimes knowledge, such as customer profile, personal network connections or client-base, is technologically grounded in knowledge bases or embedded in tacit knowledge held by employees. This sort of technology is

difficult to quantify or measure in any precise manner. This knowledge capital, that can also include system knowledge, may make a potential alliance partner extremely attractive. This can lead to organisational learning opportunities and improved productivity improvement through access to an organisational culture where double-loop learning[1] is evident (Barlow *et al.* 1998). Access to information technology and advanced and effective systems and methods of communicating rapidly and effectively are becoming increasingly important. For example, the use of web-enabled project management systems is expected to change the business of construction organisations permanently. Firms that have expertise and experience in these technologies will be highly prized potential alliance partners (Roe and Phair 1999).

Managerial motives

Managerial motives include reducing the number of suppliers to simplify the procurement process and better define accountability and responsibility. The automotive industry has moved a long way in this direction and has added value by not only reducing the number of suppliers and subcontractors but also by working more closely with them to seek and achieve continuous improvement (Womack *et al.* 1990; Womack and Jones 2000). This simplifies the supply chain management process, helps to stabilise inbound and outbound movement of physical raw materials and components and also enhances supply chain loyalty.

Strategic motives

Strategic motives include maintaining, defending or expanding market share. The industry requirement in a global economy demands rapid response to new or existing client/customer needs. Many organisations simply cannot maintain the required pace of organisational learning, staff development or business improvement and may rely on alliances with others to gain exposure to new markets or support to maintain presence in existing markets through outsourcing or periodic re-invention or re-framing (Hamel and Prahalad 1994; Doz and Hamel 1998). A core skill that alliancing partners need to develop is to understand their own core competences better and to learn how to build on alliance partners' expertise and to select partners with complementary assets – intellectual, physical, financial or virtual – that fill gaps in a package of services offered to clients/customers. Competitive advantage is derived from extracting leverage from combining organisational assets in a synergistic manner to achieve customer focus.

The strength and preference order of the above motives will vary with the nature of the alliance. For example in a survey of 180 respondents involved in a manufacturer–material supplier alliance it was found, not surprisingly, that different priorities and strength of motivation for components of the above four motivation groups varied between three types of alliance. The top five motives are ranked by importance in Table 4.1 to illustrate these differences in motives.

---

[1] Double loop learning moves beyond single loop learning where corrective action is focused on task improvement to one where systems are improved – the difference between being efficient at doing the *wrong* thing rather than being effective through doing the *right* thing (Argyris, C. and Schon, D. (1996).

Table 4.1: Top five manufacturer alliance formation motives

With material supplier	With customer	With service supplier
Reduced cycle time/lead time	Increased customer service	Increased customer service
Reduced inventory	Reduced cycle time/lead time	Reduced cycle time/lead time
Stabilised supply/demand	Improved quality	Improved quality
Improved quality	Increased customer loyalty	Internal cost savings
Increased customer service	Increased customer involvement	Achieved core competency

Source: Whipple and Gentry 2000, p. 311, with permission MCB University Press.

This helps to illustrate from the point of view of manufacturers taking part in the survey what they wanted to achieve from alliances with materials suppliers, customers or service suppliers. It would not be surprising to find that each of these three alliance partner groups might have a different priority of needs required of manufacturers in an alliance. The main purpose for presenting Table 4.1 was not the ranking order *per se* but that there are differences in priority that must be understood if trust and commitment is to be established, developed and maintained. Expectations are also important. If one alliance member has unrealistic or unrealisable expectation of other members then difficulties may well arise that weaken trust and commitment and the usefulness of the alliance. A comment made by Ian Stuart in a paper relating to supplier alliances based on research of 88 supplier–buyer alliance relationships is relevant here (Stuart 1999).

> We tentatively conclude, therefore, that buying firms that wish to establish alliances for the long run can maximise the probability of success by continually tempering the perceived benefits that both parties will achieve through the collaborative approach. Overly optimistic projections of market share improvements with potential use on other product lines, corporate image enhancements, etc. apparently serve to raise expectations and eventually lead to weakening of the bond.

Clearly understanding partner motivation is crucially important.

4.1.2   The airline industry (primarily service focused)

One of the industries that has seen a proliferation of alliances being formed and consolidated is the airline industry. The motivations here relate to the impact of regulation on business entry, customer service, cost efficiency, and market power amongst other reasons normally ascribed to alliancing. The airline industry is a useful one to investigate, as it is a truly global industry that has faced some interesting challenges that have led to alliancing. Readers interested in further information can find a wealth of detail in the book *Globalization and Strategic Alliances – The Case of the Airline Industry* (Oum *et al.* 2000).

The need for alliance formation in the airline industry stems from changes of the view of many governments during the first three quarters of the twentieth century that national airline carriers held a unique and strategic role in most countries' national identity. During that period national carriers such as QANTAS, BA, Air France, etc. were viewed by their government owners as helping define national identity as well as being a valuable resource to be dragooned in times of military conflict. That view was undermined by the later part of the twentieth century with the rise of global enterprises. Some elements of national identity persist with fear of business rather than national strategic needs prevailing. This is particularly true with the rise and importance of the tourism industry and its reliance upon air traffic. However, the USA government was particularly instrumental in forcing change towards an 'open skies' approach to airline policy relating to market access (the ability to sell tickets, be able to over-fly countries and land aircraft in countries). Oum *et al.* maintain that as of January 2000 the USA had signed open skies agreements with 41 countries (Oum *et al.* 2000, p. 2 footnote). The sheltered industry approach prevailing through most of the twentieth century had quickly developed into a scramble for access to rapidly growing airport facilities and airline management systems by the 1990s. The major motivations of airlines to form global alliances such as *Oneworld* or *Star* or for companies to form a transport and logistics alliance can be summarised as following the previous four motivation clusters – financial, technology, management, and strategic.

Financial motives

The National Semi-Conductors (NSC) and Federal Express (FedEx) alliance allows NCS to concentrate on its own core competencies and for FedEx to provide a global logistic and support role. This provides NSC with distinct financial competitive advantage. The NSC and FedEx example illustrates how a different competency can be outsourced through a strategic alliance. Within the airline industry feeder alliances allow national carriers to 'feed' passengers onto regional carriers in a transparent and badged manner that allows the national carrier, the regional carrier and the customer to benefit from this arrangement. This has led to financial advantages of limiting the cost and effort of providing a full seamless service to customers wishing to travel to destinations unattractive to national carriers. This has also led to development of transport hubs, which have generated further alliancing for delivering other ranges of associated travel industry and support services. 'Code sharing' whereby, for example, QANTAS and BA sell tickets for travel between potentially competing destinations has enabled far higher passenger load factors to be achieved leading to reduced duplication and increased revenues.

Technological motives

Airline passengers prefer to purchase tickets to travel as one transaction step. Airline alliancing has allowed travel agents and passengers to purchase a ticket leaving, for example, Heathrow in London for return travel via Hong Kong one way with a stop over in Beijing to Melbourne and returning via Buenos Aires. The Oneworld alliance permits this to happen as one booking transaction easily and seamlessly. Airlines such as QANTAS have developed websites allowing transactions to be paid online on a 24-hour, 7-day-a-week basis using a credit card. Sophisticated shared technology booking systems offered by alliances make this possible. Without the alliance the development of this technology would be unlikely (given reliance on investment by individual airlines within the alliance). Further, shared use allows cross-alliance knowledge acquisition of lessons learned, improvement potential and training and development so that enhancement to technology is more likely to occur and is less resource consuming to achieve. Many of the high impact and high expense technologies required such as booking, maintenance and other operational systems are high-risk investments. Alliances of oganisations are better placed to bear these risks jointly.

Management motives

Airline alliances offer a number of advantages attractive to alliance members. Through sharing IT and other operational systems greater efficiency and effectiveness can be achieved than would otherwise be generally the case. For example, partners can better coordinate flight schedules to minimise customer waiting times for connections, it is easier to off-load passengers to other flights if a cancellation of a flight is necessary. Apart for ticketing management advantages there are also efficiency gains in baggage handling and security. Marketing campaigns can be better coordinated for mutual partner benefit and frequent flier programs (FFP) can offer greater customer choice for redeeming points. Many of the alliance benefits gained from FFPs such as car hire, accommodation and hospitality can be extended through the alliance links. Benchmarking across partners is also a valued feature that ensures that not only 'brand' image is not harmed but that also continual improvement is facilitated.

Strategic motives

The particular nature of the airline industry and its history of being perceived as strategically important to various national governments has placed it in a unique situation. This history has inhibited merger and acquisition activity. The response to this has been pseudo-mergers in the form of airline alliances such as Oneworld or Star. Oum *et al.* (2000, p. 23) argue that '... the current alliance race will likely continue unless foreign ownership laws and nationality clauses in bilateral agreements change'. They also point to difficulties with legal, political, regulatory and institutional constraints being set to continue into the foreseeable future. They expect that alliance airline partnerships will focus on groups of airlines residing in different continents. This builds on strategic strengths of links with local and regional providers of support services as well as providing advantages of political influence to minimise strategic threats posed by potential regulation and political action.

Before leaving discussion of the airline industry, it is worth briefly discussing the features that alliance partners are engaged upon as it clearly illustrates examples of some of the motives discussed above. Oum *et al.* (2000, p. 33) indicate the joint activities undertaken by three classes of alliances. A *Type 1* airline alliance is a simple route-by-route alliance in which code-sharing or joint operations on only several routes. *Type 2* involves substantially linking two partners' networks to feed traffic to the partners' hubs. *Type 3* represents the most advanced form of alliance with equity and cross-ownership as well as operation and strategic collaboration taking place. The range and extent of collaboration and joint activities indicates that airline alliances vary considerably in the extent or expectation of involvement. These include:

(1) Coordination in ground handling and joint use of ground facilities;
(2) Shared frequent flier facilities;
(3) Code-sharing or joint operation including coordination of flight schedules;
(4) Block space sales;
(5) Joint development of systems, advertising and promotion; and
(6) Joint maintenance and purchase of aircraft/fuel.

For item 1, for example, the extent of coordination is close to very high for both Type 2 and Type 3 alliances, and low for a Type 1 alliance. In contrast, for items 5 and 6 the extent of cooperation is very low for Type 1 and Type 2 and only low/ moderate for Type 3 (Oum *et al.* 2000, p. 33).

Issues of: why form an alliance? and: what are the motives to do so? have been highlighted for two important industry groups. The first – manufacturing – had examples drawn from continuous relationships associated with production including the automotive industry. The second – airlines – was drawn from a global service industry group.

4.1.3 The engineering and construction industry (product and service focused)

A third type of alliancing arrangement that differs in some significant ways to the two previous examples is the alliancing arrangement found in the engineering and construction industry. The case studies discussed in Chapter 3 (section 10, page 63) indicate an alliance between the client, and what would be the principal contractor and major subcontractors using conventional procurement paths. In a useful study of nine building civil and engineering projects studied in the UK (Bresnen and Marshall 2000a, 2000b) the following four motivation categories used in this chapter could be seen as having motivation drivers as follows.

Financial motives
Clients were alliance members and all shared, to varying degrees, in painsharing and gainsharing arrangements with target costs developed with a view to produce a best value solution. The leverage of combined skills and expertise provided access to cost efficiencies though value engineering and constructability exercises. They all involved a high degree of contractor input into the development of the project cost plan, which provides for parties to better manage cost risks. The results of the nine UK projects were that all but one was constructed on or below cost and all on or under time.

Technological motives
While Bresnen and Marshall do not indicate in their papers what specific motivation for access to technological advantage by partners may have been evident they do observe that 'Another common theme was the perceived benefit of being able to build upon long-standing relationships and carry across core teams and workforces from project to project.... This continuity led to considerable familiarity with the technical specifications and working environment, encouraging the

direct transmission of lessons learned' (Bresnen and Marshall 2000a, p. 825). They also observe that sophisticated IT seemed to have been adopted to a limited extent and so there did not appear to be any specific desire to gain technological advantage from IT on those cases. In the cases outlined in Chapter 3 (section 10) of this book the client seemed to be the alliance member gaining most advantage from access to technological expertise of other partners of the alliance in that more options could be considered and rigorously examined and in that process the client could gain benefit from these insights when developing other projects at a later date.

Management motives

There was considerable emphasis on continuous improvement and benchmarking on the nine projects reported upon by Bresnen and Marshall (Bresnen and Marshall 2000a, p. 829). Systematically linking management requirements into other incentives provided benefits of improved management practice for all nine alliance members on each project. In all the projects reported upon as well as the four cases discussed in Chapter 3 (section 10), the experience of working on a project where there was genuine commitment and trust provided a management incentive.

Strategic motives

The main strategic motive evident on all 13 projects here was that it provided non-client alliance partners access to significant projects, to develop the potential for continuing streams of work (and thus cashflow and profits), and to do so with a chance of developing rewarding relationships with other alliance members in an environment where contractual disputes are reduced if not eliminated. The hidden costs of dispute resolution through adversarial approaches is well recognised by alliance partners and the forming of the alliance is indeed a strategy for avoiding this unproductive kind of working atmosphere. A strong sense evident from reports of these projects is that working on alliance projects is a strategic experiment to gain experience in a manner of working that could be the path to a more sustainable and rewarding future for many of the participants.

4.1.4   The National Museum of Australia project

Thus far, alliance member motives have been analysed from three industry sectors. It is now relevant to focus on a specific project as a case study to examine the motivation of alliance partners and to follow how these were translated into a selection process for assembling an alliance partnership for a major project. The case study used is the National Museum of Australia. This recently completed project (opened on 11 March 2001, one day ahead of schedule) was selected because the authors had a unique opportunity to research the project closely from commencement of construction through to its completion. We, therefore, have our own personal insights, observations, and interpretation of what we learned first-hand about this project as well as access to a wealth of documentation and other literature. One of the important resources available is the report of the Auditor-General for the Commonwealth of Australia, which was published a little over half way through the construction phase.

To deal with this section coherently we will discuss the motivation of the client who instigated the National Museum of Australia project. We use a quote from a key proponent of this project whose influence was crucial in determining that an alliance approach should be used. 'What needs to be understood is that … what is being bid is not price but quality of performance' (Service 1999). Thus trying out a new procurement approach that was considered to be possibly 'the way of the future' motivated the alliance partners to join in the experiment. There were some very interesting features of the alliance approach that seemed to attract attention. Five basic concepts of alliancing indicates how the motivation for achieving best value may be achieved by building trust through transparency and a visible, workable reward system:

(1)   The process is absolutely transparent, open book in respect of all costs.
(2)   Transparency applies to overheads and profit margins for all players. In the National Museum of Australia case the profit margins offered by each player were independently tested against industry norms and *audited* against margins included in other tenders by the relevant players.
(3)   The client having established its own budget and obtained independent pricing advice, the Alliance team, which includes the client, established what is called a Target Out-turn Cost (TOC).
(4)   The TOC is the monetary basis for determining rewards for exceptional performance and the way in which pain is borne for unsatisfactory performance.
(5)   There are extensive written quality benchmarks to measure performance: quality in this sense is not just physical quality of construction but everything that bears upon meeting the objective (Service 1999, p. 10).

The National Museum of Australia design and construction project was delivered using an alliance arrangement. This arrangement was chosen because it promised a fast delivery vehicle for a highly complex project of national (Australian) cultural significance requiring very high quality of construction, unique and significantly innovative design and high value for money.[2] There was a great deal of debate within the institutions responsible for developing the brief and project concept as to the best procurement method for its delivery. Design and construct (D&C) was considered, with and without novation, but this proposal was rejected because of the time and cost of several D&C groups each preparing complex designs within the very short time period necessary to ensure that the project would be opened on 11 March 2001 (the decision to proceed on an alliancing basis was made in July 1998). This was considered unlikely and further concerns were raised about design integrity of D&C projects. The National Museum of Australia project is a landmark, iconic project and so the decision to anchor down a design through a design competition was made and administered (Auditor-General of the Australian National Audit Office 2000, p. 43).

---

[2] The authors interviewed Jim Service (ex-chair of the National Museum of Australia Board) and Dawn Casey (director of the National Museum of Australia) who were early project champions highly influential in encouraging an alliancing approach for the project.

Other issues influencing the motivation of the procurement choice concerned the project being subject to strict probity and transparency processes that provide best government procurement practices and offer a model to the construction industry for high levels of ethical relationships between the design and delivery sectors of the industry. There has been much concern expressed relating to ethical and governance issues in Australia as well as the relationships between client and builder, and builder and subcontractor. The 'No Dispute' report called for improved relationships to be forged (NBCC 1989) with other reports and publications calling for similar actions (CIDA 1993; Office of Building and Development 1997; Clayton Utz 1998; KPMG 1998).

The Commonwealth of Australia's procurement guidelines have six core principles: value for money; open and effective competition; ethics and fair dealing; accountability and reporting; national competitiveness and industry development; and support for other Commonwealth policies (Auditor-General of the Australian National Audit Office 2000, p. 45). The alliancing concept met these objectives with the constraints of time ruling out the traditional design-bid-build approach and many other D&C related options.

While the motivation of commercial alliance members prior to commencing the project was unclear, evidence observed suggests that financial and management motivation was strong during the project's execution phase. *Financial* motivation was evident because of a 'fair' rate being struck for the at-risk profit contributions of each commercial alliance member and there was also strong evidence of motivation to achieve and exceed all performance measures. *Management* motivation was evident because all members seemed to enjoy the opportunity to learn lessons from the alliance approach with a view to being successful on future alliance projects. In this sense their motivation could be said to be *strategic*. For example, several design team leaders made the point that the removal of litigation as a likely dispute resolution method allowed design data to be directly transmitted as digital models to the building teams. Key suppliers, and what in other circumstances would be described as subcontractors, formed a sub-alliance for undertaking construction work with Bovis Lend Lease. The use of 3D design models and database information for manufacturing enabled the CAD/CAM experience to realise a *financial* motive (reducing production costs for the very complex design elements) and a *technology* motive (delivering technically complex solutions and learning new skills). For further discussion on this innovation see Chapter 9. Additionally from a *technology* and *strategy* motive point of view, the extensive use of an intranet web-based tool (ProjectWeb) on the project for all information exchange such as drawings, documents, emails, requests for information (RFIs), etc. in the context of the alliance, provided a highly prized learning and experimentation opportunity unique to this project. ProjectWeb is discussed in more detail in Chapter 6.

4.2   The selection process for the National Museum of Australia project

The National Museum of Australia project houses approximately 175 000 items and documents relating to three integrated Australian cultural and heritage themes. It was constructed to a total project budget of A$155.4 million and

achieved the fixed scheduled opening date of 11 March 2001. The historical context of this project is profound. It is an emblem of Australian heritage and a
repository of much of the priceless cultural and heritage artefacts that defined
Australia as an independent nation. The design of the building shares a unique
setting in the landscape of the National Capital Territory (ACT). It is positioned in
a large lakeside precinct that includes Parliament House, the National Gallery and
the High Court of Australia. All these are high profile, institutional icons. The
design of the finished project was required to be distinctive, unique and reflecting
the cultural heritage of the approximate 50 000 years of indigenous people's settlement and the past 200 plus years of European settlement and influence upon
nation building. It was also considered a 'flagship' project for Australia's centenary
of Federation celebrations.

4.2.1 The design selection process

A decision was made that a process of open design competition should determine
the design of the project. This was undertaken between mid-to-late 1997. During
that process 110 eligible registrations of interest were received from the design
competition advertisement of 7 June 1997, of these 76 entries were received. The
Construction Coordination Committee (CCC) of the Department of Communications, Information Technology and the Arts (DCITA) were responsible for the
selection process. An evaluation committee was appointed that included:

- The Chair of the National Museum of Australia (NMA).
- The Program Director of the Indigenous Cultures Program of the Museum of
  Victoria.
- The Deputy Chief Executive of the ACT Department of Business, the Arts,
  Sport and Tourism.
- The Chief Executive of the National Capital Authority (NCA).
- The Deputy Secretary of DCITA.
- Advisors for design implications, landscape and heritage issues and urban
  design issues (Auditor-General of the Australian National Audit Office 2000).

A Stage 1 assessment culled the entries down to 20 during early August 1997
then after receiving comments from the advisors, 11 designs were chosen for discussion and of these, five were shortlisted by mid-August 1998. Two briefing
sessions were held and detailed briefs were given to all five teams who made their
presentations for the second stage of the architectural design evaluation process
during October 1997. An announcement of the successful team was made on 29
October 1997.

It is interesting to note that the winning team was selected on design excellence
and the selection criteria rather than the lowest fee structure being submitted –
though the winning team had offered the lowest fee proposal.

4.2.2 The alliance construction team selection process

In October 1997, the Australian Commonwealth Government Parliamentary
Standing Committee of Public Works (PWC) held public hearings in December

1997 and February 1998 and gave approval to seek project alliance partners to design the project and construct it. This was after commissioning the architectural consultants so they were encouraged to form part of the project alliance team during mid-1998 and took part in the selection process for the successful general contractor and key services contractor alliance group (Auditor-General of the Australian National Audit Office 2000).

The design consortium agreed to take part in an alliance – this was an act of faith and commitment to the project goals through accepting potential risks by entering an alliance where *all* parties share painsharing and gainsharing arrangements together as a single entity. The design consortium leadership group had representatives sitting on the alliance selection panel and thus had considerable influence in decision making that affected their liability and reputation.

The selection process for the successful alliance team for building and services contractors that would include the Commonwealth Government (client) and the successful design competition team was undertaken with an evaluation team comprising (Auditor-General of the Australian National Audit Office 2000, p. 120):

- A legal advisor (Chair).
- First Assistant Secretary of Department of Communications, Information Technology and the Arts (DCITA).
- The Deputy Chief Executive of the Australian Capital Territory Department of Business, the Arts, Sports and Tourism.
- The acting Chief Executive of the National Capital Authority.
- The Architects – Principal.
- A former member of a construction company.
- Advisers: project managers, cost consultants, the architects, DCITA, alliance facilitators, probity adviser, probity auditor.

The strategic partner selection process is also significantly different to any means adopted by traditional or even partnering procurement arrangements. Figure 4.1 illustrates the process of selecting the building contractors and services contractors for the alliance. This model was presented at the first National Museum of Australia dissemination seminars held in early 2000 by the project manager for the alliance, Peter Wright, in venues in Melbourne, Canberra, Sydney and Brisbane. It was based on an earlier version of this model (see Figure 3.10).

The call for proposals was initiated on 16–18 May 1998 though advertisements in the national press. An industry briefing was conducted on 22 May 1998 at which approximately 70 people attended. The call for proposals addressing the 12-point criteria outlined in Chapter 3 (section 8) closed on 29 June 1998 and 10 proposals were received. The evaluation team undertook a rigorous and structured assessment of proposals to ensure that they met the requirements of addressing the criteria and four companies were shortlisted for interview that lasted three-and-a-half hours for each consortium. All four consortia contained major construction companies with both a national and international presence. The committee were initially inclined to follow the KPMG (see Figure 3.9) model and reduce the number of proponents to two for the two-day workshop; however, the assessed scores for the second and third ranked proponents were very close and so it was

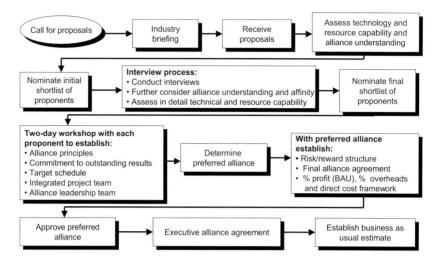

Figure 4.1:   Alliance selection process
Source: Peter Wright, Project Manager

decided that three would proceed to that stage of the process. The process went well beyond the rigour of a several-hour interview.

The purpose of the workshops was to experience working together on the key alliance principles in order to select a final alliance member of the Acton Peninsula alliance. More specifically:

- To experience working together on alliance principles.
- To develop the alliance principles, which will be used as guiding principles for the members of the alliance.
- To discuss and identify outstanding results that could be accomplished by the alliance team.
- To create the proposed alliance leadership team, its members and the proposed role for this team.
- To gain an appreciation for how it would be to work together in an alliance relationship.

Proponents sought and received coaching and simulation prior to the activity training with their consultants who were expert in this approach. After the workshops were completed the Evaluation Committee reassessed the consortium against the selection criteria and notified the preferred alliance team and they were invited to attend the Finalisation Workshop.

One interesting comment passed to us was that while it may be possible to maintain a company 'front' for a couple of hour-long interviews, it was impossible to hide real intentions, sympathies or prejudices during the two-day intense workshop with dinner and drinks sessions designed to break down inhibitions and highly formal behaviour. The two-day workshop appears to have effectively revealed true intentions, vulnerabilities and obvious strengths of all proponents.

The Finalisation Workshop was used to determine the commercial and contractual issues that included risk and reward. Due to commercial sensitivities these details can't be released. In general, it can be revealed that matters of finalising the alliance roles, responsibilities, the risk/reward structure, profit contribution, overhead arrangements, and direct cost framework details were established. It should be noted, as indicative of the atmosphere and philosophy of alliancing, that profit margins to be placed at-risk as part of these arrangements were based on an open-book auditing of profit margins achieved over a representative period. The alliance members did not negotiate their margins as such but agreed that their average business profits achieved over the representative period or the previous three years would be used. This is a radical departure from a negotiation where one party seeks to gain advantage over another. This action established the tone of ethical dealing, 'fair play' and established the foundation for trust and commitment amongst alliance members. The message was clear, the remuneration would be based on a market-based return so that alliance members would neither be disadvantaged by committing their most talented and valued staff to the project, nor seek to extract a profit level above that expected under normal business conditions. The risk/reward structure supplied the vehicle for rewarding the delivery of excellence that the alliancing objective sought to achieve.

Once the team was established they worked together and established the business-as-usual (BAU) estimates which were independently checked against cost plans and other cost data. This ensured that a competitive Target Out-turn Cost (TOC) was developed. In this process with the client and design team as alliance partners with the building and services contractors working each organisation became a single team focused on 'best-for-project'. The TOC was developed from the winning design scheme and refined within the boundaries of the design integrity of the winning submission. The contractors brought with them their experience in value analysis and other techniques for ensuring value for money not just for capital expenditure but also considering the whole-of-life costs of the project. In a sense this established some of the advantages of a BOOT scheme in ensuring that operating and maintenance costs were fully considered. Construction of the project commenced in February 1999.

The interviewing process to derive a shortlist of potential alliance members requires sophistication and judgement of a client as do the facilitated workshops. Clearly, the alliance approach is difficult for novice clients. Once a successful alliance team is established, the final alliance agreement can be formulated including the alliance charter, the target costs and time, and other performance requirements and the risk/reward agreements. Once the alliance conditions are approved by the project funding body the project can be executed.

An important issue that must not be ignored is that the construction alliance group was required to include not only the general building contractor but also the services contractors. This addressed a generally under-recognised fact that often the services component of building works can account for a substantial and significant proportion of capital costs and contribute to overall running costs through the life of a project. It makes best-value sense to draw the services contractors' potential decision-making contribution into the detailed design process and general production phase. Each contender needed to deliver a coherent and complimentary presentation in which the combined organisations would address each criterion. The interview process placed a great deal of pressure on contenders achieving a coherent and complimentary submission.

After selection of the successful alliance partner, each of the unsuccessful contenders was given the opportunity for a debriefing meeting with representatives of the selection panel to provide feedback on reasons why their submission and presentations were not successful. Several of the contending firms declined the opportunity. Of those who took the opportunity, all approached the debriefing in a positive manner and sought to learn from the exercise. Feedback to the debriefing panel indicated that there was unanimous support for the use of alliancing, that traditional contracting does not produce a good outcome for everyone and that there was a desire to develop alliancing further in the building construction industry. It was also interesting that one of the contenders made the point that their company had not been accustomed to maintaining records that were necessary for many of the above criteria. The innovation, EMS, and community development criteria in particular require not only a new mindset in contractors but also the willingness to maintain records of achievements in these areas.

Several feedback comments concerned surprise that the process was so rigorous, transparent and professional. This placed acute demands upon participants to achieve very high levels of communication, presentation and persuasive skills related to factual matters of substance rather than sales and marketing rhetoric. One interesting comment made was that the cost of the presentation and its preparation by aspiring alliance teams was considered to be less than the cost of tendering for an equivalent project. This was primarily because there was no tendering based on cost estimate preparation or detailed planning. While considerable effort was required in preparing proposals based upon the 12 criteria outlined below, this was developed from existing presentation and management information resources and it was limited to explaining how this kind of project could be delivered without detailed reference to design details.

The assumption governing the process was simply that by having the best qualified people involved working together with the best interests of the project uppermost in their minds, the best and most effective and appropriate solutions would emerge.

This is a substantially different mindset to that which prevails in other project delivery strategies including those that utilise partnering where mechanisms for cooperating and managing disputes is pivotal.

4.2.3 Selection criteria

The project's design is highly innovative, complex and unique and makes significant demands upon a construction team well beyond that normally to be expected for an institutional building project. This placed severe quality and buildability demands upon the construction management team and the workforce. As a 'flagship' project with highly sensitive political significance and a fixed scheduled opening date, the project presented a potential industrial relations nightmare. It was necessary that workplace relations would be managed to deliver industrial harmony, enable innovation, and ensure that quality and time project objectives be met. Moreover, the project's opening was the first 'gift to the nation' that began the national celebrations marking Australia's birth as a federated independent nation. A traditional design-bid-build or design and construct procurement approach was, as indicated earlier in this chapter, not considered the optimal approach to meet the government's exacting requirements.

Selection of the construction alliance members was based upon 12 criteria.

(1) **Demonstrated ability to complete the full scope of works including contributing to building, structural, mechanical and landscaping design**. The focus was on proof of performance on complex projects similar to the Museum project where the proponents had actively contributed to a design process to improve project outcomes. At least three examples were requested. Evidence of buildability or constructability was sought that proved that the contractor was capable of contributing positively to detailed analysis of design to improve productivity while maintaining design quality integrity.

(2) **Demonstrated ability to minimise project capital and operating costs without sacrificing quality**. At least ten examples were requested of projects brought within or below budget demonstrating evidence of understanding of life cycle costs. Value analysis would be an appropriate technique to have been used here to demonstrate lifecycle costs savings as well as smarter ways of achieving a quality outcome. Again, sophistication in the use of tools and techniques that challenge design assumptions and obtain a smarter solution was sought.

(3) **Demonstrated ability to achieve outstanding quality results**. A minimum of three examples of what was considered to be of outstanding quality was required. Testimonials, industry and professional association awards provide suitable evidence supplemented by a formal presentation with photographs and/or other clear forms of evidence to convey quality performance. The successful contractor consortium had to demonstrate that they could produce excellent quality.

(4) **Demonstrated ability to provide the necessary resources for the project and meet the project program**. At least three projects greater than the A$50 million level were required to be used to demonstrate this capacity. The organisation chart and CV of key staff was also a source of evidence together with mobilisation plans and global method statements of how the work was planned and organised. The client-nominated auditor also assessed financial capacity. This requirement was necessary to ensure that only financially sound and capable partners would be selected. The financial systems used by the alliance partners may also be an issue to ensure that they meet requirements of an open-book approach and a capacity for transparency to the auditors.

(5) **Demonstrated ability to add value and bring innovation to the project**. At least three examples were required of process improvements introduced over the past three years. This required a demonstrated commitment to continuous improvement, innovation and/or breakthrough invention. The construction industry is perceived to be a generally poor performer at introducing innovation as compared to other production and manufacturing

industries (Lenard 1996; Lenard and Bowen-James 1996). Bringing an innovative mindset was assumed to be a precursor to helping develop a target outcome cost that represents better value for money and good construction time performance as opposed to merely reducing scope or cost cutting in a reactive manner. Innovation is also highly important in overcoming unforeseen problems when they arise.

(6) **Demonstrated ability to achieve outstanding safety performance**. This required at least an example of a past safety plan from a previous project and presentation of supporting data such as lost claims/million man hours over the past three years, corporate OHS safety policy and how policy was translated into action. OHS is an especially sensitive issue for a government client as government represents all stakeholders involved in projects including the public.

(7) **Demonstrated ability to achieve outstanding workplace relations**. At least three years of data and statistics of performance on disputes and how they were managed. Corporate workplace policy and action plans and evidence of the nature and experience of workplace agreements over the past three years. The government was also especially sensitive to past poor workplace relations and the effect on construction time performance (Ireland 1983; Ireland 1988). The project opening date and its symbolic significance was an objective that could not be compromised.

(8) **Successful public relations (PR) and industry recognition**. At least three examples of successful PR and industry recognition from previous projects such as proactive community involvement, previous track record of managing community expectations and credible stakeholder involvement. Examples of where a potential PR disaster may have been turned around. This requirement is interesting in that one aspect is about 'spin doctoring' – an ability to represent a poor performance in its best light. A second aspect is that of being part of a showcase exercise that generates kudos not only for the project and its delivery team but also for the government as project sponsor.

(9) **Demonstrated practical experience and philosophical approach in the areas of developing ecological sustainability and environmental management**. A minimum of one environmental management system (EMS) plan developed and implemented was required. There was also a focus on nominated people having good understanding, experience, and qualifications to formulate and manage EMS plans. This requirement demonstrates greater concern and awareness by government for the environmental impact that the construction industry has upon society in general. This requirement is interesting as it indicates greater interest in the concept of a triple bottom line[3] (Elkington 1997).

(10) **Demonstrated understanding and affinity for operating as a member of an alliance**. Each of the participating companies was required to provide

---

[3] To demonstrate a positive contribution to sustainable development, any company/project must not only deliver sound financial success but must also contribute beneficial social as well as environmental outcomes.

examples of working in a non-adversarial and collaborative manner as well as demonstrate their views on participating in risk/reward schemes. The willingness wholeheartedly to support and embrace the alliance philosophy was required. There was a focus on ideas, team working, sound past relationships and general knowledge about the alliancing concept. This criterion was about a demonstrated capacity for building, developing and maintaining trust and commitment.

(11) **Substantial acceptance of the draft alliance documented for the project including related codes of practice, proposals for support of local industry development, employment opportunities for Australian indigenous peoples.** There was a focus on outstanding record of working with government, local communities and accepting broader responsibility for an ethical and socially responsible manner of working. The government was very keen to be proactive in supporting the local indigenous community. Further, part of the project included a new education and administrative facility for the Australian Institute of Aboriginal and Torres Strait Island Studies (AIATSIS) so this organisation formed part of the client group. There had been a natural desire, therefore, to ensure that benefits would flow to indigenous peoples and that this project could provide a model for demonstrating how improved support for indigenous peoples could be achieved.

(12) **Demonstrated commitment to exceed project objectives.** This required a demonstration that the proposed alliance partnership was truly committed to the project ethos with highest level corporate championing and an understanding of the calibre and qualities that differentiated the project needs from a business-as-usual case where conflict and adversarial actions prevail.

Similar principles were adopted in developing sub-alliances between the main alliance through the building and services engineering contractor. These were with the steel fabricator, the glass and aluminium fabricators, the landscapers and the Audio Visual and Information Technologies specialists. There are some interesting features of the above criteria. They appear to reveal three clusters of criteria each with distinctive motives. These are tabulated in Table 4.2.

Traditional motivational criteria relate to the business-as-usual situation of cost, time, quality and financial capacity to undertake the work. This does not serve to encourage any innovation beyond profit maximisation through working in a clever or effective manner. The second cluster headed stakeholder recognition is interesting as it not only recognises an obligation for excellence in occupational heath and safety (OHS) and environmental management systems (EMS) but also goes beyond protecting workers and the general public. The focus on excellence in workplace relations could be seen as a protective measure to minimise risks associated with industrial disputes being associated with large or high-profile projects that has plagued the Australian construction industry in the past (Ireland 1983; Ireland 1988; Walker 1994, p. 89–103). However, this cluster also contains criteria that could be viewed as addressing triple bottom line issues where not only financial considerations are reported against and considered but also environmental sustainability and social sustainability (Elkington 1997). The third cluster

Table 4.2: Motivation clusters for the National Museum of Australia selection criteria

Traditional	Stakeholder recognition	Attitudinal issues
(1) Ability to complete the full scope of works	(6) Achieve outstanding safety performance	(5) Add value and bring innovation
(2) Minimise project capital and operating costs without sacrificing quality	(7) Achieve outstanding workplace relations	(10) Understanding and affinity
(3) Achieve outstanding quality results	(8) Successful public relations (PR)	(12) Commitment to exceed project objectives
(4) Provide the necessary resources for the project and meet the project programme	(9) Ecologically sustainability and environmental management	
	(11) Support of local industry development, employment opportunities for Australian indigenous peoples	
These are the standard performance measures for most if not all projects	(6) OHS and (9) EMS are becoming more important due to duty of care. (7) and (11) in combination reflect interest in triple bottom line	These reflect sophistication in creating value for money through a real contribution to an intelligent way of working collaboratively

relate to issues concerning a mindset that supports the alliance concept and working collaboratively.

It was important to identify what motivates each of the members in the alliance. These motivators are pivotal in establishing successful risk and reward relationships that are based on agreed targets. Rewards, however, include intrinsic factors such as job satisfaction, recognition, access to learning and other less obvious factors that enhance the experience of working on projects such as participatory decision making. These rewards were not made explicit, yet as others have argued, these should be recognised in designing and evaluating partnering and alliancing relationships (Bresnen and Marshall 2000b). A risk and reward graph indicates the gain/pain share of the overall success of the project measured against key performance indicators. Risk and reward provisions encourage but do not guarantee cooperative behaviour between alliance members, including the client (Bresnen and Marshall 2000a).

The common key performance indicators are time and cost. Depending on the particular project other key performance indicators may be included, such as quality, the environment and safety. Figure 4.2 illustrates the risk and reward relationships. The risk and reward structure for the National Museum of Australia is made up of cost, time, design integrity and quality.

All components have a down side (pain) and an up side (gain) except for time. There is no reward for finishing early but there is very significant financial pain for even one day's delay. Before risk and reward can be established there must be an agreed Target Out-turn Cost (TOC). The TOC is the agreed amount the building would cost to deliver, as determined by the entire alliance leadership team (ALT).

It should be noted that there was no reward for early completion and a heavy

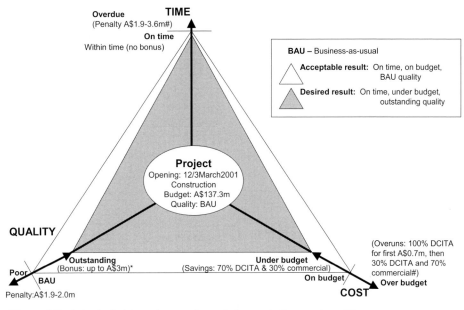

Figure 4.2: A typical risk and reward graph for cost, time and quality
Source: Auditor-General of the Australian National Audit Office (2000, p. 103)

penalty of $1.9 million for handing over the project even one day late rising to a maximum of $3.6 million on a sliding scale for a three month delay. Cost over-runs for the first $0.7 million would be accepted by DCITA and thereafter shared on the basis of 30 per cent of DCITA and 70 per cent the commercial alliance members. Quality monitoring is complex and will be further explained later. The aim was to encourage excellence in quality and so incentives to plough cost savings back into improving quality was factored into the reward formula. There are no rewards for BAU quality and a penalty of between $1.9 and $2.0 million for failing to meet the BAU benchmark measure. A quality pool of $3.0 million provides the incentive to achieve an outstanding result.

An independent Quality Panel was established and maintained surveillance and review of the project throughout construction and for 12 months after the project handover. Quality measures reach beyond the required norm, such as concrete strength and building code regulation measures. From 1998 to mid-1999 a Quality Working Group comprising members from each of the alliance partners, the alliance facilitator, an independent technical advisor and specialist advisor with direct experience of developing quality measures for another alliance project (the Sydney Water alliance). DCITA appointed three consultants to form the panel and their backgrounds include architecture and construction, exhibitions and museum design, and construction project management (Auditor-General of the Australian National Audit Office 2000). Table 4.3 provides the detailed break-down of the quality measures and their weighting factors. For greater detail on this aspect see Chapter 5, section 5.2.4.

Each component was scored on a scale of −10 (poor) to +10 (outstanding) with 0 representing BAU. Some measures could use statistical data, for example lost

Table 4.3: Quality measures for the project		
Category/sub-category	**Quality measure**	**Weighting factor**
Buildings and exhibitions		
Buildings	• Quality of build finishes	5
($0.9 million quality pool)	• Non-conformances	2
	• Defects	2
Exhibitions	• Design quality	3
($0.9 million quality pool)	• Use of content	3
	• Integration of technology	1
	• Accessibility	1
	• Visitor experience	2
Construction phase		
Environment	• Environmental management	1
($0.3 million quality pool)	• Waste management	2
	• Water quality	2
	• Air quality	1
	• Energy efficiency/greenhouse gas emissions	3
	• Ecologically sustainable development	1
Indigenous employment opportunities	• Enhancing opportunities in construction period	1
($0.3 million quality pool)	• Enhancing opportunities beyond construction period	1
	• Training	2
	• Employment	2
	• Supportive workplace	1
Public relations	• Promoting the site	1
($0.3 million quality pool)	• Industry recognition of alliancing	2
	• Stakeholder image	2
Safety	• Management processes	1
($0.3 million quality pool)	• Safety outcome	3
	• Individual attention	2

Source: Auditor-General of the Australian National Audit Office (2000, p. 100).

time injury (LTI) measures for OHS where BAU is, say, 1:21,000. Qualitative judgements comprised a large proportion of the measures and survey information was also used (for example for public relations or customer satisfaction surveys). Examples from the National Museum of Australia Draft Quality Measures indicate how these measures can be linked to selection criteria via a quality pool concept, see Table 4.4 and Table 4.5 below. The approach taken on this project provides a best practice model for alliance member selection that is rigorous, well thought through from an implementation perspective as well as providing an opportunity to address the performance needs of a wide range of stakeholders.

An alliance leadership team (ALT) was established for the National Museum of Australia alliance to manage the project much like a board of directors ultimately manage a company. This supports member motivation and inclusion in decision making and 'having a voice'. All alliance partners were represented on that team. A policy of devolving dispute resolution as far as possible was instigated much as is the case with many partnering charters. The ALT was responsible for making

Table 4.4: Water quality measures – use of statistical measures

Poor	Business as usual	Outstanding
Discharge greater than 10% above mean background samples	Discharge to Lake Burley Griffin has suspended solids level of not more than 10% above mean background samples	Discharge to Lake Burley Griffin has suspended solids level of not more than 5% above mean background samples
OR	OR	OR
Discharge to Lake Burley Griffin greater than 60 mg/L of suspended solids on more than one occasion	Discharge to Lake Burley Griffin to be less than 60 mg/L of suspended solids	Discharge to Lake Burley Griffin to be less than 30 mg/L of suspended solids
Weighting 1		

Source: National Museum of Australia Draft Quality Measures.

project policy decisions and where necessary, resolving disputes or issues needing that high level input. These were reached by consensus of ALT members based upon what was best for the project rather than for any individual party. This supported and maintained commitment so that effectively each ALT meeting (generally monthly) became an alliance reinforcement exercise. The alliance contract virtually bans litigation. One of the authors (Renaye Peters) attended many meetings as part of her research role and observed that opinion and argument was resolved at this level by the ALT in a professional manner by ALT members in a way that power differential seemed to play no discernable part. There seemed to be a genuine attempt to form a collegial atmosphere where diverse views were valued. Further, heated discussion and debate surrounding many issues and diversity of views appeared to be welcomed, encouraged, expected and managed on a 'what is best for project' basis.

Figure 4.2 illustrates the alliance as it emerged. The company and organisation names are presented below in abbreviated form. Names are less important than the nature of team groups that coalesced, gaining critical mass, to enable them to be considered as alliance members. It is interesting that no single design team could fulfil the criteria and so each organisation had first to form a syndicate or sub-alliance to have the opportunity to submit a credible proposal. Similarly, the building contractors needed the services contractors in their group. The museum

Table 4.5: Supportive workplace measures – use of qualitative measures

Poor	Business as usual	Outstanding
No effort is made to support indigenous employees on the site. Indigenous employees feel unwelcome on the site	Reactive approach to any issues concerning indigenous employment. Indigenous employees feel neither particularly welcome nor unwelcome on site	Proactive approach to foster a supportive culture for indigenous employees on site. Indigenous employees feel welcome on site
Weighting 1.4		

Source: National Museum of Australia Draft Quality Measures

exhibition designer procurement process has not been discussed in this chapter though this was undertaken in parallel with the building and services contractor alliance selection process. Figure 4.3 illustrates the complexity of the alliance and illustrates how sophisticated a client body must be to contemplate a process such as that embarked upon on this project.

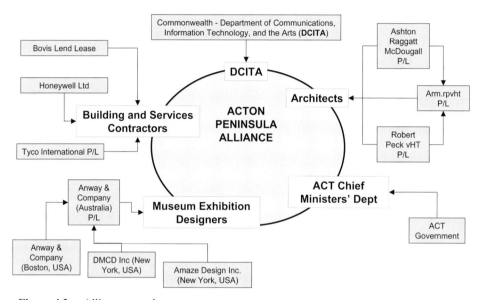

Figure 4.3:   Alliance members
Source: Auditor-General of the Australian National Audit Office (2000, p. 38)

4.3   Chapter summary

The important distinction drawn between partnering and alliancing is that with partnering aims and goals are agreed and dispute resolution and escalation plans are established but partners still retain independence and may individually suffer or gain from the relationship. The contractual relationship between the client and contractor is similar to a traditional contract. With alliancing, the parties form a cohesive entity, which jointly shares risks and rewards to an agreed formula. Thus if the project is delivered one day late for example, *all* partners jointly *share* the penalty. Rewards are likewise awarded for successfully exceeding expectations. The contractual arrangements are significantly different.

This chapter began with a substantial section that explored the question: why form an alliance? We saw that three important purposes are co-option, co-specialisation, and learning and internalisation. Alliance members cannot forget that they may be supporting the elevation of a newcomer or follower into a leadership position in an industry segment or business area. This is often desired when organisations seek to re-invent themselves through a transformation of potential value that they can contribute to their customer base. Other times, maintaining a status quo or blocking competitive pressure is a real issue to an alliance relationship. We saw that emerging B2B ventures are forcing sophisticated organisations to form alliances to present customers and clients with a coherent, seamless, and effective means to meet their needs. E-commerce activity has shown us that alliances are not only desirable but also pivotal in moving towards a future where innovation and effectiveness contribute to an improved way of conducting business to satisfy society and organisational needs. We also saw through the work of Wise and Morrison (2000) that alliancing in the B2B world of the twenty-first century may facilitate the re-invention of organisations to unearth value in their competencies and capacities. The model for future B2B that they present is fundamentally rooted in the concept of alliancing and motivation for creating better value through *financial, technological, managerial* or *strategic* means. An interesting link to these motivations is that they represent change from a state of complacency to one of alert intent to compete globally, to focus on customer needs and to provide an intellectually challenging and otherwise rewarding environment in which people wish to contribute their energies, knowledge and spirit of adventure.

One of the most significant implications that can be drawn from the first section of this chapter, in which the rise in alliances in the e-commerce area was briefly discussed, is that e-commerce is ripe for re-configuring the supply chain. This potential cannot be underestimated. Business is already being radically challenged by the need for e-commerce solutions and a greater focus on networking organisation to achieve this. The traditional business enterprise is about to be assailed by new competitors with new approaches in which alliances will form a significant feature (Kanter 2001). The construction and engineering industry, for example, tends to be composed of a majority of traditional organisations that are ill-prepared for this kind of shock. Extensive re-engineering of the construction industry has been shown to have the potential to deliver substantial benefits to customers and the construction industry (T40 1994; Koskela 2000). It is unclear whether the implication of this will be that increasing use of alliancing will spur greater focus on e-commerce or vice versa for the construction industry and other similar business sectors. The authors' view is that these two factors will drive industry changes hand-in-hand.

This chapter moved from consideration of traditional business to project-based enterprises. The way that alliance partners are selected on the basis of service provision first, and cost considerations later is a novel project delivery approach.

Rigorous selection criteria and process is an important issue for clients who have either little inclination to place so much emphasis on a transparent and ethical process or do not have, or wish to commit, the required resources to perform a selection process in this manner. The approach adopted in the National Museum of Australia project was highly time and resource consuming from the client's viewpoint, but necessary for public accountability to demonstrate that the client's risk acceptance by not first selecting on the basis of lowest price was well founded.

The alliance team's formation of a management group as a true joint management team with democratic membership ensured that trust and commitment was truly encouraged and manipulation discouraged by the system of alliancing was an important feature. This supported trust and commitment and resulted in enhanced perceptions of the desirability of working on the project. Results from a survey of trust and commitment issues (unpublished at this point of writing) indicates that questions such as 'We *share* technical and commercial *information* relating to our projects without the need to protect ourselves' returned a response of double the confidence measure than a business-as-usual measure. Another indicative question asked was 'I feel part of the project's community' which also had an almost double score response comparing the business-as-usual situation with the alliancing project[4]

Risk/reward arrangements also encouraged a team approach to innovative problem solving. The National Museum of Australia project adopts practices followed by other alliance projects in the energy and mining industries, for example the Andrew Drilling Platform project – North Sea UK (KPMG 1998) and the East Spar Development (ACA 1999). The success of transferring this form of project procurement from heavy engineering construction to building construction is a focus of this Australian milestone project.

The National Museum of Australia project provides a useful illustration of alliance member selection in practice. The selection process was discussed and a detailed model of the process presented. The key selection criteria were also presented and discussed in detail and the link between motives, selection criteria and performance measurement was explored. Vitally important distinctions were drawn between partnering and alliance approaches. Teams or partners in partnering arrangements are often selected with project price determination as being a key component. With alliancing, the philosophy was to first select the best possible team of project partners and then develop the design in line with its original integrity but using the intelligence, skills and abilities of the alliance team to seek efficiencies – in buildability and administration – as well as to mould the project outcome to meet the budgeted out-turn cost[5] representing best value for the project's scope and quality.

The philosophy for the alliance concept can be summarised as follows. When the best available people are hired to work in a truly collaborative and cooperative way, then the project outcome will represent best value. This focus on gaining access to a project opportunity through demonstrating that an organisation has the best available people has far reaching consequences. There are implicit implications in this about attracting, hiring and retaining talented key people. The success of organisations wishing to be alliance members will be increasingly dependent upon the quality of their employees, the tacit knowledge embedded in employees' beings and the capacity of people to work constructively in alliance teams in which they develop trust and commitment.

Best value primacy may be the defining element of an alliancing approach. While the National Museum of Australia project presents only one case, findings from this case study cannot be generalised. However, it presents a most useful example of alliancing experience for building construction projects.

Finally, the selection process and the behavioural characteristics required of both client and the design/construction teams indicates that this approach should be limited to clients and team partners who share a sophisticated understanding of how true collaboration may be established, maintained and undertaken. Alliancing requires a sophisticated and involved client to drive and benefit from the

---

[4] It is beyond the scope of this chapter to discuss the various surveys and data gathered on this case study project. Several key papers will emerge over the coming year that will deal with these in depth.
[5] Literally the cost that the project would 'turn out' to cost.

process. Clearly, novice or unprepared team partners might find this approach too challenging to fully reap benefits that can be gained including the vital one of project success in terms of the quality of inter-team relationships.

References

ACA (1999) *Relationship Contracting – Optimising Project Outcomes*, Sydney, Australian Constructors Association.

Argyris, C. and Schon, D. (1996) *Organizational Learning II: Theory, method, and practice*. Reading, MA, Addison-Wesley.

Auditor-General of the Australian National Audit Office (2000) *Construction of the National Museum of Australia and the Australian Institute of Aboriginal and Torres Strait Islander Studies, Audit Report*. Canberra, Australia, Australian National Audit Office.

Barlow, J., Jashapara, A. and Cohen, M. (1998) 'Organisational Learning and Inter-firm 'Partnering' in the UK Construction Industry.' *The Learning Industry Organization Journal*. **5** (2): 86–98.

Bresnen, M. and Marshall, N. (2000a) 'Building Partnerships: Case studies of client–contractor collaboration in the UK in construction industry.' *Construction Management and Economics*. **18** (7): 819–832.

Bresnen, M. and Marshall, N. (2000b) 'Motivation, Commitment and the Use of Incentives in Partnerships and Alliances.' *Construction Management and Economics*. **18** (5): 587–598.

CIDA (1993) *Building Best Practice in the Construction Industry – a practitioner's guide*. Sydney, Commonwealth of Australia.

Clayton Utz (1998) 'Alliance Contracts: A glimpse of the future.' *Australian Construction Law Newsletter*. August/September: 7–8.

Doz, Y.L. and Hamel, G. (1998) *Alliance Advantage – The art of creating value through partnering*. Boston MA, Harvard Business School Press.

Elkington, J. (1997) *Cannibals with forks*. London, Capstone Publishing.

Hamel, G. and Prahalad, C.K. (1994) *Competing for the Future*. Boston, Harvard Business School Press.

Ireland, V. (1983) *The Role of Managerial Actions in the Cost, Time and Quality Performance of High Rise Commercial Building Projects*. Sydney, University of Sydney.

Ireland, V. (1988) *Improving Work Practices in the Australian Building Industry – A comparison with the U.K. and U.S.A., Research report*. Sydney, Master Builders Federation of Australia and University of Technology Sydney.

Kanter, R.M. (2001) 'The Ten Deadly Mistakes of Wanna-dots.' *Harvard Business Review*. **79** (1): 91–100.

Koskela, L. (2000) *An Exploration Towards a Production Theory and its Application to Construction*. VTT Technical Research Centre of Finland. Helsinki, Finland, Helsinki University of Technology.

KPMG (1998) *Project Alliances in the Construction Industry*, Literature Review. Sydney, NSW Department of Public Works & Services.

Lenard, D.J. (1996) *Innovation and Industrial Culture in the Australian Construction Industry: A Comparative Benchmarking Analysis of the Critical Cultural Indices Underpinning Innovation*. Newcastle, University of Newcastle, Faculty of Architecture.

Lenard, D.J. and Bowen-James, A. (1996) *Innovation: The Key to Competitive Advantage*. Adelaide, Australia, Construction Industry Institute Australia, University of South Australia.

NBCC (1989) *Strategies for the Reduction of Claims and Disputes in the Construction Industry – No dispute*, Canberra, National Building and Construction Council.

Office of Building and Development (1997) *Partnering and the Victorian Public Sector*, Melbourne, Australia, Office of Building and Development, Department of Infrastructure, Victorian Government.

Oum, T.H., Park, J.-H. and Zhang, A. (2000) *Globalization and Strategic Alliances – The Case of the Airline Industry*. Oxford, Pergamon.

Porter, M.E. (1990) *The Competitive Advantage of Nations*. New York, Free Press.

Porter, M.E. (1998) 'Clusters and the New Economics of Competition.' *Harvard Business Review*. **76** (6): 77–90.

Roe, A.G. and Phair, M. (1999) 'Connection Crescendo.' *Engineering News Review*. May: 22–26.

Sawhney, M. and Parikh, D. (2001) 'Where Value Lives in a Networked World.' *Harvard Business Review*. **79** (1): 79–86.

Service, J. (1999) Alliancing for richer, for poorer. *The Chartered Building Professional* 8–10.

Stuart, F.I. (1999) 'Supplier Alliance Success and Failure: A longitudinal dyadic perspective.' *International Journal of Operations & Production Management*. **17** (6): 539–557.

T40 (1994) *T40 Construction Research*, Canberra, Australia, Australian Building Research Grants Committee, Department of Industry, Technology and Regional Development.

Walker, D.H.T. (1994) *An Investigation Into Factors that Determine Building Construction Time Performance*. Melbourne, RMIT University, Department of Building and Construction Economics.

Whipple, J.S. and Gentry, J.J. (2000) 'A Network Comparison of Alliance Motives and Achievements.' *Journal of Business & Industrial Marketing*. **15** (5): 301–322.

Wise, R. and Morrison, D. (2000) 'Beyond the Exchange – The future of B2B.' *Harvard Business Review*. **78** (6): 87–96.

Womack, J.P. and Jones, D.T. (2000) From lean production to lean enterprise. *Harvard Business Review on Managing the Value Chain*. Boston, MA, Harvard Business School Press: 221–250.

Womack, J.P., Jones, D.T. and Roos, D. (1990) *The Machine that Changed the World – The story of lean production*. New York, Harper Collins.

Yin, R. (1994) *Case Study Research*. Thousand Oaks, California, Sage.

Chapter 5

Managing Risk and Crises Resolution – Business-as-Usual Versus Relationship-based Procurement Approaches

Derek Walker and Martin Loosemore

We saw in Chapter 3 that problem resolution is a key advantage of partnering and alliancing relationships. In this chapter we will extend the brief discussion undertaken in Chapters 3 and 4 relating to managing problems, risk and crises. We will compare the business-as-usual and an alliance approach and explore how an alliance or partnering approach can obviate many of the crises and problems that arise in traditional contracting situations. We address the following:

- What is the difference between risk, uncertainty, problems and crises?
- How does a business-as-usual crisis typically develop from its risk management approach and how does a relationship-based procurement experience differ from the business-as-usual approach?
- What key characteristics of a relationship-based procurement approach help teams to build solutions to problems and potential crises that they may confront?

5.1 Introduction – risk, uncertainty and crises

One of the fundamental tasks of project management is risk identification and management. While there are plenty of texts dealing with risk management, see for example Flanagan and Norman (1993) and Raftery (1994), we feel it relevant to define terms and briefly describe risk management strategies. Two terms that frequently are used in project management are risk and uncertainty. **Risk** is a measure of something likely to happen in the **future**. It is usually understood as being problematic although there is a potential positive side to all risks which is often ignored. Risks are usually expressed in terms of a ratio or percentage that something is likely to occur, i.e. a 1 in 20 or 5 per cent chance and impact in terms of low, medium, high, etc. intensity. In assessing any project we are all confident that we understand or know something about what problems might arise but are unsure about their degree or impact – this represents uncertainty.

Uncertainty is recognition that all is not known. We can anticipate some risks based upon past experience or indicated from the literature but there are other risks that we cannot or do not have the background knowledge to anticipate (i.e. that which is **unknown**) and problems that we are unable to foresee – unknown or unanticipated problems. Every project presents unknown risk and uncertainty.

Risk management is the process of attempting to understand and quantify risks and developing a contingency plan that may be implemented to avoid or mitigate them. Residual risks can be retained and insured against or absorbed into a management reserve (contingency) fund for later use, if need be. Alternatively, we can transfer risk through outsourcing arrangements to those better able to manage it or share it with willing partners in a joint venture.

Companies without effective risk management strategies are unable to predict or react effectively to problems and are vulnerable to crises. Organisational **crises** are unexpected risks that develop into decisive periods of acute difficulty, which threaten the viability of an organisation, its business units or key products (Aguilera and Messick 1986; Fink 1986). What distinguishes a crisis from a day-to-day problem is the extreme sense of urgency that hyper-extends an organisation's coping capabilities, producing stress and anxiety among organisational actors and stakeholders (Allen 1990; Pearson *et al.* 1997).

Before discussing the management of crises it is important to understand that there are at least three different types of crises: *creeping, sudden* and *periodic*, each has a different required response (Jarman and Kouzmin 1990).

Creeping crises are generally systemic – they are often something that should have been anticipated and seen as inevitable at some time or other. Refusing to acknowledge a perilous behaviour could fall into this category such as poor occupational health and safety (OHS) practices or ingrained prejudice resulting in legal action over employee discrimination. Another example could be ignoring a serious potential environmental management systems (EMS) risk such as waste liquid spills and escapes from building sites. These crises are symptomatic of poor management and/or a poor communication process and are often pre-dated by unheeded warning or ignored tell-tale warning signs. A management system that condones complacency in OHS or EMS procedures is almost sure to experience this kind of crisis sooner or later. Poor communication systems may be caused by 'corporate deafness' (not wanting to hear or acknowledge disturbing news) or poor coordination of information within or across team/organisational boundaries. An open management style that accepts and welcomes diverse views and opinion is less likely to be susceptible to creeping crises. Relationship-based procurement systems have greater propensity to obviate this kind of crisis because the very nature of relationship-based procurement systems is one of inclusion of project stakeholders and teams and the integration of these groups through an inclusive information communication system.

A **sudden** crisis is one that occurs seemingly from nowhere and often appears overwhelming. One that is all too frequent in the construction industry is the fatality of a workman or collapse of temporary support systems such as falsework or scaffolding or in some cases earth or other support walls. These crises may be subsequently found to have been creeping crises, if as mentioned earlier OHS or EMS protocols were ignored or flouted. However, sudden crises are usually the result of exceptional contributing factors. For example, unseasonable and extraordinary rainfall may undermine a foundation wall. While the weather experienced may have been a once in a 100-year event, it is nevertheless not generally expected that precautions should have been made to accommodate this kind of risk for every project as the incidence may seem too remote to warrant specific measures to be planned for.

The third type of crisis is a **periodic** crisis such as the impact of business cycles, economic cycles and other changes that ebb and flow in a predictable way but where the timing is not easy to predict accurately. In the construction industry, speculative building in the housing sector can often lead to developers being subject to financial crises and distress. Such crises not only affect the direct instigators (such as developers that speculatively over-build) but also extend to encompass their entire supply chain and possibly also their customer base.

Although crises are often perceived to be a sign of managerial failure, there is an increasing realisation that they are an inevitable and healthy part of organisational life which have to be planned for (Pascale 1991; Frazer and Hippel 1996; Furze and Gale 1996; Lerbinger 1997). While crises can destroy unprepared organisations they can strengthen those which are well prepared since a well conceived crisis management plan can harness the many potential opportunities contained in a crisis (Pascale 1991; Furze and Gale 1996). This crisis management planning should be part of an integrated and thoroughly implemented risk management process. However, evidence suggests that many construction organisations exist in a low state of crisis preparedness, having an inadequate understanding of their risk exposure, of how to mitigate those risks and of the internal systems needed to cope with, learn from and recover from their eventuality (Teo 1998).

All construction companies are prone to crisis but this varies depending upon the nature of work they are engaged in. For example, a company involved in demolition work is likely to be more prone to crises than a company involved in house building. Similarly, a company involved in complex, innovative and non-routine business activities is more prone to risk. This also goes for high-value business activities involving complex financing arrangements and large business activities because there is simply more to go wrong. However, crisis-proneness is also dependent upon the way an organisation is managed and research indicates that crisis-prone organisations have certain characteristics in common (Pauchant and Mitroff 1992; Mitroff and Pearson 1993; Pearson *et al.* 1997), many of which are associated with the business-as-usual arrangement found in the construction industry. For example, crisis-prone organisations tend to be characterised by a culture of mistrust, inequality, suspicion and a short-termism. They also tend to have inflexible, formal structures and penal, exploitative and task-oriented cultures that stress the importance of short-term profit gains over longer-term corporate goals and relationships. The naive mindset that is nurtured in such organisations is that crises are someone else's responsibility, that they are essentially negative in nature and that they happen to others. Crisis-prone organisations also operate in a reactive mode and discourage learning, precipitating crises and then mismanaging them to cause further damage. To a crisis-prone organisation, there is little justification for the re-examination of existing organisational practices in the aftermath of a crisis. This is because the short term and transitionary relationships that such organisations engage in create a perception that the circumstances that led to a crisis are unlikely to occur again.

5.2   The process of preparing for and managing crises

In contrast to crisis-prone organisations, crisis-prepared organisations have a culture of awareness, collective responsibility and sensitivity to their social and financial responsibilities to stakeholders and the wider environment (Ginn 1989; Lerbinger 1997; Pearson *et al.* 1997). These are sentiments that are more likely to be found in partnering and alliancing type projects where a longer-term attitude towards working relationships exists. This allows crisis management planning to be more easily and systematically incorporated into strategic planning processes so that it is an integral part of organisational life at all levels (More 1995). In such organisations, project managers are able to drive and support crisis management efforts by providing clear statements of fundamentally held core beliefs and attitudes relating to organisational priorities. They are also more able to 'let-go' of formal, standardised systems and procedures which serve them well in *normal* times but which become restrictive and counter-productive during a crisis (Sagan 1993). Having the confidence to do this requires trusting relationships and effective horizontal and vertical communications with external and internal stakeholders.

During a crisis, effective communication is essential and companies with a track record of effective communication as an intrinsic part of their day-to-day life are most likely to turn it to advantage (Mindszenthy *et al.* 1988; Aspery and Woodhouse 1992; Sikich 1993). Effective communication systems are particularly important in dealing with external stakeholders such as emergency services, the public, the media and existing and potential customers. The media, in particular, play an important role in constructing the public's image of events and the poor communications can result in distortions of the truth, unjustified mistrust, suspicion and irrevocable damage to customer relations. In essence, a crisis-prepared organisation has well developed and widely understood crisis management plans which keep them in constant touch with *what* type of crisis it faces, *when* they begin, *why* they occur and *who* they affect (Mitroff and Pearson 1993). It is the state of knowledge in these areas that represents the fundamental difference between crisis-prone and crisis-prepared organisations.

Crisis management can be seen as a process of several distinct phases, as depicted in Figure 5.1 (Loosemore 2000, p. 46). The first phase of the process is **detection** through monitoring potential events that could precipitate a crisis. In common risk management parlance, this phase is often called the risk identification stage. Swift reactions to potential crises are essential to mitigate potential damage and maximise potential gains and early warnings are a critical part of this process. To facilitate such warnings, monitoring activities should be preventative in focus, aimed at detecting potential problems before they arise. However, while prevention is better than cure, it is impossible to create a crisis-free environment and monitoring activities need to have a reactive focus to detect crises as soon as possible after they have arisen.

Having detected a potential or actual crisis, the next stage is **diagnosis.** This is the process of analysing the event further to see if a response is justified and if so, what that response should be. In common risk management parlance, this phase is often called the risk analysis stage.

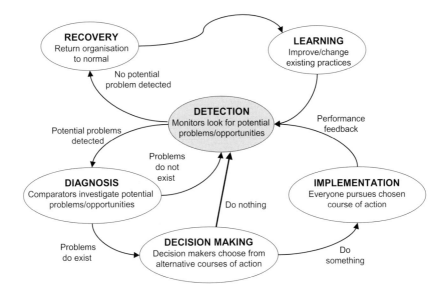

Figure 5.1: Phases of crisis management
Source: Loosemore, M. (copyright © 2000) *Crisis Management in Construction Projects.* New York, USA, American Society of Civil Engineering Press.

If the diagnosis process indicates that the implications of a detected event fall outside acceptable limits of performance deviation then an appropriate response must be formulated. In this above model, this stage is referred to as **decision making** and in common risk management parlance, this phase is often called the risk response stage. The aim of any decision should be to realign the project goals with project performance or, if the crisis has already occurred, to mitigate any damage and maximise any potential gains.

Such decisions are likely to have widespread social, technical and monetary implications and the **implementation** phase can often be problematical. To ensure that the process goes smoothly, constant monitoring and feedback is necessary to ensure that further adjustments can be made, if need be. This cycle of monitoring, diagnosis, decision making, implementation and feedback continues until the crisis is resolved.

The focus should then turn to recovery and learning. **Recovery** involves returning an organisation to 'normal' and dealing with the aftermath of the crisis, a process that takes time, specialist skills and patience. **Learning** is important to prevent future crises and to enhance the capacity to deal with future crises more effectively. As Gonzalez-Herrero and Pratt 1995 argue, crises present profound learning opportunities for organisations because they reveal weaknesses that other events cannot.

This cycle must operate proactively and reactively and each phase must be managed effectively if the potential costs of crises are to be mitigated and the potential benefits maximised. The only difference between proactive and reactive crisis management is in the timing of the cycle, proactive crisis management focusing on decisions as they are being made and reactive crisis management, as they are being implemented. The idea is that every decision made in a project which involves significant risk, is subject to continuous scrutiny from its inception to its full implementation. The following sections consider the potential inefficiencies that can arise in this cycle and explain why partnering arrangements can help to minimise them.

5.2.1   Monitoring

One of the most common reasons why people overlook potential problems is that they did not know what they were looking for. This is often the case in business-as-usual contracting arrangements where project goals are unclear due to the lack of longer-term interaction with clients and the exclusion of many players from early stages of the procurement process where the majority of project goals are established. In contrast, relationship-based procurement projects enable long-term relationships with clients to be forged, which facilitate a better understanding of their business objectives and culture. It also enables a greater proportion of the project team to be involved in critical and early goal-setting activities.

Another problem in business-as-usual contracting arrangements is confused monitoring responsibilities. This can often arise from large numbers of contracts that have to be complex and voluminous to control the lack of trust that is often generated between project participants. In contrast, the sense of mutual respect and trust engendered in partnering arrangements allows for fewer and simpler forms of contract which allow monitoring responsibilities to be more clearly understood. Most importantly, it also engenders the sense of collective responsibility between different risk takers that is necessary for effective monitoring processes. In business-as-usual contracting arrangements, confused monitoring responsibilities can also arise from the frequent and often *ad hoc* changes in project team membership. This relational instability is less evident in relationship-based procurement contracts and the result is more uniformity and stability in the monitoring of potential risks.

Another major cause of monitoring problems is conflicting objectives between members of an organisation. This has been one of the defining characteristics of the business-as-usual approach and has been one of the main driving forces behind the development of relational contracting. The problem with conflicting interests is in the selfishness and communication problems it creates, precisely at a time when effective communication is more important. Effective communication is particularly important during a crisis because most crises tend to affect a variety of project stakeholders who all need to be involved if an effective and holistic response is to be made. When conflicts of interest exist, it may be in the interests of one stakeholder to withhold information about another because they may directly benefit from any escalation of costs. Clearly, this is not in the interests of the project as a whole and is far less likely to occur in the environment of fairness and equity that pervades partnering-type contracts.

5.2.2   Diagnosis

Having detected a potential problem, it is often the case that people cover it up rather than communicating it to those who can decide what to do and act upon it. In a business-as-usual environment where competitive pressures are likely to have cut any contingency allowances to the minimum, this type of denial behaviour is particularly likely. It can also be caused by the culture of division, blame, fear and mistrust that is associated with such arrangements. This creates an organisational shadow that can haunt an organisation until the repressed problem grows to the point where it is a fully blown crisis. This type of defensive behaviour is less likely in relationship-based procurement arrangements where there is less chance of recrimination and blame and where there is a greater willingness to share the losses as well as the benefits of risky business activities

5.2.3   Decision making and implementation

When a potential problem does rise to the surface the process of allocating responsibility to deal with it is itself a problematical process with considerable potential for conflict. In the tightly resourced environment of a business-as-usual contract, there is little room for charity and to make an adequate response without the temptation of looking for ways of reclaiming losses in other aspects of the project. Furthermore, complex contracts that are a necessity in such projects make it more likely that people have not realised their responsibilities. The common response is an attempt to deny them, ironically often using the complexity of the contract as a weapon.

Assuming responsibility for dealing with a potential problem is accepted by someone who is willing and able to deal with it, then further problems due to conflicting interests because the actions of one decision maker are unlikely to coincide with the desires of another and there is likely to be no sharing of resources to deal with it. The lack of agreement over proposed solutions has the potential to create major problems in the acceptance and implementation of any solution.

5.2.4   Recovery and learning

Assuming that an effective response is made, managers must not see the return to normality as the end of the crisis management process. Crises present profound **learning** opportunities by revealing important improvements for application to future crises. Furthermore, they can contribute to improved effectiveness through the **unlearning** of crisis causing behaviours and procedures that may be ingrained within an organisation.

In many ways, learning and unlearning are arguably the most important phases of crisis management because much of the knowledge we use today to construct, manufacture, and operate engineering and built facilities has been acquired from analyses of past mistakes. Unfortunately, the time-pressured, temporary, transitionary, fragmented, and divided nature of business-as-usual construction project organisations does not encourage the far-sighted attitudes that inspire people to learn and unlearn. Evidence of this can be found in the emerging literature about facilities management that shows little evidence of the construction industry evaluating the effectiveness of its final product, let alone individual crises that occur during its production (Barrett and Stanley 1999). Despite these potential problems, a desire to learn, however difficult and painful, is the key to preventing repeated mistakes and improving performance and the continuous nature of partnering arrangements create a far more conducive environment for this to occur within.

5.3   Countering dysfunction in managing crises

Many of the problems that we have discussed above, relate to barriers that are created by the business-as-usual approach. Figure 5.2 illustrates the situation where one team member may hold the key to a solution and yet is ignored by others in the project team.

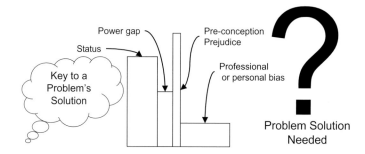

Figure 5.2:   Barriers to teamwork

The barriers are institutional and personal. From the institutional perspective, status and power distance may be entrenched as part of the contractual arrangements. From the personal perspective, in-built bias may be a core driver of this attitude based upon past experience, expectations triggered by others (colleagues, friends and the operating organisational culture of the team). Indeed, a recent study on professional stereotyping in the construction industry has confirmed widely suspected ingrained prejudices between builders, architects, engineers and other construction team professionals (Loosemore and Chin 1999). The implication of their findings is that these stereotypes influence people's initial behaviour in response to a crisis, particularly early in the project's lifecycle before people have the opportunity to get to know each other and overwrite their general prejudice with their most recent and enduring experience with the person in question. Stereotyping also can be seen from a decision theory point of view as being a person signature skill which is defined by a person's preferred way of tackling problems. We all generally have a preferred task mode or selection of what to do. We have a preferred cognitive approach on how we approach a task or solution and a preferred technology or way in which the task may be executed. These three components define signature skills (Leonard-Barton 1995, p. 63). Thus if there is a stereotypical response for builders, for example, to regard architects as generally arrogant and likely to reject offers of advice on how to simplify an unnecessarily complex and possibly hazardous design detail then they may not wish to invest the energy and social capital to propose such improvements. Further, if the architects have a stereotypical bias to regard builders as always looking for a way of claiming extra's for work to maximise their profits regardless of the justification for the 'extra', then any simplification or safety advice may be rejected out of hand. This can reinforce biases and prejudice and when crises develop such behaviours and attitudes may drive the default signature skills to be employed. This is how a creeping crisis can fester and develop.

Relationship-based procurement systems do not necessarily obviate these problems but they do provide the framework for choice to remove the barriers and for teams to better understand each other. Solutions should be collaboratively built rather than being dictated or negotiated. Building a solution means adding to stock of feasible solution options by generating them through sharing perspectives and expanding the insights, perceptions and wisdom offered by a diversity of individuals. In a business-as-usual situation, typically solutions are either pre-scribed by the hierarchy or generated through a negotiation process where an outcome is often a sub-optimal satisficing solution. In an alliancing project solutions may be built through a more rigorous process where options are generated and consequences assessed from the project and stakeholder perspective and an appropriate win–win solution settled upon.

5.3.1   Building solutions to problems

Figure 5.3 provides a model of how solutions to problems may be built. Ability requires both technical and people-management skills to be able to diagnose the problem's source as well as generate feasible solution options. High-level communication skills are required to be able to appreciate the perspective of others and a right to a perspective by others and to be able to communicate and under-

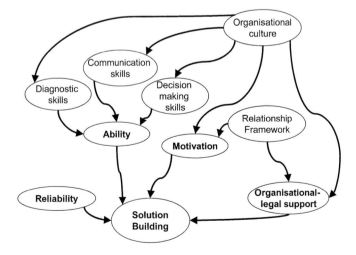

Figure 5.3:   Solution building model

stand issues and processes. Equally high-level decision-making skills are required to assess options and to evaluate outcomes and to decide upon the preferred option.

Motivation to arrive at a best-for-project solution to a problem requires an innovative organisational culture that supports decision making by allowing mistakes to be made providing lessons are also learned (Leonard-Barton 1995; Leonard and Straus 1997). Organisational culture pervades all aspects of the ability to build solutions. It also requires an operating environment that supports collaboration and rewards innovative behaviour. Those best able to manage risk should be empowered to do so.

A supporting framework is also required that provides a reward and penalty regime that supports solution building and provides the administration and system protection to allow innovative solutions to be attempted and also supports a more cooperative environment in which to build and deploy solutions.

The question of a partner's reliability will act as a filter to effective solution building. Positive evidence of experience between partners will enhance effectiveness of solution building through generating enthusiasm and commitment. Poor reliability experience either triggers defensive routines being entered into or the unreliable partner's contribution will be downgraded or ignored. Thus a sub-optimal situation will arise similar to a business-as-usual scenario outcome.

5.3.2   The National Museum of Australia project

The fundamental difference between business-as-usual and relationship-based procurement systems is the supporting role and relationship framework of the organisational culture that creates and maintains a high level of motivation. This was well illustrated on the National Museum of Australia project, which is discussed in this section.

During the two years duration of this project, no crisis appears to emerge. Problems periodically arose, some of which had the potential to build into crises. However, the problem solving approach and project organisational culture reflected the approach illustrated in Figure 5.3 and prevented this from happening.

Table 5.1 provides comments by respondents to the National Museum of Australia project case study where 22 of the senior management team were interviewed. These illustrate the extent of organisational culture as a defining feature of the relationship-based procurement systems, in particular project alliancing. We selected two sets of lessons learned. The first related to a no-blame culture. The second relates to skills perceived as necessary of team members for project alliancing. Chapter 10 of this book provides an extensive in-depth discussion and analysis of the people-side of managing projects.

We have also chosen to present evidence of the organisational culture and approach to solution building for the National Museum of Australia project to illustrate how crises may have been avoided. We will draw together some of these results to illustrate behaviours that help explain how risk and crises may be treated differently in a project alliance arrangement over a business-as-usual situation. Respondents were asked to compare their experience of negotiation in the following three situations using a scale of 1 = strongly disagree and 7 = strongly agree for:

Table 5.1: Organisational culture issues between BAU and project alliancing on the National Museum of Australia project

Page number and interview number references are provided for comments in Table 7.1

'How important was the no-blame clause in the project alliance agreement' (page 116 Interviewees I-1, 12 ... IN)	• On a couple of occasions we could have turned to litigation – and held up the works' (I-1) • 'We still tend to blame – but you just get in there and fix it' (I-8) 'No one had to prove they were right or wrong for financial gain – instead we just got things fixed' (I-22) • 'It causes a dependency and reliance on the players' (I-10) • 'We all knew that we were a team and had to work together' (I-13) 'We all had a shared destiny' (I-15) '... part of the culture which was that we are in it together' (I-20) • 'There was still political manoeuvring and pushing and finger pointing – but it was well managed and did not affect the outcome of the project' (I14) • 'The team has to get in and sort it out – it is a fundamental component of the alliance contract' (I-18) • 'It was more about achieving the end rather than who should I blame' (I-19) • 'No comment' (I2, 3, 4, 5, 12, 16, 17, 21)	• It is clear from the lessons learned comments that a general feeling of community was evident and that problems and crises should be 'fixed' • Responsibility for obviating problems was seen as shared amongst the project community of teams and individuals
'If you had to pick personnel for a project alliance – what skills would you be looking for?' (page 117)	• Interpersonal competencies – relationship building, negotiation skills, e.g. humility, respect for all levels of relationships, cooperative personality (cited by 14 of 22 interviewees) • Cognitive competencies – information seeking and analytical thinking, e.g. creativity, people who can think outside the square, people with an open mind, willing to share knowledge (cited by 10 of 22 interviewees) • Intrapersonal competencies – perseverance and self control, e.g. people management skills, take an idea and implement it, personal commitment (cited by 10 of 22 interviewees)	• High levels of commitment required • High levels of assertiveness to promote and discuss options and implement ideas • Flexibility is valued and encouraged • Very high levels of evidence of cooperation and team spirit • Supports the Figure 7.3 model

Source: Peters *et al.* (2001)

(1) Average to normal business-as-usual (BAU) – the most common situation – usually high/constant conflict.

(2) Best BAU – the occasional project where all parties to the project work exceptionally well together as a team.

(3) Project alliancing – the project delivery strategy that the parties are currently using on the Acton Peninsula Project for the National Museum of Australia – aim is to force collaboration as the only means to achieve the best outcome for the project and hence all teams involved.

A relative agreement index (RAI) was then constructed using the formula where:

$$RAI = \frac{\sum w}{A \times N}$$

w = weight given to each statement by the respondents from the 1 to 7 range described previously
$A = 7$ (the highest weight)
N = the total number of respondents

The closer the RAI is to 100 the higher the level of agreement is with the statement proposed and conversely the closer the RAI is to 0 the lower the level of agreement is with the statement proposed.

Table 5.2 illustrates the stark difference in negotiation style between BAU and Project alliancing. The low values indicates that alliancing on this project facilitated an approach where parties were able to reach a win–win outcome more quickly than with BAU. This indicates the negotiation style culture operating in the National Museum of Australia project. The study also found that principled negotiation was clearly more evident (Peters *et al.* 2001). Chapter 10 more fully explores principled negotiation and its impact upon human capital.

Table 5.2: Negotiation styles for the National Museum of Australia project

Negotiation styles	Average normal	BAU likelihood	Project alliancing	PA likelihood
Rate your responses for the appropriateness and the likelihood of your using the following negotiating tactics in the following situations:				
Make an offer or demand so high (or low) that it seriously undermines your opponent's confidence in his/her ability to negotiate a satisfactory settlement	29%	35%	17%	22%
Convey a false impression that you are in absolutely no hurry to come to a negotiated agreement, thereby putting more time pressure on your opponents to concede quickly	37%	43%	20%	26%

Source: Adapted from Peters *et al.* (2001).

Table 5.3 illustrates how the project team partners may be better equipped to diagnose problems. By sharing information and knowledge and openly cooperating while actively building trust, teams expand their capacity to recognise causes from symptoms. This is supported by results provided in the fourth and fifth rows, which indicate that team members expand their repertoire of possible solutions to consider by volunteering help and searching beyond the immediately obvious.

The results reported in Table 5.4 clearly indicate a more open and project focus rather than organisational team focus for alliancing rather than business-as-usual. The combination of close physical and electronic contact is clearly a major driver of communication supporting the ability to build problem solutions. The differ-

Table 5.3: Organisational cultural diagnostic support behaviours

Question/statement (with page reference, our bolding of text) Relative Agreement Index (RAI) scores	Average to normal BAU	Project alliancing
1. 'We **share** technical and commercial **information** relating to our projects without the need to protect ourselves' (S4 page 202)	41%	85%
2. 'We believe that by **cooperating** with our partners **openly** we reduce the likelihood of opportunistic behaviour' (S5 page 202)	42%	87%
3. 'We actively attempt to **build trust** with our partners through mutual moral and other types of **support'** (S6 page 202)	51%	93%
4. We **volunteer help** and support to our partners when they need help and we are happy to provide resources in a crisis' (P1 page 205)	34%	71%
5. We are **continually exploring options** outside the immediately obvious' (V21 page 219)	48%	85%

Source: Peters *et al.* (2001, page numbers indicated within the table).

ence between BAU and Project alliancing RAI values is in many cases close to or in excess of 100%.

Results indicate a high level of sophistication in Table 5.4. Strong diagnostics and communication skills are necessary for sound decision making. The organisational culture clearly supports the ability of team members to collaboratively build solutions. Decision-making skill measures are presented in Table 5.5 below.

Table 5.5 again clearly indicates a very high margin between scores for business-as-usual versus project alliancing for decision making. The evidence is high for a culture of a collaborative approach. Mutual adjustment appears strong with a sophisticated approach to sharing risks (first, third and fourth rows). The second and fifth row results indicate strong empowerment to facilitate decision making as does row 6. Yet again the project alliance scores are very high compared to business-as-usual.

Table 5.6 results clearly indicates that project alliancing for the National Museum of Australia project provides a culture and system that rewards participants to collaborate and cooperate. Results reported in the first and second rows indicate goals being well aligned. Mental and intellectual stimulation seems strong as indicated in rows 3 and 4. The results reported in the fifth to seventh rows indicate that personal rewards are recognised.

Organisational/legal support present on the National Museum of Australia project is illustrated in Table 5.7. These comments are taken from interviews undertaken with 22 management staff as part of a lessons learned reflection exercise.

The table indicates that project alliancing provides a supporting framework for better communication, better data, information and knowledge sharing, better collaborative and cooperative decision making and in the National Museum of Australia project case an IT system that enhanced documentation communication currency. The project used a facilitator who was involved at the start of the project

Table 5.4: Organisational cultural communication support behaviours

Question/statement (with page reference, our bolding of text) Relative Agreement Index (RAI) scores	Average to normal BAU	Project alliancing
1. 'We **trust** our partners' integrity to be able to **discuss sensitive issues** with them in order to resolve disagreements over such issues without fear of appearing a "non-team" player if these issues are important' (O8 page 202)	36%	84%
2. 'We have a strong sense of **mutual connection** with our partners – we feel like part of our partners' organisation' (A6 page 205)	29%	73%
3. 'We respond to disagreements by **rationally debating and discussing** ways to resolve conflicts rather than withdrawing or seeking formal remedies' (C11 page 205)	43%	94%
4. 'We **communicate openly** with our partners when problem solving and are not afraid to own up to mistakes' (A2 page 208)	39%	88%
5. 'We believe that **close physical proximity** to our partner organisations for extended periods of time on site is of vital importance in maintaining a good team relationship' (R6 page 208)	44%	93%
6. 'We believe that **fast and effective electronic communication** technology with our partner organisations for extended periods of time on site is of vital importance in maintaining a good team relationship' (V7 page 208)	51%	72%
7. 'I regularly **share ideas** with my colleagues from different companies' (C10 page 211)	41%	77%
8. 'I enjoy the **mental stimulation** of working in a cross-disciplinary organisation' (B13 page 211)	49%	85%
9. 'I work in close personal contact with many of my colleagues from different companies' (F7 page 211)	36%	78%
10. 'I work in **close electronic contact** with many of my colleagues from different companies' (F8 page 211)	38%	66%
11. 'The major source of **authority** that people use is related to their **expertise and knowledge** of the task at hand' (P12 page 217)	49%	85%

Source: Peters *et al.* (2001, page numbers indicated within the table).

to help establish a collaborative and cooperative culture and this was periodically reviewed. Team spirit was strong and so follow-up facilitation workshops were deemed unnecessary, however, the facilitator remained as a mentor to the senior project team throughout the project and was reported to have been a useful resource. Throughout the study, it was evident that this project had a powerfully committed culture that placed best for project as foremost in team members' minds (Peters *et al.* 2001).

The study also reported interesting examples of the environment for project alliance behaviours that supported solution building. Table 5.8 illustrates some of these.

Table 5.5: Organisational cultural decision-making support behaviours

Question/statement (with page reference, our bolding of text) Relative Agreement Index (RAI) scores	Average to normal BAU	Project alliancing
1. 'We are willing to **sacrifice** in the **short term** something to a partner to ensure that the **long-term relationship** remains intact and functioning to mutual advantage' (O10 page 202)	43%	83%
2. 'When problems arise we concentrate on solving them rather than trying to find somebody to blame' (C12 page 205)	43%	93%
3. 'We see our partners and us as **sharing risk** on the **basis of mutual competence**, whoever can best control risk volunteers to accept and manage it' (C13 page 205)	36%	89%
4. 'I am confident the majority of time that I **understand** what is **expected** of me' (C1 page 217)	56%	86%
5. 'We generally **focus** more on **solving problems** than relying on formal procedures and protocols for raising and documenting them – basically we just get on with it' (E23 page 219)	44%	88%
6. 'We are closely involved in **working out how to achieve our goals** and plans' (E28 page 219)	51%	80%

Source: Peters *et al.* (2001, page numbers indicated within the table

Table 5.6: Organisational motivation support behaviours

Question/statement (with page reference, our bolding of text) Relative Agreement Index (RAI) scores	Average to normal BAU	Project alliancing
1. 'Our **goals** and those of our partners do **not conflict**' (A8 page 205)	36%	80%
2. 'Our **goals** and those of our partners are in **harmony** with project goals' (A9 page 205)	33%	77%
3. 'I **enjoy** the **mental stimulation** of working in a cross-disciplinary organisation' (B13 page 211)	49%	85%
4. 'One of the reasons why I was attracted to this project was to be **mentally stimulated** and to learn new things' (B15 page 211)	45%	71%
5. 'I believe that I get good **recognition** for my contribution' (I34 page 218)	44%	85%
6. 'I believe that the project is **good for my career plans**' (I35 page 218)	52%	77%
7. 'I get fair reward for my contribution for the work I do relative to that of others' (E32 page 219)	52%	78%

Source: Peters *et al.* (2001, page numbers indicated within the table)

Table 5.7: Contrasting support behaviours between BAU and project alliancing on the National Museum of Australia project

Page number and interview number references are provided for comments presented in Table 7.7.

Lesson learned	Behaviour difference examples	Comments
'How important is facilitation to project success. When do you need facilitation and how much do you need?' (page 119)	• General supportive comments as a culture setting exercise for cooperative behaviour – for 10 of 22 interviewees, no comment for 7 of them, 3 had done this with the project director and 2 were unimpressed • Was seen less useful as an ongoing arrangement because team spirit was high • Recognition of a need to find a mechanism to change the win–lose culture	• Facilitation is a change management action and prepares the ground for behaviours that help solve problems and obviate the potential causes of conflict • Team spirit is vital for sound crisis and risk management
'How much do you think the delivery strategy of project alliancing contributed to the success of the project' (page 120)	• Three key factors, no blame, no litigation and no external arbitration – supported by 18 of 22, 2 interviewees with no comment and 1 saw the constructors as the problem solver being an issue • Teamwork enhancement (as one big project team) and facilitating joint action, 11 of 22 interviewees	• Clearly it provided the framework for a focus on the success of the project thus team members seek to protect the protect's interests • Facilitates wider knowledge assets and decision making
'In your experience what do you think some of the potential outcomes could have been if a different delivery strategy would have been used' (page 121 Interviewees I-1, 12…IN)	• Cost over-runs cited by 14 of 22 interviewees • Time extensions 8 with little or no time change by 4 interviewees • Collaborative risk sharing and friction over decision making and communication problems increased, 9 of 22 interviewees for a different approach • Six interviewees used language like 'would have been a disaster' (I2), 'contractual nightmare' (I12, I16)	• Appears to be strong recognition that project alliancing facilitates reduction in crisis conditions emerging • Better able to deal with problems growing into crises
'What was ProjectWeb like – has it assisted you with your work?' (page 122)	• No comment by 16 of 22 interviewees • Four highly positive comments regarding communications gains (I15, I12, I16, I21) and 1 negative comment (I10)	• Provided some useful help on communication but personnel and team issues main issue for commitment

Source: Peters *et al.* (2001, page numbers indicated within the table)

A number of negative questions were asked. These indicate low RAI results, which suggests that both business-as-usual and project alliancing rate low on these factors. In general it appears that the organisation supports the empowerment and authority to build solutions to problems without undue interference. The facilitation and legal/administrative infrastructure provided the necessary resources and culture to provide confidence of team members to act in the best interests of the project.

Having a capacity through ability and motivation to build solutions to problems

Table 5.8: Organisational support behaviours

Question/statement (with page reference, our bolding of text) Relative Agreement Index (RAI) scores	Average to normal BAU	Project alliancing
1. 'We have the **confidence and support** of our company's **top management** to act in the way we do to others' (S11 page 202)	55%	78%
2. 'My immediate supervisor firm encourages me to **develop my skills** through structured learning activities' (courses training) (F11 page 211)	44%	64%
3. 'I feel that the **physical working conditions** are **poor** for me' (I36 page 213) NOTE this is a negative question	28%	30%
4. 'I feel that I have **insufficient authority** to meet contractual or ethical obligations' (I39 page 213) NOTE this is a negative question	29%	35%
5. 'I feel that the **working atmosphere** between groups is mainly characterised by **conflict** and **point scoring**' (I40 page 213) NOTE this is a negative question	42%	27%
6. 'The hierarchy facilitates a lot of **group interaction**' (E30 page 219)	37%	82%
7. 'I find the organisational structure here **stifles initiative**' (V8 page 219) NOTE this is a negative question	31%	32%

Source: Peters *et al.* (2001, page numbers indicated within the table)

is insufficient if teams lack confidence in the reliability of their project team partners. Table 5.9 illustrates research results to illustrate the trust environment on the National Museum of Australia project. Proven reliability that builds confidence in the team members' experience in proving that they can be relied upon to do what they take responsibility for is evident from results in the first, second and fourth rows. Reliability indicators that are evident from results illustrated in the third, fifth and sixth rows suggest that team members are reliable. A track record of team members having high levels of empathy, sensitivity, and open-mindedness is evident from the seventh to tenth rows. Mutual understanding is evident from the result in the eleventh row. Results in the twelfth and thirteenth row indicate that team members under the project alliance system demonstrate joint efforts to achieve building solutions.

The above evidence supports the model illustrated in Figure 5.3 that suggests that under project alliancing on the National Museum of Australia project, in particular, solutions and responses to problems and crises are developed and built rather than imposed or sub-optimally negotiated. Building solutions means that tangible and intellectual resources as well as goodwill are jointly used in a respectful manner so that the full energy and effort of all parties are available for proposing, probing and developing solutions is more likely to occur. The power balance is evidently well considered so that intimidation on the part of those with greater contractual or institutional influence is minimised so that the voice of the less formally influential can not only be heard but also actively considered.

Table 5.9: Organisational reliability behaviours

Question/statement (with page reference, our bolding of text) Relative Agreement Index (RAI) scores	Average to normal BAU	Project alliancing
1. 'Our word is **reliable** – we do what we say' (S1 page 202)	54%	93%
2. 'We **fulfil our obligations** to our partners – we do what we have agreed to do' (S2 page 202)	54%	92%
3. 'We have the **confidence** and **support** of our company's **top management** to act in the way we do to others' (S11 page 202)	55%	78%
4. 'Our **actions towards others** (in the supply chain who are not our direct partners) reflects how we would like them to **act towards us**' (O7 page 202)	58%	91%
5. 'We have a strong sense of **mutual reliance** – we need our partners' cooperation to function effectively' (A5 page 205)	43%	88%
6. '**Achieving** what is expected of me for work-related **objectives** for the project is **very important** to me' (C4 page 217)	66%	92%
7. 'We **recognise** our **partners' contribution** as of equal importance in achieving project objectives and goals' (A7 page 205)	43%	85%
8. 'We maintain **open** lines general **communication with** our **partners** in order to prevent hesitation, reservation or other defensive behaviour' (A1 page 208)	49%	92%
9. 'We are **alert** to issues that our **partners** may find **sensitive**' (A4 page 208)	39%	83%
10. 'We know what is an **acceptable behaviour** to our **partners**' (A8 page 208)	49%	80%
11. 'Our partners know what is an **acceptable behaviour to us**' (A9 page 208)	43%	76%
12. 'We take considerable effort to **allow** our **partners** to **learn from our experiences**' (C1 page 211)	37%	76%
13. 'I feel **part** of the project's **community**' (I33 page 218)	47%	92%

Source: Peters *et al.* (2001, page numbers indicated within the table)

5.4   Chapter summary

This chapter has demonstrated that the main benefits of alliancing arrangements in terms of facilitating positive crisis management are:

- A positive, open and flexible environment where people have the freedom and confidence to respond creatively without fear of blame, recrimination or punishment;
- A culture of collective responsibility where interdependence between different people's risks are openly acknowledged and recognised;

- A culture of altruism where the losses and benefits are shared between project participants as far as possible;
- More effective communication that facilitates a more effective detection and response to potential crises;
- Continuing and established relationships which allow project participants to build trusting and efficient working relationships and to understand more clearly their different needs and goals and their crisis vulnerabilities;
- A collaborative environment where people focus on their similarities more than their differences and can build mutually beneficial solutions to problems;
- Fewer resourcing problems that can generate the tensions and conflicts often associated with crises;
- A long-term outlook that facilitates learning; and
- A supportive environment that aids recovery.

While crises hold obvious dangers for managers, they also present unique opportunities for improvement. This research indicates that the dangers are much greater and benefits far fewer managers on business-as-usual projects than they are for those managing alliancing projects. The main benefit of relationship-based procurement systems is that they transfer the traditional emphasis of risk management from reactive to proactive, resulting in fewer crises, but they also help reactive strategies deal more effectively with those that do arise. In contrast to alliancing projects, business-as-usual arrangements appear to lock people into a set of confrontational and selfish relationships and rigid performance standards that stifle the openness, confidence and freedom that enables people to take advantage of the major opportunities which crises pose. The result is pessimistic organisations that focus on risks rather than opportunities and which therefore, are unable to achieve their full potential.

5.5.   References

Aguilera, D.C. and Messick, J.M. (1986) *Crisis Intervention: Theory and Methodology*. St Louis, USA, C V Mosby Company.

Allen, R.E. (1990) *The Concise Oxford Dictionary of Current English*. Oxford, UK, Clarendon Press.

Aspery, J. and Woodhouse, N. (1992) 'Strategies for Survival.' *Management Services*. **36** (11): 14–16.

Barrett, P. and Stanley, C. (1999) *Better Construction Briefing*. Oxford, UK, Blackwell Science Ltd.

Fink, S. (1986) *Crisis Management: Planning For the Inevitable*. New York, USA, American Management Association.

Flanagan, R. and Norman, G. (1993) *Risk Management and Construction*. Oxford, Blackwell Science.

Frazer, N.M. and Hippel, K.W. (1996) *Conflict Analysis: Models and Resolutions*. New York, USA, North Holland.

Furze, D. and Gale, C. (1996) *Interpreting Management – Exploring Change and Complexity*. London, UK, International Thomson Business Press.

Ginn, R.D. (1989) *Continuity Planning: Preventing, Surviving and Recovering from Disaster*. Oxford, UK, Elsevier Science Publishers Ltd.

Gonzalez-Herrero, A. and Pratt, C.B. (1995) 'How to Manage a Crisis Before – Or Whenever – It Hits.' *Public Relations Quarterly*. **40** (1): 25–29.

Jarman, A. and Kouzmin, A. (1990) Decision Pathways From Crisis – A Contingency Theory Simulation Heuristic for the Challenger Space Disaster (1983–1988). In: *Contemporary Crisis – Law, Crime and Social Policy*, ed Block A. Netherlands, Kluwer Academic Press: 399–433.

Leonard, D. and Straus, S. (1997) 'Putting Your Company's Whole Brain to Work.' *Harvard Business Review*. **75** (4): 110–121.

Leonard-Barton, D. (1995) *Wellsprings of Knowledge – Building and Sustaining the Sources of Innovation*. Boston, MA, Harvard Business School Press.

Lerbinger, O. (1997) *The Crisis Manager: Facing Risk and Responsibility*. New Jersey, USA, Lawrence Erlbaum Associates.

Loosemore, M. (2000) *Crisis Management in Construction Projects*. New York, USA, American Society of Civil Engineering Press.

Loosemore, M. and Chin, T. (1999) 'Occupational Stereotypes in the Construction Industry.' *Construction Management and Economics*. **18** (5): 559–567.

Mindszenthy, B.J., Watson, T.A.G. and Koch, W.J. (1988) *No Surprises: The Crisis Communications Management System*. Tororonto, Ontario, Canada, Bedford House Publishing Ltd.

Mitroff, I. and Pearson, C. (1993) *Crisis Management: A Diagnostic Guide for Improving Your Organization's Crisis Preparedness*. San Francisco, USA, Jossey-Bass Publishers.

More, E. (1995) 'Crisis Management and Communication in Australian Organizations.' *Australian Journal of Communication*. **22** (1): 31–47.

Pascale, R.T. (1991) *Managing on the Edge*. Harmondsworth, UK, Penguin Books.

Pauchant, T.C. and Mitroff, I. (1992) *Transforming the Crisis-Prone Organisation: Preventing Individual, Organizational and Environmental Tragedies*. San Fransisco, USA, Jossey-Bass Publishers.

Pearson, C.M., Misra, S.K., Clair, J.A. and Mitroff, I. (1997) 'Managing the Unthinkable.' *Organisational Dynamics*. **26** (2): 51–64.

Peters, R.J., Walker, D.H.T., Tucker, S., Mohamed, S., Ambrose, M., Johnston, D. and Hampson, K.D. (2001) *Case Study of the Acton Peninsula Development*, Government Research Report. Canberra, Department of Industry, Science and Resources, Commonwealth of Australia Government: 515.

Raftery, J. (1994) *Risk Analysis in Project Management*. London, E&FN Spon.

Sagan, S.D. (1993) *The Limits of Safety: Organizations, Accidents and Nuclear Weapons*. Princeton, NJ, USA, Princeton University Press.

Sikich, G.W. (1993) *It Can't Happen Here: All Hazards Crisis Management Planning*. Oklahoma, USA, PennWell Publishing Company.

Teo, M.M.M. (1998) An Investigation Into the Crisis Preparedness of Australian Construction Companies. B.Sc. Sydney, Australia, University of New South Wales.

Chapter 6

Enabling Improved Business Relationships – How Information Technology Makes a Difference

Bruce Duyshart, Sherif Mohamed, Keith Hampson and Derek Walker

We argue in earlier and later chapters of this book that effective information and communication technologies are pivotal in supporting relationship-based procurement strategies. This chapter provides an insight into how the use of advanced IT and relationship-based approaches are critical in enhancing communication effectiveness on construction projects. We draw upon evidence presented by us more fully in a research report that investigated the National Museum of Australia project as a case study of the project alliancing approach. The case study compared business-as-usual and relationship-based approaches and included a substantial investigation into the role IT played in improving project-based business relationships and vice versa (Peters *et al.* 2001, Part B).

6.1   Introduction

For centuries, the unique combination of site positioning, design requirements, materials selection, budget constraints and the availability of specialised skills have led to the development of buildings that are essentially prototypes to themselves. Today, the majority of construction business processes are still heavily based upon traditional means of communication such as face-to-face meetings and the exchange of masses of paper documents such as technical drawings, specifications and site instructions. The need to improve the timeliness and increase the efficiency and effectiveness of these information exchanges, at a relatively low cost, has been long recognised by the industry. However, the use of IT in construction has not progressed to the level that can be seen in other industries such as manufacturing. This is due to a number of historical, industrial and market forces that perpetuate the industry's culture. This in turn affects the extent of IT adoption in day-to-day business processes. It is not surprising that research shows that about two-thirds of construction problems are caused by inadequate communication and exchange of information to facilitate effective decision making (Cornick 1990).

To address this incessant need for effective communication, various technologies can be used in the briefing, design, construction and operation phases of any built asset. This approach can help to improve communication, increase client satisfaction, reduce coordination errors in construction, provide a greater understanding of designers' work by construction workers, create fewer ambiguities and discrepancies in documentation and generally increase awareness and recognition of issues and requirements by all project participants.

Each day, new information technologies are emerging that can be used to raise productivity and reduce costs. Although many industries have embraced information technology to achieve these fundamental benefits, its use by the construction industry in many countries, including Australia, is limited. Paradoxically, the slow take up of computer technology by the construction industry is inconsistent with its specialised and intense information management requirements.

The construction industry is well known to foster highly fragmented and competitive environments. This diversity naturally limits efficiency and effective communication. A consequence of this effect is that communication problems can lead to expensive delays with costs potentially escalating dramatically while the various discrepancies are resolved and/or rework is undertaken during the construction phase. Accordingly, a better use of information technology can lead to improved communication and money saving (DPWS 1998). Defining the scope and limitations of the use and performance of IT in the construction industry is difficult due to the relatively limited amount of detailed research that has been carried out in the field (Bjork 1999). To many people, IT in construction encompasses the use of all electronic means of information transfer (computer networks, local area networks (LANs), internet, mobile phones, faxes, etc.) Others see IT as the use of the latest technology, such as knowledge-based systems, computer-based decision support systems, object-oriented CAD, concurrent engineering and just-in-time production. Alternatively others also see IT as part of an overall management strategy that requires significant process re-engineering. This diversity has led to a number of different IT definitions. For the purposes of the research presented here, this chapter adopts an information-centric definition that encompasses *the use of electronic machines and programs for the processing, storage, managing, transfer and presentation of information*. This definition is applied to demonstrate the key role IT plays in improving the effectiveness of communication and information exchange in the context of managing a construction project.

This chapter highlights these issues and illustrates how they may be addressed. A brief introduction to IT in the construction industry is provided to highlight the limitations and extent of its current use at the strategic as well as operational levels. This is followed by a case study of the National Museum of Australia project that introduced many new IT tools into the design, construction and project management processes. A detailed analysis is then provided of the effectiveness of these new methods of project communication. An overview of the lessons learned and the extent to which the alliancing agreement has facilitated the uptake of IT tools is also examined. Finally, the likely industry impact of IT innovations is then discussed with a view to facilitate the future development of more effective construction project information and communication processes.

6.2   IT in the construction industry

Information is the cornerstone of any business process. It is not surprising there-fore, that information technology has emerged as a key enabler to change the way business is conducted. During the last decade or so, significant productivity improvements experienced by a wide range of industries have been associated with IT implementation. IT has provided these industries with great advantages in speed of processing, consistency of data exchange, vastly improved accessibility and more flexible information exchange. Compared with its counterparts, the construction industry seems to be less efficient in its core business, the develop-ment of built assets. In the area of effective use of IT data, the Australian con-struction industry lags well behind other industries in its uptake of key IT processes (Australian Bureau of Statistics 1999). See Table 6.1 and Figure 6.1.

Table 6.1: Percentage of companies using aspects of IT by Australian industry

	PCs	Networked PCs	Internet access	E-mail access	Web access
Mining	80.4	38.0	46.8	46.3	42.4
Manufacturing	70.8	20.1	31.5	29.9	25.3
Construction	**55.9**	**8.2**	**20.2**	**18.9**	**16.2**
Wholesale trade	75.8	31.1	40.9	40.1	36.3
Retail trade	53.0	17.7	16.9	16.7	13.6
Accommodation/cafes/restaurants	46.9	8.6	16.4	13.7	14.9
Transport and storage	52.9	15.8	20.0	19.8	16.4
Communication services	72.7	16.5	44.5	44.5	37.7
Finance and insurance	72.8	22.7	39.6	38.3	34.6
Property and business services	78.6	31.4	46.4	44.7	40.9
Health and community services	67.8	19.9	31.7	31.0	29.9
Cultural and recreational services	64.3	17.2	35.0	34.8	30.8
Personal and other services	43.8	13.5	21.8	21.0	19.0
Business average	**63.9**	**20.4**	**30.1**	**29.1**	**26.0**

Source: Australian Bureau of Statistics 1999. Catalogue Number 8133.0 reprinted by permission.

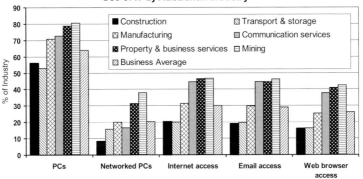

Figure 6.1:   Use of IT by the construction industry compared to associated industries in Australia
Source: Australian Bureau of Statistics 1999. Catalogue Number 8133.0 reprinted by permission.

6.2.1   Impact of IT on the construction industry

Research has shown that the principal causes of delays, disruptions, increased costs and poor workmanship are a direct result of problems with documents (NPWC 1993). Typically these problems relate to:

- errors in documents;
- contradictions and ambiguities in documents;
- late supply of documents;
- outdated information;
- inconsistent formatting;
- unnecessary duplication of documents; and
- difficulties in coordinating document standards from different disciplines.

Many of these issues can be attributed to the fact that the majority of documents traditionally use paper as a substrate, which does not facilitate easy transfer and building of content (Duyshart 1997). Accordingly project documents need to be managed with an effective use of time and resources through the use of in-built mechanisms for easy review. Given the inherent problems associated with paper-based management systems and their distribution, the majority of industry evidence points towards a case for the adoption of computer-based document management systems (Duyshart 1997). The systems proposed in this report support this purpose.

Reasons for the slow uptake of IT have been investigated and are well documented (Tucker and Mohamed 1996; Marsh and Finch 1998; Betts *et al.* 1999). Contributing factors relate to the industry norm of one-off projects, industry fragmentation, lack of client leadership, low levels of technology awareness and training, required up-front investment, ongoing maintenance costs and resistance to change. Surveys have also shown that IT is still too restricted to administrative functions of an accounting nature, or for highly specific and technical functions within the construction disciplines (Betts and Clark 1999). These findings demonstrate that the use of IT in construction has not progressed to the depth observed in other industries.

A number of surveys have been conducted in recent years to investigate the status of construction IT applications in major developed parts of the world. For example, Betts *et al.* (1991) surveyed the use of IT in construction in Singapore; Aouad and Price (1994) reviewed its use in planning in the UK and the US; O'Brien and Al-Soufi (1994) surveyed construction data communications in the UK; Aouad *et al.* (1996) conducted a survey to identify the priority IT topics likely to be of benefit in the future; Shen (1996) assessed the impact of IT on the construction and real estate industry in Hong Kong; and Samuelson (1998) examined the current situation of construction IT applications in Sweden. A recurring theme in these surveys is the huge but untapped potential IT has to modernise construction business processes.

Unfortunately, the construction industry is known for its reluctance to embrace innovation, especially in procurement and technology. This reluctance originates from the relatively high uncertainty surrounding many construction business processes. A business process, in the context of this chapter, refers to any technical or administrative activity that adds value to given input. This generic definition is adopted to cover the majority of design, construction and project management activities. From initial concept to procurement and final handover, hundreds or thousands of activities and transactions are conducted and communicated among project participants. The amount of information generated during the execution of these activities is enormous, even for a small-size construction project. For example, research by Crotty (1995) shows that approximately 1000 technical documents are produced for every A$2.0 m project cost.

Throughout the world, the construction industry in general is highly fragmented and is associated with a number of significant negative impacts, namely: perceived low productivity, cost and time over-runs, conflicts, disputes, resulting claims and time-consuming litigation. These have been acknowledged as the major causes of performance-related problems facing the industry (Mohamed and Tucker 1996; DETR 1998; Walker *et al.* 2000). Key consequences of this fragmentation problem include the lack of coordination between various project participants and the cumbersome task of handling data generated by different parties, in many diverse forms, throughout the entire project lifecycle (Evbuomwan and Anumba 1998). Accordingly, the ability to communicate and network using enhanced IT environments is regarded as a most important opportunity for the future (Aouad *et al.* 1996).

For example, recent software for customising web browser portals such as Microsoft's Digital Dashboard allows users, through a relatively simple drag and drop web portal page generator/editor, to completely customise their web browser screen. This allows them to gain access to a wide range of system software, utilities, electronically stored information or knowledge bases and other IT resources such as e-mail, discussion boards, etc. Moreover they can establish their digital dashboard portal page to place these tools and IT assets in precisely the order and appearance they want. This potential to develop a highly customised IT environment provides a highly useful psychological effect in helping users accepting and using these tools by giving then a sense of control over their IT environment. The Bovis Lend Lease ProjectWeb is an example of this kind of technology.

Strategically speaking, IT has the potential to change the landscape of the construction industry. It is an industry, however, that generally makes slow progress towards capitalising on the opportunities IT offers to gain competitive advantage and enhance performance. As the bulk of construction activities is carried out by relatively large number of small and medium organisations, and due to highly competitive business environments, only leading organisations tend to be in a position to plan strategically for IT uptake. Betts and Ofori (1999) point out that the current use of IT for strategic purposes in construction is rather limited. They also advocate the promotion of more effective strategic exploitation of IT in construction by leading organisations, national agencies and professional institutions.

Betts *et al.* (1999) analyse the current strategic role of IT and the way it is used for business efficiency in construction. In doing so, they identify a number of key competition and business strategy drivers which influence the process of strategic exploitation of IT. These include the organisation's core competencies, competitive behaviour, vision of the role of IT, and formation of national and/or international strategic alliances. It is worth pointing out that the role of IT should not only be viewed in terms of automating information handling and processing, rather in terms of adding value by redesigning traditional business processes. Mohamed (1996) highlights the inappropriateness to invest in IT simply to reinforce existing processes and expect different results.

Although the lack of progress of construction organisations in adopting IT is frequently blamed on the nature of the industry and the lack of investment in this area, Betts *et al.* (1999) suggest that the main problem lies in the way that construction organisations allow IT to be managed. They argue that the management of IT in construction organisations is in need of more attention, indicating considerable room for improvement. Until now, the conventional IT focus has been aiming strategically to reduce 'cost drivers' in the industry by eliminating unnecessary costs, requirements, processes and duplication. In addition to these cost-improvement schemes, it is argued that organisations need to have a long-term strategic framework focusing on changing business relationships from adversarial to an integrated mindset and creating value through flexibility and customisation (Alsagoff 2000).

In addition to strategic IT planning, Pena-Mora *et al.* (1999) advocate that organisations need also to develop sound strategic IT implementation plans. This is to allow greater accuracy in planning; provide data on the return of IT investment; reduce the risk associated with such a strategic IT investment; and monitor the benefits of this investment over the IT life cycle. Alsagoff (2000) stresses the importance of strategic IT implementation by considering a variety of implementation issues such as organisational structure, management style and human resource policy. Unfortunately, many organisations appear to approach the management of IT in an unstructured or *ad hoc* manner throughout the technology's lifecycle (Irani *et al.* 2001). Therefore, a formal holistic approach to the management of IT should be adopted to assess costs, benefits and risks associated with IT implementation (Stewart and Mohamed 2001).

In assessing the costs, organisations typically consider the direct costs as those attributed to the implementation and operation of the IT tools. Indirect costs typically comprise human and organisational factors. One of the largest indirect human costs is that of management time (Irani *et al.* 2001). A significant amount of resource is usually used to investigate the potential of the IT tools, and in experimenting with new information flows and modified reporting structures. Wheatley (1997) suggests that a further indirect human cost, which is often overlooked, is that of system support and trouble-shooting. From this, it could be argued that to have a successful IT implementation, the organisation needs to have a business vision to invest in IT and continued investment in training and support for its employees. However, this prerequisite may not be perceived to be economically feasible for many organisations. Research, however, shows that IT enables small and medium-sized organisations to enter new markets and enables their efforts to be integrated with the rest of a project team on an equal footing (Atkin *et al.* 1999).

In evaluating the benefits of investing in IT, any potential IT tools should have identifiable benefits for the organisation. Recognising these benefits usually requires an understanding of the work processes of the organisation and its clients. However, many benefits are primarily strategic or tactical in nature and their financial rewards are difficult to predict. Operational benefits are more tangible in nature and are more likely to display direct financial relationships. Accordingly a long-term investment viewpoint must be adopted to realise such benefits.

At the project level, information is usually considered as the processed and presented data in a given situation, and is the data that enables effective action (Fisher and Shen 1992). Information produced by many sources, retained by the creator, at many levels of abstraction and detail contributes to the fragmentation of the industry (Froese *et al.* 1997). Therefore, timely and accurate information is important for all project participants as it forms the basis on which decisions are made and physical progress is achieved. Most wasted time and cost in construction projects are traced back to poor coordination caused by less-than-optimum information handling and exchange, i.e. inadequate, insufficient, inaccurate, inappropriate, inconsistent or late information or a combination of them all.

Traditionally, project information exchange between designers and constructors has been mainly based on paper documents (Luiten *et al.* 1998). These documents come in the form of architectural and engineering drawings, specifications, and bills of quantities and materials. This practice is far from being satisfactory, with research showing that about two-thirds of the construction problems are caused by inadequate communication and exchange of information and data (Cornick 1990). Research (Kagioglou *et al.* 1998) has also noted that 85 per cent of commonly associated problems are process related, and not product related. These findings explain the growing awareness of the value of IT to bring together the major parties in the construction process, and share project as well as industry information in a meaningful way.

The need for improved project communication is widely acknowledged in the construction industry. To facilitate the management of project information and address project communication requirements, a number of IT tools have been used aiming to maximise benefits and reduce cost for the entire project team. The key to project information management, though, is the information flows associated with inter-organisational communications (Tucker *et al.* 2001). As a result, a core issue is the effective management of information, both in the form of information flows that permit rapid inter-organisational transactions between project participants, and in the form of information accumulated, coded, and stored for future use in the organisations' database structures.

The relationship between construction project participants is normally complex and involves many parameters that extend across technical, functional, business and human dimensions. As a result, attention and focus must be given to the intensive collaboration among project participants to synchronise both the input and output of the supply chain. Undoubtedly, a key enabler to successful collaboration is the ability to communicate, and share and exchange project information in a timely and accurate manner. A recent European survey (McCaffer and Hassan 2000) has highlighted the need for electronic sharing of information between Large Scale Engineering (LSE) clients' information systems and those of:

- Funding bodies in the areas of finance and accounting;
- Consultants in the areas of modelling and calculations;
- Project Managers in the areas of project planning and QA systems and document control;
- Contractors in the areas of CAD drawings, materials procurement, project

planning, QA systems and document control, and communication systems; and

- Suppliers in the area of materials procurement.

The construction industry has for many years suffered from difficult-to-access, out-of-date and incomplete information (Shoesmith 1995). Until very recently, it would have been inconceivable electronically to control and direct information flows in construction projects. Documents can now be produced and transmitted instantaneously by digital transmission at fractions of their previous costs and delivery speed. Electronic Data Interchange (EDI) permits computers and information systems to communicate directly with other computers, strengthening joint operations among organisations.

Unfortunately, the effectiveness of using IT in a construction project could be hindered by the inability to share data in electronic form and build in added value between project partners. Although it is not practical to expect compatibility between all information systems in the short term, there should be more focus on standardisation of interfaces between the different systems. IT tools should be able to exchange digital information with other applications/systems using appropriate data exchange standards (Hannus *et al.* 1999). Since the internet is a worldwide system for exchanging and distributing freeform information, it is regarded an ideal platform on which to base information systems (Adcock 1996).

Today's clients are more demanding and the construction industry is no exception. Increasingly in order to limit whole-of-life costs, government and institutional clients want a comprehensive package of construction services including maintenance, finance and management. Developers are becoming more involved in the detailed planning and design of their projects to ensure they earn a reasonable return on capital in a competitive market.

In general, clients are seeking cooperative relationships through short- and long-term partnerships and strategic alliances with one-stop service providers, requiring greater cooperation between contractors, designers, subcontractors and suppliers. These new business relationships require more open communication and a more rapid exchange and sharing of information. The need for trust, cooperation and honesty is particularly important where open information systems are used.

Reporting on findings from the European Large Scale Engineering Wide Integration Support Effort (*eLSEwise)* project, Hunter *et al.* (1999) conclude that the required attributes for the LSE contractor to be competitive are best supported by the concept of a 'Virtual Enterprise', where different organisations with supportive skills set, appropriate to meet the specific project demands, form a cohesive team, but without the need for co-location of the team members.

Cooperative business relationships that provide clients with single source solutions are known as 'virtual enterprises'. The client has a relationship with what they believe is one organisation whereas in reality they are dealing with a group of separate organisations working together. The business arrangements of a virtual enterprise can take many forms, some are formal, others informal. Usually they do not involve changes in ownership or loss of identity of the individual participants. The organisations within a virtual enterprise can be in different locations with varying time zones. Information technology can enable them to work together effectively and efficiently. For example BP's Virtual Teamwork Project is a case in point where scarce knowledge workers communicate globally via videoconferencing equipment, multimedia, e-mail, application sharing, and other ICT tools to share knowledge and significantly enhance the productivity of their organisations as well as provide wider and deeper capacity for complex problem solving (Davenport *et al.* 1998; Davenport and Prusak 2000, p. 20). The strength of these business partnerships and inter-organisational links lies in: the recognition of each individual member's complementary strengths plus shared values and objectives (DPWS 1998).

From the above, it is clear that the construction industry could make greater use of IT to gain a strategic competitive edge. According to Walker and Rowlinson (1999), this can be achieved in three ways. First, IT can reduce the cost of communication transmission through productivity gains and reduction of the need to multiple-handle information in the supply chain. Second, IT tools have the ability to allow clients already using the appropriate technology, to communicate easily with the industry. Third, if the use of these IT tools is well thought through then there could be a quality of service advantage because it offers the ability to maintain online current status information of projects.

6.2.2 Project collaboration websites

The ubiquitous presence and increasing maturity of the internet, is gradually leading to its validation as a legitimate business medium. For example, in the US the number of companies that have implemented a dedicated internet connection has almost doubled to 97 per cent in 2001 from just 51 per cent four years ago (ZweigWhite 2001).

As a result of this trend, many technology vendors are now taking existing construction industry IT tools such as computer aided design, cost planning/estimating, scheduling, accounting, project management and site management systems and internet-enabling them. This internet-enablement may range from a simple capability to transfer files over the internet up to a complete re-write or development of a system to be delivered using a web server and web browser client.

In the closing decade of the twentieth century and in recent years, information and communication technologies has been noted to be a significant enabler of knowledge management facilitating organisational learning. The literature indicates that IT is only an enabler contributing about one-third of the knowledge management effort with the remainder being the way in which human interaction allows the generation, transfer and use of knowledge (Zack 1999; Davenport and Prusak 2000; Kluge *et al.* 2001). The interest shown, particularly in the USA, by IT facilitation of knowledge management has been sparked by advances in access to knowledge assets through web browsers.

One of the fastest growing areas of web-based development for the construction industry has been in the area of project extranets. Also known a 'project collaboration website', a project extranet is simply a secure website used for the centralised management of project information by all project participants. By posting all project information to a web server it becomes instantly accessible to authorised team members regardless of their location. Project participants can then easily view, mark-up, approve and distribute these documents. Throughout the entire process all changes and revisions can be automatically tracked and archived in an activity file. This type of IT support facilitates development of communities of practice that share and build knowledge (Wenger and Snyder 2000; Davenport 2001).

In October 2001, over 230 project extranet systems have been identified (see www.extranetnews.com) emphasising the rapidly expanding demand and experimentation with these systems. Each of these systems has a particular emphasis on the types of information they manage and for which industry. For example, solutions exist for construction, architecture, manufacturing, GIS and others and can be used for CAD management, office documents, general collaboration, and electronic commerce or general reference.

Perhaps one of the most significant and successful examples of a web-based application currently to be used in the Australian construction industry has been the development and implementation of a project collaboration system called 'ProjectWeb' by Bovis Lend Lease. This company has been effectively utilising the internet since 1997 and has recognised the requirement for project members to be able to connect, coordinate and manage projects while faced with increasing numbers of geographic, logistical and time-based constraints. Utilised on over 300 projects with participants in over 15 countries, this system has proven to be an invaluable tool for Bovis Lend Lease to manage the project information generated by the thousands of companies that they collaborate with on a daily basis.

This particular system is discussed in the following section to highlight its role in enabling improved project information management capabilities and business relationships, compared to the common business-as-usual approach.

6.3 Case study – implementation of an IT solution for the delivery of the National Museum of Australia project

With the combination of increasingly complex building types, increasing numbers of outsourced project participants, tightly programmed fast-track construction methods and a globally competitive marketplace, the ability to deliver profitable projects on time and within budget is becoming a significant challenge. Without the effective use of information technology (IT) to facilitate the process of information management amongst project teams, it is unlikely that *any* major improvements to the delivery process will occur by continuing to use traditional processes (Duyshart 1997).

The National Museum of Australia project involved the design and construction of a new museum in Canberra, Australia for the Commonwealth Department and Communications and the Arts. One of the key challenges for this project was that the complexity of the building type and the challenging design produced by the architects. In addition to the challenging design it was a project that, by the nature of the alliancing delivery model, had additional imperatives to be delivered on time and under budget. An additional geographic constraint was added by the fact that the exhibition designers were located in Boston, USA, the architects were in

Melbourne and the project was in Canberra. From the onset, there was a high potential for IT to add value in supporting the project team.

6.3.1   Developing an IT return on investment (ROI) model

The decision for the National Museum of Australia project to make extensive use of IT to assist in the delivery of the project was not a straightforward process. Prior to the commencement of the project it was necessary to develop an extensive business case that:

- identified the drivers for the use of IT;
- identified the potential technologies that could be used;
- specified the actual technological components and their cost;
- calculated a positive ROI for the project.

This business case had to be presented to a number of board meetings of the alliance team and had to be independently audited by the project alliance auditors KPMG.

It wasn't until these milestones had been achieved that a budget for the use of IT was established for the National Museum of Australia project. One of the conditions of the approval was that a parallel study would be conducted that analysed how effectively these technologies were used and the findings then disseminated to industry.

The ability to develop sound ROI calculations for expenditure on IT has been an ongoing issue since the development of the first computer. In particular, this dilemma has been particularly prevalent in the conservative construction industry. Accordingly, on the National Museum of Australia project an extensive ROI study had to be performed to justify expenditure on IT. The following outlines the basic principles that were used in the calculation of this ROI.

The question of whether there is a direct relationship between IT expenditure and productivity is one that has been widely debated. In the early 1990s the so-called 'productivity paradox' emerged as a hypothesis that productivity levels had stagnated despite massive investments in IT. Extensive research by MIT between 1987 and 1991 on 367 large firms, which generated a total $1.8 trillion dollars output to the US economy, proved that expenditure on information systems (IS) made an average gross ROI of 81 per cent. This result was significantly higher than for other types of capital expenditure, which averaged at about 7 per cent thus dispelling the 'productivity paradox' debate (Brynjolfsson and Hitt 1994b).

It became evident from that research that there are large numbers of factors that contribute to productivity apart from savings in labour cost. The essential question to arise is whether it is possible to measure productivity accurately as a result of expenditure on IT.

Even though there are extensive calculations that can be performed such as ROI, IRR, and NPV, these techniques have often been demonstrated to be incapable of capturing the entire benefit of investment in IT. The reason for the failure of traditional economic measurement techniques is that they don't take into account the so-called *intangible* or *soft* factors such as accuracy, quality, convenience, variety, timeliness, flexibility, functionality, reliability, usability, user satisfaction, utilisation, relevance, and security. These factors have been estimated to account for the majority of the benefits resulting from investments in IT and can result in vastly improved customer service and client satisfaction.

Of the companies analysed in various research projects on information systems expenditure, it was found that the delivery of *customer service* had the single greatest impact on relative productivity and that customer-focused companies were likely to be 7 per cent more productive than their competitors (Brynjolfsson and Hitt 1994a). By contrast, those companies which used **cost cutting** as the sole reason for investing in information systems performed 4 per cent *worse* in financial productivity than their competitors.

In summary, it is now widely acknowledged that there are a large number of contributing factors that should be considered in the evaluation of IT expenditure. A simple method of representing this is illustrated in the following formula:

> Value of IT expenditure = Tangible benefits (cost savings)
> + Intangible benefits (e.g. service,
> quality, timeliness)

Since Brynjolfsson's original 'productivity paradox' findings (Brynjolfsson and Hitt 1994b), difficulties in estimating the benefits to derive from expenditure on IT have continued. In particular, it has also been highlighted that 'proven ROI models for industry-specific extranet-based business-to-business electronic commerce applications will not be in use until at least 4Q 2002 (0.8 probability)' (Terhune 1998). Accordingly, for the purposes of ROI calculations on the National Museum of Australia project, it was only possible to quantify tangible benefits and provide a conservative estimate of intangible benefits that should also be taken into consideration.

The methodology used included the calculation and provision of the following supporting material:

An estimate of tangible savings including:
(1) Cost savings on key overhead preliminaries costs; and
(2) Cost savings from more efficient document handling processes using information management systems.

A highly conservative estimate of intangible savings including:
(1) Improved customer satisfaction;
(2) Improved quality;
(3) Improved service;
(4) Improved accuracy;
(5) Reduced rework;
(6) Improved convenience;
(7) Improved security;
(8) Improved flexibility;
(9) Improved functionality;
(10) Improved reliability;

(11)   Improved usability;
(12)   Improved user satisfaction;
(13)   Improved utilisation;
(14)   Improved relevance of information;
(15)   Improved security.

Provision of an extensive list of additional benefits categorised according to:
(1)   Technical infrastructure benefits;
(2)   General business benefits;
(3)   Project management benefits;
(4)   Client, consultant, subcontractor, supplier and authority benefits;
(5)   Document management benefits.

In the calculation of tangible savings a best practice scenario was calculated. This figure represented the savings that *could* be achieved if adoption of the proposed systems was 100 per cent utilised. Realistically, this figure was unlikely to ever be obtained on this project due to most participants' lack of prior experience with new IT systems. Accordingly, a much lower target of attainability (40 per cent) was set for the project that had a higher probability of being achieved. Having set this figure, however, it was also important to present the best practice targets as it helped the senior project management team to realise the potential upside of supporting and fully implementing the IT systems to best practice industry levels.

Tangible savings – reduction of identified overhead preliminaries costs
The preliminaries component of the project budget typically identifies the costs associated with the delivery of a project. On the National Museum of Australia project, seven major areas of cost were identified that could be directly affected through the use of information technology initiatives. These cost areas were:

- **Plan printing** – reduced numbers of prints through centralised management and user-demand printing.
- **Paper supplies** – reduction of overall paper consumption.
- **Photocopier/fax hire** – reduction of hire charges based on reduced number of copies.
- **Couriers** – reduction of the numbers of physically exchanged documents.
- **Photography** – elimination of traditional chemical based processing with use of a digital camera.
- **Interstate travel** – reduction of travel requirements due to improved communication and access to information.
- **Telephone/mobile** – reduction of number of calls due to improved communication.

For each of these cost areas, the actual project preliminaries budget was identified and then cost reductions targeted that could be realised through the use of IT initiatives. Table 6.2 illustrates the targets that were set. Actual ROI calculations used the National Museum of Australia target reductions, which were quite conservative.

Table 6.2: Cost reductions on preliminaries

Preliminaries cost	Best practice target reduction	National Museum of Australia target reduction
Plan printing	60%	20%
Paper supplies	50%	10%
Photocopier/fax hire	40%	10%
Couriers	90%	50%
Photography	95%	80%
Interstate travel	25%	5%
Telephone/mobile	20%	10%

Tangible savings – improved efficiency in handling of technical documents
In the calculation of savings that could be realised from improved efficiencies in document handling, it was important to provide an accurate estimate of the number of project documents that would be produced. Independent research indicates that approximately 1000 technical documents are produced for every A\$2 m of project cost (Crotty 1995). Therefore, based upon a A\$135 m project cost for the National Museum of Australia:

Total technical documents expected to be created = 67 500

With the introduction of the project collaboration system, a wide range of documents types would be managed in a far more efficient and cost effective manner. If a conservative estimate of 15 minutes saving per document, over the project life of that document (including all revisions), can be assumed then the following savings could be realised, based upon an average charge-out rate of A\$70/hr.

Potential savings from 67 500 documents $\times$ 0.25 $\times$ \$70/hr = A\$1.18 m

This figure does not take into account the savings that are achieved in search times to retrieve documents, to verify issue and other standard procedures. This amount is therefore a highly conservative saving of time and cost, but would represent best industry practice if 100 per cent utilisation of all IT systems were achieved. Accordingly, this figure was discounted to a utilisation target of only 40 per cent in order to increase the probability that the identified savings could be realised:

Target utilisation from handling 67 500 documents = 40% $\times$ \$1.18 m = A\$472 500

On the National Museum of Australia project, the total tangible savings are indicated in Table 6.3.

Intangible savings – estimating a conservative figure
As previously illustrated, the ability to calculate the monetary value of intangible savings is not currently available. Accordingly, on the National Museum of Australia project a highly conservative estimate was used of approximately 20 per cent of the best practice tangible savings.

Table 6.3: Estimate of tangible savings on case study project

Tangible benefits	Best practice (@100% utilisation)	National Museum of Australia project target (@40%)	Remaining potential (BP – target)
Estimated reduction of preliminaries overhead costs	$484 950	$194 340	$290 610
Estimated improved efficiency in handling documents	$1 181 250	$472 500	$708 750
Total tangible benefits	$1 666 200	$666 840	$999 360

Total IT ROI – final calculations

The total return on investment for this project was then calculated as:

> Total target tangible benefits
> + conservative estimate of intangible benefits
> − budget to implement IT initiatives
> = Estimated return on investment

Using this methodology, it was possible to:

(1) Demonstrate a positive return on investment for the project that was esti-
    mated as being realistically achievable (based on the conservative estimates
    that were used); and
(2) Highlight that it was possible to achieve an additional A$1 m worth of tan-
    gible benefit savings if industry best practice was sought.

Other key ROI considerations also needed to be considered and are detailed
under the following headings.

6.3.1.1   *Project information management solutions*

The need for improved project communication is a widely documented issue in the
construction industry. In this industry today, a key area of opportunity is to ensure
that project team members can effectively communicate without being over-run by
transaction costs. These costs are well recognised as tending to increase exponentially as the number of participants on any project increase. To address this
issue, project team members need to be able to participate in a project as if they
were a large, single, 'virtual' enterprise (Crotty 1995).

In order to facilitate the management of project information and to address
specific project communication requirements towards this aim, it was proposed to
implement a number of IT initiatives. These initiatives were aimed at maximising
benefits and reducing costs for the entire project team. The following areas were
addressed on this project.

6.3.1.2   PC provision

Traditionally, PC usage in a construction site office has been very low and limited to a number of key functions such as accounting, cost planning and secretarial support. With the lowering costs of PCs, affordability has increased resulting in an improved cost–benefit case for computers to be used by all project participants requiring them as a part of their job function. With the subsequent introduction of computer-based information management tools on this project, there was also an increased need for all project participants to have ready access to a computer to be able to communicate effectively on the project. Accordingly, this project significantly increased the availability of computers for the entire project site team.

6.3.1.3   Network LAN

A functional requirement for computers needing to share services such as file servers, printers, and internet access, was to build a local area network in the project site office. To accommodate these requirements a local area network was established with 50 network points. Connectivity to the internet was provided via a 256 K ISDN link.

6.3.1.4   Video conferencing

To improve communication and reduce travel costs, video conferencing facilities were occasionally used for the project. This system involved the use of desktop PictureTel equipment, compatible with most industry-standard video conferencing facilities. Rather than purchase high quality equipment and additional ISDN lines specifically for the project, the nearby facilities at the Bovis Lend Lease Canberra Office were used.

6.3.1.5   Digital handheld and web cameras

To improve the recording, distribution and communication of construction progress a handheld digital camera and a web camera connected to the internet were used. These cameras were used for internal reporting purposes as well as for a public marketing website used to promote the development of the museum.

6.3.1.6   Project collaboration websites

One of the key initiatives for information management on this project was the implementation of a project collaboration website or project extranet (as described in the previous section). Bovis Lend Lease (as the appointed project managers) offered the use of an internally developed tool called 'ProjectWeb' to be used for this purpose. ProjectWeb provided a wide range of tools for management of drawings, documents, images and many workflow documents such as Requests for Information (RFI), and Site Instructions. The ProjectWeb collaboration system provided by Bovis lend Lease was the primary tool used for information management on the National Museum of Australia project.

As a new medium for communication, the internet has surpassed the adoption rate of all previous media such as newsprint, radio and television with breathtaking speed. The internet's worldwide penetration and high level of standardisation has contributed to increasing levels of globalisation. In 1993, the introduction of the world wide web has led to an explosion in internet usage, paving the way for extensive commercial use due to its capability to provide a graphically-based tool for sharing information using text, hyperlinks, full-colour graphics, photos, audio and video images through computers (Tam 1999).

The approach in using web-based technologies means that business partners can easily adopt its standards and obtain access to it from any single point of entry available through the internet. The use of the web does not require any proprietary software or hardware systems. The minimum user requirement, therefore, is a connection to the internet and an industry standard web browser.

The concept of using a web-based information management system is to use the internet as the communication channel and the widely available technologies associated with the web as a publishing medium for global distribution of information. Using this technology, a project team member was able to provide project information easily for any other participants located anywhere in the world. In particular, the application of the Bovis Lend ProjectWeb system, used on the National Museum of Australia project, was argued to provide significant qualitative as well as quantitative advantages to the construction management process (Hampson *et al.* 2001, p. 242 and p. 482). These can be explained as follows:

- Electronic linkages can fundamentally change the nature of inter-organisational relationships by raising trust and lowering risk. Baldwin *et al.* (1996) suggest that such linkages are essential to establish alliance relationships between project participants. The basis of all trust is an acceptance by those involved in the process that each is telling the truth about their knowledge domain. One of the key advantages of a project collaboration system is that it not only provides the electronic linkages between project participants, but also the necessary levels of business process transparency that supports alliances. It is a key inter-organisational requirement for better performance of each participant, i.e. to achieve a win–win outcome.
- One of the recent systems developments aimed specifically at utilising the ubiquitous nature of the world wide web is ProjectWeb. Bovis Lend Lease designed and developed a suite of secure, internet-based interactive business tools for the centralised management of project information via a web browser. This system was used instantly to share, visualise and communicate project information between any project participant including staff, clients, consultants, subcontractors, suppliers and authorities. It also improved communications amongst the entire project team not just the contractor. Bovis Lend Lease specifically developed the system as a project information management system for projects on which it has been engaged as the Project Manager and uses it to improve the value of their services. All project participants had access to ProjectWeb, to the appropriate level required for their respective project role.
- ProjectWeb has a number of modules to manage project documents. The modules include day file, help, project administration, project directory, project information, document library, drawing register, file transfer, image

library, project calendar and transmittals. The types of documents which are typically used during a construction project and which can be managed, stored, transmitted and viewed include incoming documents (once scanned), legal documents, drawings and specifications, requests for information, site instructions, memos, faxes, cost reports, meeting minutes and agendas, monthly reports, presentations and programmes. The overall system also has: a 'day file' daily activity log, application help, support feedback, project administration functionality, a project directory and project overview information. The digital dashboard approach allows users to have access to these modules through a user-friendly portal. Figure 6.2 illustrates a typical screen interface for a user.

Some of the key benefits to result from the implementation of this system for the National Museum of Australia in particular were:

(1)   The achievement of improved communication and coordination of information between all project participants resulting in increased efficiencies, better-facilitated decision making and improved project control;
(2)   The ability to increase flexibility for project participants by easily becoming a part of the project team through the use of ubiquitous, low-cost internet and web technology;
(3)   The ability to facilitate the strategic alliances established for this project;
(4)   The ability to increase savings in standard operating costs and approaches resulting from improvement in delivery processes, responsiveness, reporting and turnaround; and
(5)   The opportunity to provide an innovative business-to-business information technology capability for the construction industry in delivering a high profile government project.

With the above functional features in mind, time requirements of transferring and checking information were significantly decreased. Research has shown that the availability of the current and valid information has greatly decreased the response time required to handle any unforeseen situation that may arise (Tucker *et al.* 2001). The application of ProjectWeb enhanced the general process of collecting, gathering and reporting different types of data and information during the design and construction phases. There was a strong feeling on the National Museum of Australia project that by providing participants with information on time, information that was correct and that did not require later changes, the whole culture in relation to project information exchange would change (Peters *et al.* 2001). It was common practice for late changes to result in the need for rework, which can be directly linked to a reduction in the level of quality being achieved or increase of non-productive work time.

6.3.1.7   *3D CAD modelling*

To optimise information flow and communication of design intent between the architects and other project participants, 3D CAD modelling software was extensively used throughout the entire project.

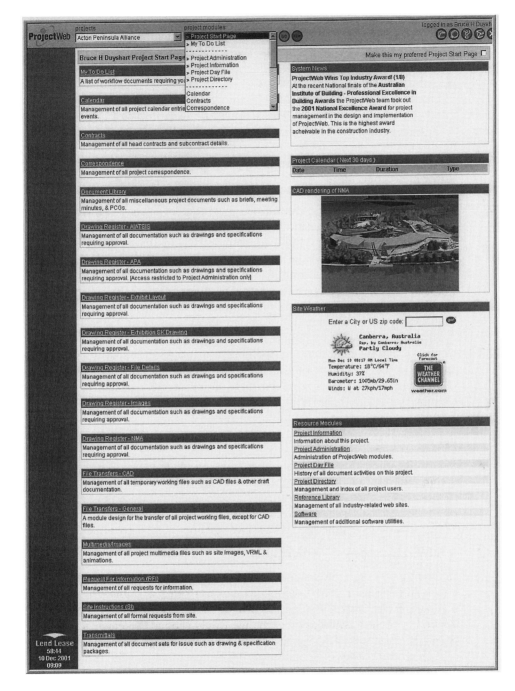

Figure 6.2: Typical screen for ProjectWeb

The relatively complex form of the National Museum of Australia building provided a huge challenge for the entire delivery team. Principal amongst these was the problem of communicating the design intent to a built form. To achieve this ultimately required the 3D model data generated by the architects to also be made available for other disciplines using a range of different computing systems.

The initial 3D model developed by the architects was used to communicate the design intent of the museum. This required the use of spatial analysis techniques such as walkthroughs and fly-overs, combined with numerous computer renderings.

Once the concept model had been built, the architects further developed it through a number of iterations eventually to develop a model detailed enough to be used by the cost planner to take off accurate quantities. The façade contractors were also able to generate exact panel layouts using special software that 'unfolded' the 3D surface of the model into cut-out templates.

Within the 3D envelope that had been defined by the architect's CAD model, the structural engineers were able to model all of the structural members accurately. The model of the structural members could then be transferred to a number of steel detailing companies. The transfer of the 3D modelling data was able to eliminate a significant amount of steelwork shop drawings that would have traditionally been drawn in 2D from scratch.

The entire process of modelling in 3D and using it to communicate design intent, resolve design issues, accurately estimate cost and generate cutting templates was able to reduce significant amounts of time by avoiding intensive rework. More importantly, however, it removed large areas of risk associated with communicating the design integrity of a complex building form.

6.3.1.8   Marketing website

A nominal allowance was made in the budget towards the provision of a public marketing website. This site augmented an existing site of the client, but was able to make extensive use of the drawings, images and other documentation generated by the project team, using the resources identified in this chapter.

6.3.1.9   IT technical support

In order to procure, install and maintain all of the IT systems that were provided as a part of the site office setup, it was necessary to employ a full time technical resource in the early stages of the project and then periodically thereafter. The extent of systems implemented (server, network, and approximately 40 PCs) more than justified a full-time role for this position during the first half of the project.

A significant benefit to result from this support was that the IT equipment was able to kept fully operational at all times with no major downtime or lost productivity.

6.3.1.10   IT training support

Perhaps the most significant of all initiatives was the provision of a full-time IT trainer or facilitator as Bovis Lend Lease referred to the role.

Traditionally, it is always the hardware and software that receives focus in an IT budget with little or no attention paid to how the end-users will be able to cope with the introduction of these new technologies. Anecdotal evidence suggests that a significant contributor to the failure of many IT-led initiatives is this lack of user support. This support is required at both a technical literacy and business process level.

The facilitator or 'coach' influenced not only the immediate project site team but also all other related project participants including clients, consultants, sub-contractors and suppliers. Considering the range of new systems and the number of project participants, this full-time role was well justified.

6.4   An assessment of the application of IT on the National Museum of Australia project

In additional to the development and delivery of the National Museum of Australia on the Acton Peninsula, a part of this study investigated the theory, history, process, cost and benefit and cultural change involved in the development and implementation of an integrated information technology system (Peters *et al.* 2001, Part B).

This part of the study provided a framework for comparing the use of information technology with the business-as-usual activities associated with the design, construction and project management of the Acton Peninsula alliance team. This framework identified the role of information technology as a tool for cultural change that also facilitated the creation of an integrated and efficient team.

6.4.1   Defining an evaluation framework

The development of an evaluation framework for this case study has been primarily based on a study of the potential benefits of the IT tools used on the National Museum of Australia Project (Peters *et al.* 2001, Part B). The framework, shown in Figure 6.3, examines the implementation of IT from seven different but interconnected perspectives:

(1)   **Information technology**: this perspective is the centre of the framework. It focuses on the IT tools used and addresses their technical aspects. It covers issues such as tool performance, reliability, availability, security, accessibility, user friendliness and suitability to the process.

(2)   **User utility**: this perspective is concerned with user satisfaction and perceived value of IT use. From the user's perspective, the value of the tool is based largely on the extent to which it helps them perform their tasks more efficiently and effectively. This perspective covers usage-related issues when interacting with IT tools. These issues include IT acceptance, utilisation, availability of training and technical support, satisfaction with the IT tool, and the quality of its output.

(3)   **Project organisation**: this perspective deals with the role that IT plays in facilitating the integration of project participants with the introduction of IT.

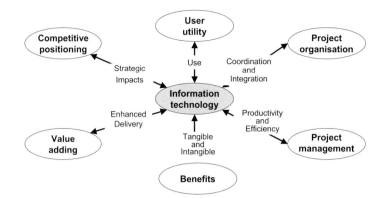

Figure 6.3: Seven perspectives of the IT analysis framework
Source: Hampson *et al.* (2001, p. 478)

Typically IT use increases inter-organisational cooperation between project participants.

(4) **Project management functions**: this perspective examines the impact of IT on project management functional goals. In particular, it focuses on information needs, project quality and delivery timeliness for design, construction and project management functions. The measurement and evaluation of project management functions focus on the impact IT has on the productivity and efficiency of these functions.

(5) **Benefits**: this perspective investigates links between IT implementation and any project-related short-term benefits. This perspective includes both tangible and intangible benefits. Tangible benefits such as time and cost savings are expected due to the reduction of paper-based workload, faster response times and less rework. Intangible benefits may include process flexibility in generating, handling and manipulating data, ease of workload, and the ability to more easily detect errors or inconsistencies.

(6) **Value-adding**: value is a much broader concept than benefits. This perspective captures the relationship between IT implementation and the overall project delivery process. It examines the perceived value-added aspect of the process in terms of generating business value to the client (delivering a project through a more robust delivery process) as well as to all project stakeholders (cultural change and extended partnerships).

(7) **Competitive positioning**: in addition to evaluating IT use on a particular project, there is also a need to measure and evaluate IT contribution to the strategic capability of the organisation. This perspective assesses the IT impact has on the organisation at the strategic level. This area of study is concerned with how lessons learned on this project are disseminated and hence contributed to the overall strategic positioning of the organisation. For example, this examines how the benefits achieved can be translated into an increased organisational capability with an ability to view IT as being an integral part of future business and project activities.

The questionnaire developed for the survey of project participants reflected the seven perspectives of the framework described previously. The seven perspectives and the topics that were addressed are shown in Table 6.4.

Table 6.4: IT evaluation framework perspective topics	
Perspective	**Topic**
Information technology	• Reliability • Secureness against unathorised use • User-friendliness • Appropriateness for the application/function • Suitability for site conditions (if applicable)
User utility	• The IT tool/system used most frequently • Level and frequency of training provided • Level and frequency of technical support provided • Accuracy and quality of the tool/system output
Project organisation	• Enhancement of coordination between project participants • Reduction in response time to answer queries • Establishment and support of the project alliance • Empowerment of participants to make decisions
Project management functions	• Facilitation of document transfer and handling • Keeping and updating records • Enabling of immediate reporting and receiving of feedback • Identification of errors and/or inconsistencies
Benefits	• Time savings (e.g. processing, responding, etc.) • Cost savings (e.g., rework, travelling, overheads) • Improved document quality • Decrease in number of design errors • Decrease in number of RFIs
Value-adding	• Leading to a more satisfied customer • Leading to a more streamlined process • Enabling a cultural change among project members • Improvement in computer/IT literarcy
Competitive positioning	• Enhancement of organisation's image in the industry • Attraction of more sophisticated clients • Increase in the capability for global cooperation

6.4.2   Survey methodology

The survey was administered three times to a range of project participants between February 2000 and January 2001 as illustrated in Table 6.5. The survey forms were delivered individually to the participants who were then given instructions on how to complete the survey. Three days after distributing the survey, respondents were visited again to collect completed surveys and to assist with any further queries they might have had. The majority of the survey responses were collected after three days. A week after the initial survey distribution all remaining surveys were collected.

6.4.3   Survey results

The survey questionnaire consisted of 29 statements that addressed the seven perspectives of the IT evaluation framework. Each survey participant scored each

Table 6.5: IT survey details			
	Survey date	**Surveys sent (number)**	**Surveys completed (number)**
Survey 1	February 2000	45	41
Survey 2	October 2000	35	30
Survey 3	January 2001	35	31

response by assigning a value, for example a 'Low' or 'Strongly Disagree' response counted as one and a 'High' or 'Strongly Agree' as five (Peters *et al.* 2001).

The main findings of the survey can be summarised as follows:

- Three professional groups (managers, designers and administrators) agreed that the predicted time savings resulting from the implementation of the project collaboration system, ProjectWeb, were actually realised;
- Most users (in particular the designers and administrators) were satisfied with the accuracy and quality of the output of information;
- The designers and administrators also believed that the ProjectWeb system assisted them in enhancing coordination between project participants, whereas the management users felt that its real value lay in its ability to help establish and supporting the project alliance. Users in administration roles felt that IT did not empower participants to make decisions, rather it acted as a communication and decision support tool.
- Managers felt that the level and frequency of training was below average for their requirements;
- Managers and administrators acknowledged ProjectWeb's role in being able to facilitate document transfer and handling;
- As predicted, the ability to provide immediate reporting was not rated highly by designers. These users were more interested in having streamlined processes;
- Overall, the managers acknowledged that there were noticeable improvements in IT literacy, but all three groups were neutral about whether this achieved a more satisfied and sophisticated project outcome in terms of decision making.

The average scores of each of the three surveys were plotted on a spider diagram as illustrated in Figure 6.4, Figure 6.5 and Figure 6.6. As illustrated in these three figures, the differences between the three surveys are minimal but do decline slightly from the first to the third survey. The results of both the second and third surveys are very similar. The largest declines observed were in the perspectives of Project organisation and Strategic positioning. Drops of 7 per cent and 11 per cent respectively could be due to:

- relatively high initial expectations;
- becoming more familiar with the occasional shortcomings of the web-based systems; and
- becoming increasingly aware of alternative systems.

Despite these particular areas, overall users appear to have adapted very well to the use of a web-based collaboration system with the worst acceptance score being

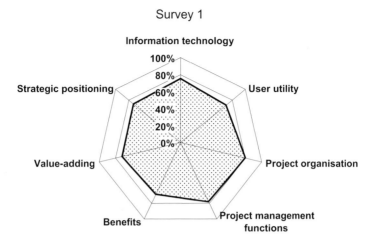

Figure 6.4:   Average scores of the seven IT perspectives for Survey 1

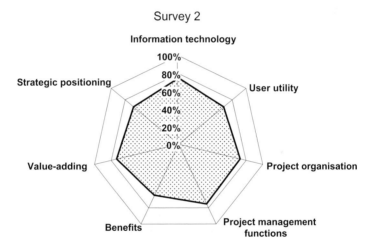

Figure 6.5:   Average scores of the seven IT perspectives for Survey 2

62 per cent and the best 80 per cent. Typically, in a survey of this type, any score above average is very positive. The evenness of the scores also indicates that there seems to be no major shortcomings felt by users and that the web-based system they have used is very effective in assisting them to carry out their tasks.

6.4.4   Analysis of project collaboration data

The analysis of data from the project extranet system, ProjectWeb, concentrated on three main areas:

(1)   **Users**: to analyse user activities in the context of the alliance form of contract and their ability to take up use of the system;

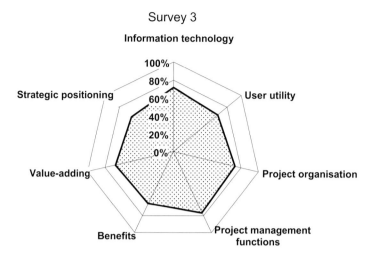

Figure 6.6:   Average scores of the seven IT perspectives for Survey 3

(2) **Communications**: to verify that the integrated and cooperative nature of alliancing did have some impact on the way project participants operated;

(3) **Performance indicators**: to investigate the potential for performance indicators to be extracted from the available system data using Requests for Information (RFIs) as an example.

6.4.4.1   Users of the system

Users of the ProjectWeb system were given an account that provided restricted access to the project based upon a user-identification and password. A ProjectWeb administrator would create an account for every team member when they joined the project and these accounts could either be retired or reactivated over the course of the project depending on their ongoing involvement in the project. Figure 6.7 below shows the total 'team' size and also the addition and retirement of members by week of the project.

After the initial 'sign-on' of over 70 users, there was an average of 5 new additional users per week for most of the project. After the halfway mark of the project, new additions began to taper off, as might be expected.

Retirement of user accounts that were no longer actively working on the project was carried out intermittently throughout the project. In particular one large group of users was retired about 80 per cent of the way through the project. The number of users continued to rise until about three-quarters of the way through the project when a plateau was reached.

6.4.4.2   Communications

With the implementation of a project collaboration system, the ease of communication is greatly enhanced compared to traditional methods of communication. One of the advantages of having all project users on one integrated system is that the specific activities and activity levels can be accurately recorded. On a major

ProjectWeb team members

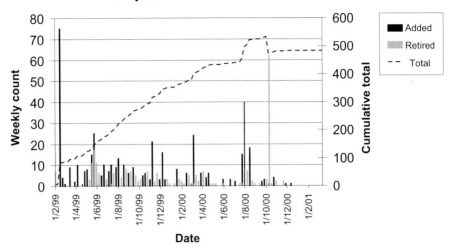

Figure 6.7: Team size as indicated by ProjectWeb users over time

project like the National Museum of Australia, the intensity of communication traffic can be readily appreciated. To the authors' knowledge, there is no comparative information available on any similar projects that have used the same medium.

In investigating the activity logs of the project collaboration system, the intensity of document traffic can be presented in two different ways:

- The number of messages of a particular type over time (message traffic); and
- The number of interactions between any two participants (participant inter-actions).

When presented on a weekly basis, the message traffic numbers reflect the demands of the project at that moment of time as well as the significance of that individual form of communication. Specific project communication forms include: Requests for Information (RFIs), Site Instructions (SIs), amendments, variation orders, delay notifications, correspondence, action items, submittals and document transmittals.

Interactions between participants were observed to be partly dependent and influenced by the effectiveness of information and communication technology resources applied to linking teams on the construction project. This was reflected in the communication traffic patterns.

6.4.4.3 *Traffic*

When analysing the volume of document traffic the actual numbers of documents being sent can be measured, however, similar to e-mail, this system allows documents to be addressed directly to people using the *To* field and circulated to other recipients for information using the *CC* field. For the purposes of this study, traffic is defined as the total number of electronic messages sent, i.e. the product sum of all *To* recipients and *CC* recipients for all items sent. The ratio of items sent compared to the total traffic is an indication of the multiplicity of communications.

Figure 6.8 shows the weekly traffic for all memo documents raised over the entire project. Memo traffic continues to rise until about three quarters of the way through the project. The *To* traffic is typically always greater than *CC* traffic especially towards the end of the project.

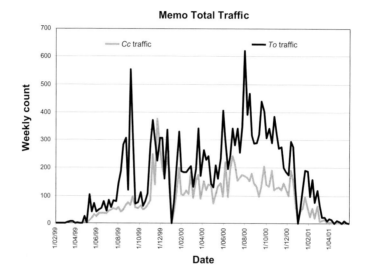

Figure 6.8:   Total traffic count of memos over time

This indicates the extent to which memos are used for broadcasting to the project team. After a couple of large-scale broadcasts early in the project, the ratio settled to a steady value of about 4 to 1 before falling away towards the end of the project. One explanation for this observation is that in the early stages of a project, specific project roles, responsibilities and confidence in certain individuals are less clear and hence more people tend to be notified to ensure that people are aware that a specific communication has occurred. For example, rather than just address the specific architect that a memo should be sent to, the memo is also sent to all members of that company on the system. The slowdown towards the end of the project is a reflection of this declining need to provide additional information and a lessening role of the design consultants. As design issues are resolved, the amount of correspondence to design consultants should decrease and construction corre-spondence should increase.

Figure 6.9 shows the weekly traffic for Transmittals that were sent over the entire project. In contrast to the Memo traffic, Transmittal traffic had a peak about a quarter of the way into the project and continued to decline until the end of the project. At all stages of the project the *To* traffic was always greater than *CC* traffic. In contrast to the observation for Memo traffic, the trend of the ratio remained stable (and slightly higher than the traffic volume of memos) throughout the entire project. Towards the end of the project a number of larger fluctuations

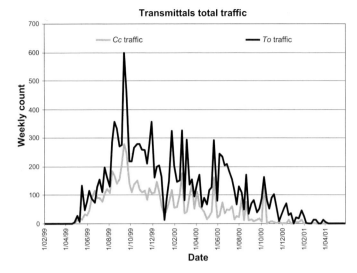

Figure 6.9:   Total traffic count of transmittals over time

occurred. This may have been caused in part by reissue of drawings for as-built documentation or for the few construction problems that required redesign.

Figure 6.10 shows the weekly traffic for RFIs over the entire project. Similar to the Transmittal traffic, the RFI traffic had an early peak about a quarter of the way into the project and then began to decline until about three quarters of the way through the project. In contrast to both the Memo and Transmittal traffic, the trend of the ratio was very unstable with large fluctuations throughout the entire project. The average ratio was only slightly greater than one, indicating a strong

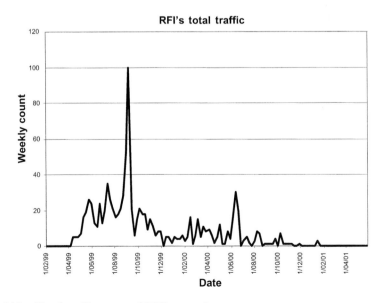

Figure 6.10:   Total traffic count of RFIs over time

degree of specific direction to individuals. The study results also indicated strong face-to-face interactions facilitated by co-location and the alliancing approach: 44 per cent rank agreement index for business-as-usual compared to 93 per cent for project alliancing.

Participants also indicated that they believed that fast and effective electronic information and communication technologies between teams for extended periods of time on site were of vital importance in maintaining good team relationships. Their rank agreement index for business-as-usual was 51 per cent compared to 72 per cent for project alliancing on the project (Peters *et al.* 2001, p. 208).

6.4.4.4 *Interactions*

Interactions between the project participants (users of ProjectWeb) as measured by the number of memos, transmittals and RFIs, illustrate the complexity of the communication requirements of a large construction project. The range of inter-actions also reflects the form of resource procurement system used on this project. Some traditional forms of contract ensure that most information exchanges (at least formal exchanges) only occur between a limited number of organisations or people. Using the data from ProjectWeb it is possible to view a map of the sender, receiver, size and direction of project interactions. This ability can potentially lead to a greater understanding and significance of the impact of the alliancing process.

The interactions where at least 50 exchanges took place during the first quarter of the project are shown in Figure 6.11. The thickness of the lines joining the various participants shows the intensity of interaction on a logarithmic scale so that a line of double the thickness of another indicates that the traffic was 100 times bigger.

Figure 6.11 shows that communications across the alliance members and many of their contractors were well established during the first quarter of the project.

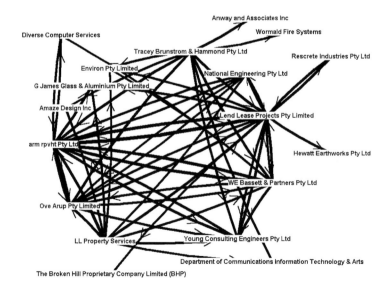

Figure 6.11:   Communications channels up to 1 September 1999

There were also many small contractors and subcontractors exchanging information, but on a much smaller scale in both number and recipients. Even at this stage of the project, the complex web of information exchanges was well established.

Figure 6.12 illustrates that during the second quarter of the project, communications increased considerably in both number and participants. The number of participants with over 50 communications (averaging almost one per working day) increased with many more of the participating companies appearing in the overall traffic. The overall structure changed very little. The changing intensity of communications reflected the changing demand for contractors' services as the project progressed.

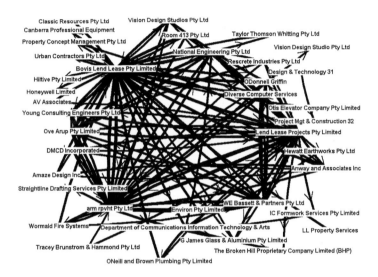

Figure 6.12: Communications channels from 2 September 1999 to 1 March 2000

A similar view of communications occurred during the third quarter as illustrated in Figure 6.13.

By the time the project reached the fourth quarter, the intensity of communication had decreased and the number of participants with over 50 communications during that quarter had declined as illustrated in Figure 6.14. As in the previous mappings, companies that had limited exchanges between themselves and/or the alliance team are not shown. At this stage of the project, the majority of communication was between the contractor and individual companies.

The star like shape of the communication channels illustrated in Figure 6.14 at this stage of the project is similar to that expected on more traditional contracts. One explanation for this pattern is that there was a reduced need to interact between participants, as the majority of them were nearing completion of their contracts and had acquired the required information to complete their tasks. A further possibility was that many communications were at that stage less formal since only limited detail was required that could be acquired by a simple telephone call or face-to-face interaction. The build up in trust that was evident (see Chapter 10 for more details) between participants also appeared to have been an

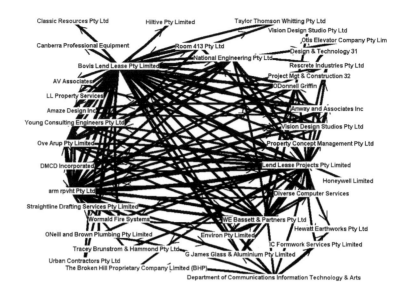

Figure 6.13: Communications channels from 2 March 2000 to 1 September 2000

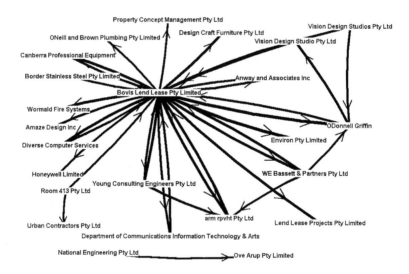

Figure 6.14: Communications channels after 2 September 2000

influencing factor. It may also be possible that as the project was in the construction phase, most interaction was directly related to the contractor with construction queries.

6.4.4.5 *Performance indicators*

Requests for Information (RFIs) are a common communication document type on a construction project. They are used specifically to ensure that unknown issues about the project are not overlooked and they provide a mechanism to seek the missing information and obtain a response in timely manner. The ProjectWeb system allowed three possible workflow actions, to 'reply' to the RFI, to 'forward' it to another ProjectWeb user for action or to close the RFI out once the author was satisfied with the response.

One distinct advantage in the use of a web-based system is that it ensures all RFI details are readily available and transmitted to the intended recipient quickly. However, instant notification of the RFI does not necessarily assure the recipient will provide an immediate response – some responses take time to resolve and at other times the person who can supply the required information is not available.

On the National Museum of Australia project the number of 'on-time' responses was very low. Initially this observation was accentuated by the fact that the default time to respond was set at 1 day so that it was very common for an RFI response to be 'late' as defined by the system. The default date problem was modified during the course of the project and many inadvertent late RFIs were also closed off. An additional factor in the statistics was that, according to the system, an RFI is not considered 'closed' until the user undertakes a specific action to close off the RFI, even though the issue may have been resolved verbally.

In comparison with the limited data available from other construction projects the actual number of RFIs is much less than usual probably. This was possibly a result of the alliance procurement method. The system did record the purpose of RFIs – errors, omissions, design changes, clarifications, etc. – which are useful for performance indicators focusing on where improvements could be obtained. Alliancing meant that design changes happened throughout the construction stage and may have contributed to a lower number of RFIs than expected. An RFI delay performance indicator is illustrated in Figure 6.15. This indicator shows a typical fluctuating start before settling down to a project dependent level. A performance

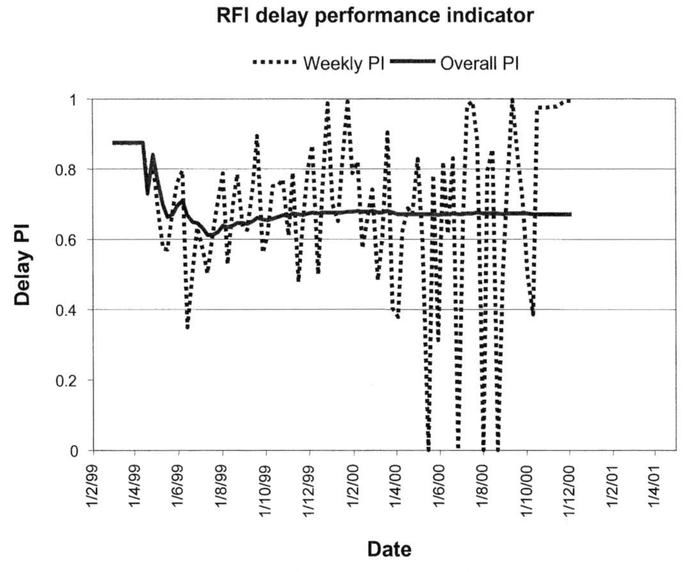

Figure 6.15:   Delay performance indicator for Requests for Information (RFIs)

indicator similar to this could be used on a weekly or monthly basis to assist in ensuring that delays are minimised. (Note: Bovis Lend Lease had already used this technique on some projects using this system.)

6.4.5   IT study conclusion

In analysing these results, it should be made clear that this project was the earliest example in Australia to be making use of a building project alliancing approach. Accordingly, there were a number of system and behavioural, training and support issues that needed addressing over the course of the project. In context, the National Museum of Australia project was one of the first 10 projects that Bovis Lend Lease had initiated using this technology, a figure which had risen to over 300 by mid-2001.

Whilst it is impossible to draw comprehensive conclusions from the study of just one project, there are still a number of clear benefits in the approach to project information management that can be supported through this study. In particular, it can be stated that:

- Users were generally satisfied with the performance and benefits of using a project collaboration system such as ProjectWeb;
- Users were satisfied that genuine savings in time could be realised;
- There were significant numbers of users on a project that could benefit from a project collaboration system;
- In general, project communications were analogous to the use of e-mail where they could be addressed '*To*' and circulated '*CC*';
- Not all document types were handled the same way by users;
- Similar to e-mails, documents were quickly created and distributed, but it took considerably longer for the recipient to respond. Typically, users have a high expectation of the recipient's ability to respond in the time frame they had allocated to the task;
- If required, Key Performance Indicators (KPIs) could be built around the performance of a project team once a single system was used to manage all project information; and
- Given the observable and numerous relationships that exist between project participants, there was a clear case for the need for a centralised project information management system to facilitate this process.

6.5   IT and project alliancing

Project alliancing is a relatively new project delivery method in the Australian construction industry. The main principle in project alliancing is the concept of 'driving decisions which are best for the project' through shared and transparent risk–reward mechanisms. In essence, the alliance drives project participants to work within a more collaborative framework and effective team environment than is traditionally used in project delivery methods (Walker *et al.* 2000).

The premise of this section is to explore the relationship between project alliancing and the utilisation of the IT tools used on the National Museum of

Australia project, especially the use of the Bovis lend Lease project collaboration system ProjectWeb. On the National Museum of Australia project, project alliancing allowed participants to synergistically operate and interact with each other, free of the normal non-productive contractual behaviours seen in the industry (Peters *et al.* 2001) – also see Chapter 10.

The decision to adopt ProjectWeb as an information management system on the National Museum of Australia project was a business-driven and strategic decision supported by senior management. Strategic effects, however, are difficult to quantify – however, the Government client realised the importance of benefits to this approach if widely adopted in industry. At an operational level, ProjectWeb was able to increase collaborative behaviour and facilitate ease of data capture, whilst significantly reducing costs associated with paper-based reproduction and transmittal.

A number of management challenges are always created. The National Museum of Australia project presented a major challenge to ensure that implementation of the IT system was successful. One of the most important success factors in the use of collaboration tools is to ensure that *all* project participants are committed to using IT tools. This in turn forms a critical mass of both users and content that should be seen as an immediate advantage for all participants. The National Museum of Australia project participants, for example, commented that they saw immediate benefits from use of the IT collaboration system for both routine and process oriented communication, hence its successful adoption.

Traditionally, the 'technology-organisational structure relationship' has been regarded as a major issue in organisational theory, with most technological innovations disregarding the issue of how people in organisations go about acquiring, sharing and making use of information (Davenport 1994). In essence, they glorify the use of IT and ignore the human psychology factor. Kraut *et al.* (1998) argue that despite the advances in IT, the establishment and maintenance of personal relationships is the glue that holds collaborative efforts together. This theory is supported by comments made by several construction contractors who recognised the need for intimacy required for trust and relationship building that is often lacking when using video-conferencing facilities (Walker and Rowlinson 1999).

The impact of project alliancing and the use of IT tools on the National Museum of Australia project is explored and discussed in the following section across four key dimensions, namely: mutual cooperation, conflict, benefits and satisfaction.

6.5.1   Mutual cooperation

Mutual cooperation refers to the ability to have cooperative working relationships with business partners that are supported by collaborative activities, transparent interfaces, and implicit trust whilst sharing the risks associated with a project. Project alliancing requires parties to form relationships and work cooperatively to provide a more complete service to the client. This cooperative working style requires that parties increase their knowledge of each other.

On the National Museum of Australia project, use of a project collaboration system (such as ProjectWeb) allowed parties to become more familiar with the project participants and to understand each other better. They also learnt how to openly interact with each other (without the need to over-protect themselves) and share project technical and commercial information.

Research indicates that introduction of IT typically increases inter-organisational cooperation between project participants as a result of better communication and coordination (Atkin 1999). Research undertaken on the National Museum of Australia project supports this finding and highlights that the use of the ProjectWeb system has facilitated the interaction between all project participants – as evidenced by the mapping of communication and interaction patterns during the construction phase (Peters *et al.* 2001). Compared to business-as-usual levels, this partially explains the significant improvement in the effectiveness of project participant communication. The results of a survey administered three times in February 2000, October 2000 and January 2001 showed that the key users (designers, managers and administrators) adapted very well to the ProjectWeb. Feedback from users indicated that designers and administrators believed that ProjectWeb enhanced coordination between project participants, whereas managers feel that its real value lay in establishing and supporting the project alliance. Nevertheless, they were all in agreement that it played a key role in the integration of project participants.

6.5.2   Conflict

Conflict in the context of the construction industry can be conceptualised as being the frequency and depth of disagreement between project parties. If the business relationships match strategically then there tend to be relatively high levels of mutual knowledge, resulting in less conflict. In a traditional project delivery process, each project participant brings to the project interaction process a different set of expectations and performances. Ultimately, these differences can result in conflict. Conversely, project alliancing encourages a clear understanding of individual and collective responsibilities and accountabilities. Notwithstanding this encouragement, there is little value in creating a strong project alliance relationship early on in the process and then neglecting to consider the ways in which the relationship can be actively maintained throughout the course of the project.

Building and maintaining solid relationships with all parties is important in alliances because there are many synergies that can be capitalised upon. For example, sharing information more broadly about customer needs and providing feedback and creative ideas from parts of the supply chain, are a key feature of alliances (Doz and Hamel 1998). Using a project collaboration system such as ProjectWeb, proved to be advantageous in sharing project information. Users of the system acknowledged its role in facilitating document transfer and handling as well as enabling immediate reporting and receiving feedback.

Project alliancing adopts a no-blame no-disputation principle. This implies that parties need to put their energies together to solve problems, rather than to attribute blame. Anecdotal evidence suggests that project alliancing has the potential to change the adversarial culture of the construction industry and to reduce the amount and negative impact of conflict (Peters *et al.* 2001).

6.5.3   Benefits

In the context of managing construction projects, the generic term 'benefits' goes beyond traditional time and cost savings to encompass non-monetary or intangible value derived from IT implementation. Innovative use of IT minimises procedural delay of interaction, reduces the transaction costs among project parties and encourages the business relationship to be more efficient. As illustrated in the National Museum of Australia case study, a number of principal benefits were clearly realised in the design phase of the project. These included successful negotiations regarding buildability of the design, value engineering exercises and the ability for CAD/CAM data to be applied to the fabrication of a highly complex structure and building envelope (Peters *et al.* 2001).

Project participants have also acknowledged, via a survey, the role that IT played in realising other benefits during the construction phase (Peters *et al.* 2001). For example, ProjectWeb is believed to have enabled streamlining of a number of administrative processes as well as improving both document quality and staff computer literacy level. In addition to these project-focused benefits, participants also identified a number of long-term benefits, such as enhancing their organisa-tion's image in the industry, and having acquired a demonstrated capability for national and global collaboration.

6.5.4   Satisfaction

Satisfaction can be regarded as the net of benefits (rewards) minus the cost of maintaining the business relationship. Accordingly, both rewards and cost can be measured in terms of economic and social exchange. As interactions between project participants develop, participants continually assess their relationship (i.e. the outcomes of interactions). These assessments involve rewards and costs associated with the relationship. Ultimately, this leads to the basis for affective reactions of the parties toward their relationship – such assessments determine the degree of satisfaction or dissatisfaction each participant feels toward that rela-tionship. This traditional view of satisfaction was challenged in the National Museum of Australia project due to the adoption of project alliancing. Project participants proved willing to sacrifice – in the short term – something to a partner to ensure that the long-term relationship remains intact. The level of willingness to do so was found to have a rank agreement index value of 43 per cent for business-as-usual compared to 83 per cent for project alliancing (Peters *et al.* 2001, p. 202, response O10)

When mapping the interactions between project participants (via ProjectWeb) and measuring the range and number of transmittals, memos and requests for information, the intensity of interactions among participants throughout the pro-ject can be truly reflected. Due to the increase in the number of participants, the intensity of communication continued to increase reaching its peak during the second and third quarters of the project. Compared to business-as-usual levels, project participants indicated that fast and effective electronic communication technology with their partner organisations was of vital importance in maintaining a good team relationship. However, they also noted that ProjectWeb is highly effective for explicit knowledge transfer and development but not nearly as effective for tacit knowledge transfer (Peters *et al.* 2001, part B).

6.6   Chapter conclusions and future directions

In summary, the key observations and conclusions from the National Museum research project were:

- Users were able to adapt very well to a web-based collaboration system and have generally rated the system used on the National Museum of Australia project very highly from several perspectives;
- Use of the web-based collaboration system began slowly. With the alliancing approach, it might have been expected that all participants should have begun earlier in the project. However, there were many first-time users of this technology and one possible explanation could be the associated learning curve. This learning phase may not be required in later projects. Typically, it takes time to adjust to the detailed capabilities of any new technology as was indicated by the fact that traffic only stabilised at about 25 per cent of the way through the project;
- The changing traffic patterns after an initial surge was indicative of behaviour that users tended only to use system functionality that was considered helpful, more efficient or essential to their job;
- Throughout the project there were considerable numbers of additions to the project participants and these changes in staff required ongoing training and support to maintain effective use of the various IT systems;
- A web-based system collects a significant amount of data that could be further used to assist in monitoring ongoing performance of the project team. Timely feedback on identifiable and predetermined KPIs via appropriate indicators would enhance the capabilities of a web-based system and offer further value to the project team;
- The assessment of IT implementation declined a minor degree from an initial very high position in early stages of the project. Despite the excellent performance of the system this was ongoing familiarisation with the system and identifying that further improvements could be made, i.e. there were rising expectations of the capabilities of the system and a rising number of potential improvements identified;
- The number of RFIs generated was considered to be low for a project of the size of the National Museum of Australia project. This could be attributable to the alliancing process. If there were a wider range of performance indicators on this study for this aspect, there would have been more opportunities to assess a wider range of implications of using an alliancing approach;
- Lateness of RFI response was considered as high. This measurement rose as the project progressed, but this was considered to be mostly due to a system default value of one day and the fact that the RFI module functionality did not require a formal close-off until towards the end of the project. The significance is not in the lateness of RFI responses, but in the difference between the expectations of how users would operate the system and their actual usage;

- ProjectWeb initially generated large quantities of recipients for documents both directly and via the *CC* function. As the project progressed, the total traffic declined but not the number of electronic memos, etc. sent. There are possibly two reasons for this observation. At the start of the project, it was felt that everyone should know as much as possible about what was going on (keeping all participants informed) and then as trust and awareness of specific relevant team members became known communications became far more focused in the numbers of people to whom they were addressed; and
- Preliminary indications are that many of the characteristics identified through this study on the National Museum of Australia project, also appear to be similar to another project that uses the same web-based system for communications and document management.
- These results strongly support the hypothesis that the use of a web-based collaboration system facilitates the interaction between all participants.

Clearly the construction industry needs to move away from being highly risk-averse in terms of embracing IT to one in which it matches its client's expectations of IT use. One of the reasons for this is that many clients now wish to better integrate reporting and tracking systems and to align their information flows to build information value at each stage in order to achieve better decision making. Much of the supply chain literature supports this view (MacMillan and McGrath 1997; Kim and Mauborgne 1999; Womack and Jones 2000).

Since the conclusion of the research on the National Museum of Australia project, there have been a number of industry advances in the availability and use of project collaboration systems. These second and third generation systems all provide similar functionality to that found in ProjectWeb, but are similarly in the early stages of their full development cycle.

In many ways, the development of these systems is analogous to the introduction of CAD systems that were initially used as replacements for manual drafting. Similar to CAD systems, project collaboration systems are likely to go through a phase of rapidly accumulating functionality. There is also an increasingly divergent range of product vendors, each with their own set of value propositions and approaches. It is anticipated that the system features will eventually standardise and consolidate into more clearly defined core sets of functionality such as for project management, site management and design management capabilities.

Over the next decade, it is expected that the functionality and reliability of project collaboration systems will improve considerably and that they will become standard software used on all projects in the way that word-processing, CAD and project scheduling systems are used today. By this time, there should also be a clearer industry standard in the functionality and interoperability of these systems. Concurrently, advances in the availability and affordability of high-speed internet bandwidth will mean that performance and accessibility to these systems will also continue to improve.

However, despite all of these technical advances, the underlying fundamental successful factor in the use of a project collaboration system will be the ability of an organisation to understand and apply their own business processes to it. In addition, they will also have to understand the processes of their business partners if they are successfully to maintain their business relationships. As this case study illustrated, our current business understanding of project collaboration systems is still in its infancy and it will take many years before we better understand the consequences and implications of using what is fundamentally a new direction in the manner in which project information is managed.

The whole thrust of this book argues that the value created through a relationship-based approach to project procurement lies with building synergies and sharing information and knowledge that can be used to improve both effectiveness and efficiency. IT is thus seen as an enabler or facilitator. To fully capture this value there needs to be a more sophisticated and committed approach to project procurement than has been witnessed during the twentieth century. We argue that clients and more forward thinking contractor/consultant alliances will drive this new direction and we have seen the National Museum of Australia project as an exemplar project where this approach has been manifested. IT support has been shown in this chapter to be pivotal to a relationship-based procurement approach. Figure 6.16 illustrates how IT may enable greater project and team benefits and value to be realised.

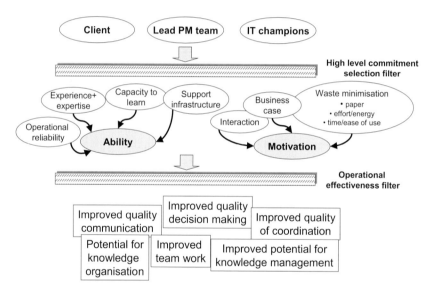

Figure 6.16:   A model of IT infrastructure support for projects

This chapter has clearly identified the need for high-level commitment of the client and the project management team to provide sufficient IT infrastructure to enable information to flow unhindered between project participants. We have argued that electronic representation of information allows a greater capacity to build solutions to problems and to translate plans into action than the use of paper-based systems. This is because at each stage paper-based systems have to re-create existing information and this process leads to leakages, distortion and error. When electronic forms of information transfer are used the re-creation process is redundant. Any errors or problems that occur do so because of poor information interpretation or misunderstanding in the information value chain linking one member of this chain to the next. Additionally, there needs to be a high-level commitment manifested by IT champions who can drive a strategy for effective IT support.

For IT operational effectiveness to occur there needs to be both IT enabling ability and motivation. Ability includes a capacity to deliver IT support infrastructure that is operationally reliable. This includes human support in terms of people to help who have sufficient experience and expertise to deal with and solve the daily snags, problems and can address the frustrations of linking complex systems that leads to user disenchantment with IT. Ability requires a capacity for all users and facilitators of IT infrastructure to learn and build competence in IT use. IT is a technology that is rapidly advancing and changing. Thus IT infrastructure requires both hardware and humanware.

Project team members must be motivated to apply the IT tools at their disposal. Issues relating to human capital assets and establishing a people-based approach to enabling relationship-based procurement systems to be effectively applied is discussed in more detail in Chapter 10. Motivation is required at both the individual/ team level and at the business level. This chapter contains, in section 6.3.1 and its sub-sections details of benefits and considerations that formed the basis for a business case for IT infrastructure support on the National Museum of Australia project. Motivation has also been shown from evidence presented in that research study (Peters *et al.* 2001) to relate to a desire to minimise waste in paper use, managerial energy and effort, and IT systems enabling tasks to be more easily undertaken. Personal benefits are important.

Potential outcomes of an effective IT supporting infrastructure when combined with a truly cooperative approach to information sharing and collaboration in solution building are illustrated in Figure 6.16. These include improved quality of communication, coordination and communication. This leads to improved teamwork. With the right environment for sharing knowledge by gaining further value from information through adding in contextual insights, improved knowledge management can be achieved and a knowledge/learning organisation can be developed. It has been argued elsewhere that clients and project team members miss the opportunity to capture knowledge from projects and that this represents a hidden waste that costs society dearly (Walker and Lloyd-Walker 1999).

This chapter set about introducing the importance of IT to relationship-based procurement systems and traced the current use of IT in the construction industry. The National Museum of Australia project case study data and results provided a useful focus for demonstrating what can be achieved and is now one example of current best practice. The case study also served as a useful framework for discussing implications for the future of relationship-based procurement in the construction industry. We have stressed in this chapter that IT hardware and software alone provide only a part of the solution to the problem of driving out management waste. Chapter 10 addresses the vital human related issues that underpin successful application of IT infrastructure.

6.7   References

Adcock, M. (1996) *EDI in construction: A guide to introducing electronic data interchange between companies.* London, Business Round Table.

Alsagoff, S.A. (2000) *IT in Construction Engineering: Are We Chasing Our Tails?* 4th Asia-Pacific Structural Engineering and Construction Conference, Kuala Lumpur.

Aouad, G., Alshawi, M. and Bee, S. (1996) 'Priority Topics for Construction Information Technology.' *Construction Information Technology.* 4 (2): 45–66.

Aouad, G. and Price, A. (1994) 'Construction planning and information technology in the UK and US construction industries: a comparative study.' *Construction Management and Economics.* **12**: 97–106.

Atkin, B. (1999) *Measuring Information Integration in Project Teams.* CIB W78 Workshop, Vancouver, Canada.

Atkin, B., Davies, P., Dubois, A.M., Gravett, J., Jagbeck, A., Smith, D. and Walker, S. (1999) Benchmarking Projects Information Integration. In: *Strategic Management of IT in Construction,* eds Betts, M. and Smith, D. Oxford, Blackwell Science.

Australian Bureau of Statistics (1999) *Business Use of Information Technology – Preliminary, Australia 1997–98,* Catalogue number 8133.0, Canberra, Australian Bureau of Statistics.

Baldwin, A.N., Thorpe, A. and Carter, C. (1996) The Construction Alliance and Electronic Information Exchange: A Symbolic Relationship. *The Organisation and Management of Construction – shaping theory and practice,* eds Langford, D.A. and Retik, A. London, UK, E&F Spon. **3**: 23–32.

Betts, M., Cher, L., Mathur, K. and Ofori, G. (1991) 'Strategies for the Construction Sector in the Information Technology Era.' *Construction Management and Economics.* **9**: 509–528.

Betts, M. and Clark, A. (1999) *Strategic Management of IT in Construction.* Oxford, Blackwell Science.

Betts, M., Jarrett, M. and Shafaghi, M. (1999) Current Strategic Practice. *Strategic Management of IT in Construction,* eds Betts, M. and Smith, D. Oxford, Blackwell Science: Chapter 9.

Betts, M. and Smith, D. (1999) *Strategic Management of IT in Construction.* Oxford, Blackwell Science.

Bjork, B.C. (1999) 'Information Technology in Construction: Domain Definition and Research Issues.' *International Journal of Computer Integrated Design and Construction.* **1** (1): 3–16.

Brynjolfsson, E. and Hitt (1994a) 'Paradox Lost? Firm-level Evidence of High Returns to Information Systems Spending.' *http://ccs.mit.edu/CCSWP162/ccswp162.html.* MIT.

Brynjolfsson, E. and Hitt (1994b) 'Productivity Without Profit? Three Meaures of Information Technology's Value.' *http://ccs.mit.edu/CCSWP190.html* MIT.

Cornick, T. (1990) *Quality management for Building Design.* London, Butterworth Architecture Management Guides.

Crotty, R. (1995) *Advanced Communications in Construction – Broadband Communication in Construction,* London, Bovis Construction Ltd.

Davenport, T.H. (1994) 'Saving IT's Soul: Human Centered Information Management.' *Harvard Business Review.* **72** (2): 119–131.

Davenport, T.H. (2001) Knowledge Workers and the Future of Management. *The Future of Leadership – Today's Top Leadership Thinkers Speak to Tomorrow's Leaders,* eds Bennis, W., Spreitzer, G.M. and Cummings, T.G. San Francisco, Jossey-Bass.

Davenport, T.H., Delong, D. and Beers, M. (1998) 'Successful Knowledge Management Projects.' *Sloan Management Review.* **39** (2): 43–57.

Davenport, T.H. and Prusak, L. (2000) *Working Knowledge – How Organizations Manage What They Know.* Boston, Harvard Business School Press.

DETR (1998) *Rethinking Construction,* Report. London, Department of the Environment, Transport and the Regions.

Doz, Y.L. and Hamel, G. (1998) *Alliance Advantage – The Art of Creating Value Through Partnering.* Boston, Harvard Business School Press.

DPWS (1998) Information Technology in Construction – Making IT Happen, Sydney, NSW, Department of Public Works and Services (DPWS).

Duyshart, B.H. (1997) *The Digital Document*. Oxford, Butterworth-Heinemann.

Evbuomwan, N.F.O. and Anumba, C.J. (1998) 'An Integrated Framework for Concurrent Life-Cycle Design and Construction.' *Advances in Engineering Software.* **29** (7–9): 587–597.

Fisher, N. and Shen, L.Y. (1992) *Information Management Within a Contractor*. London, Thomas Telford.

Froese, T., Rankin, J. and Yu, K. (1997) 'Project Management Application Models and Computer-Assisted Construction Planning in Total Project Systems.' *Construction Information Technology.* **5** (1): 39–62.

Hannus, M., Watson, A., Luiten, B., Degiune, M., Sauce, G. and Van Rijn, T. (1999) 'ICT Tools for Improving the Competitiveness of the LSE Industry.' *Journal of Engineering and Architectural Management.* **6** (1): 30–37.

Hunter, I., Mitrovic, D., Hassan, T., Gayoso, A. and Garas, F. (1999) 'The eLSEwise Vision, Development Routes and Recommendations.' *Journal of Engineering and Architectural Management.* **6** (1): 51–62.

Irani, Z., Sharif, A.M. and Love, P.E.D. (2001) 'Transforming Failure into Success Through Organisational Learning: An Analysis of a Manufacturing Information System.' *European Journal of Information Systems.* **10**: 55–66.

Kagioglou, M., Cooper, R., Aouad, G., Hinks, J., Sexton, M. and Sheath, D. (1998) *Final Report: Process Control*. Salford, UK, University of Salford.

Kim, W.C. and Mauborgne, R. (1999) 'Strategy, Value Innovation and the Knowledge Economy.' *Sloan Management Review.* **40** (3): 41–54.

Kluge, J., Stein, W. and Licht, T. (2001) *Knowledge Unplugged – The McKinsey & Company Global Survey on Knowledge Management*. New York, PALGRAVE.

Kraut, R., Galegher, J. and Ewdigo, C. (1998) 'Relationship and Tasks in Scientific Research Collaboration.' *Human Computer Interactions.* (3): 31–58.

Luiten, G.T., Tolman, F.P. and Fisher, M.A. (1998) 'Project Modeling to Integrate Design and Construction.' *Computers in Industry.* **35** (1): 13–29.

MacMillan, I. and McGrath, R.G. (1997) 'Discovering New Points of Differentiation.' *Harvard Business Review.* **75** (4): 133–145.

Marsh, L. and Finch, E. (1998) 'Attitudes Towards Auto-ID within the UK Construction Industry.' *Construction Management and Economics.* **16**: 383–388.

McCaffer, R. and Hassan, T.M. (2000) *Changes in Large Scale Construction Arising from ICT Developments*. Proceedings of the Millennium Conference on Construction Project Management, Hong Kong.

Mohamed, S. (1996) *Information Technology and Construction Process Re-Engineering: A Recursive Relationship*. Proceedings from the International Conference on Urban Engineering in Asian Cities in the 21st Century, Bangkok, Thailand, Asian Institute of Technology.

Mohamed, S. and Tucker, S.N. (1996). 'Options for Applying BPR in the Australian Construction Industry.' *International Journal of Project Management.* **14** (6): 379–385.

NPWC (1993) *Integration of Documents: Quality Management of Documentation for Construction*. Canberra, National Public Works Council.

O'Brien, M.J. and Al-Soufi (1994) 'A Survey of Data Communications in the UK Construction Industry.' *Construction Management and Economics.* **11**: 443–453.

Pena-Mora, F., Vadhavkar, S., Perkins, E. and Weber, T. (1999) 'Information Technology Planning Framework for Large-Scale Projects.' *Journal of Computing in Civil Engineering.* October: 226–237.

Peters, R.J., Walker, D.H.T., Tucker, S., Mohamed, S., Ambrose, M., Johnston, D. and Hampson, K.D. (2001) *Case Study of the Acton Peninsula Development*, Government

Research Report. Canberra, Department of Industry, Science and Resources, Commonwealth of Australia Government: 515.

Samuelson, O. (1998) *A study of the Use of IT in the Construction Industry, the Life Cycle of Construction IT Innovations.* Proceedings of the CIB W78 Conference, Stockholm, Sweden.

Shen, Q. (1996) *The State of IT Applications in the Construction and Real Estate Industry in Hong Kong and their Impact on the Industry.* Proceedings of the Construction Information Technology Conference, Sydney, Australia.

Shoesmith, D.R. (1995) 'Using the Internet as a Dissemination Channel for Construction Research.' *Construction Information Technology.* **3** (2): 65–75.

Stewart, R.A. and Mohamed, S. (2001) 'Using Benchmarking to Facilitate Strategic IT Implementation in Construction Organisations.' *Journal of Construction Research.* **2**: 25–33.

Tam, C.M. (1999) 'Use of the Internet to Enhance Construction Communication: Total Information Transfer System.' *International Journal of Project Management.* **17** (2): 107–111.

Terhune, A. (1998) 'ROI in the Extranet World.' *http://www.gartnerweb.com.* GartnerGroup Interactive.

Tucker, S.N. and Mohamed, S. (1996) Introducing Information Technology in Construction: Pains and Gains. *The Organisation and Management of Construction – shaping theory and practice*, eds Langford, D.A. and Retik, A. London, E&F Spon.

Tucker, S.N., Mohamed, S., Johnson, D.R., McFallen, S.L. and Hampson, K.D. (2001) *Building and Construction Industries Supply Chain Project.* CSIRO Confidential Report: 56.

Walker, D.H.T., Hampson, K.D. and Peters, R.J. (2000) *Relationship-Based Procurement Strategies for the 21st Century.* Canberra, Australia, AusInfo.

Walker, D.H.T. and Lloyd-Walker, B.M. (1999) Organisational Learning as a Vehicle for Improved Building Procurement. *Procurement Systems: A guide to best practice in construction*, eds Rowlinson S. and McDermott, P. London, E&FN Spon.

Walker, D.H.T. and Rowlinson, S. (1999) Use of World Wide Web Technologies and Procurement Process Implications. *Procurement systems A Guide to Best Practice in Construction*, eds Rowlinson S. and McDermott, P. London, E&FN Spon.

Wenger, E.C. and Snyder, W.M. (2000) 'Communities of Practice: The Organizational Frontier.' *Harvard Business Review*, **78** (1): 139–145.

Wheatley, M. (1997) 'Hidden Costs of the Humble PC.' *Management Today*, January: 52–54.

Womack, J.P. and Jones, D.T. (2000) From Lean Production to Lean Enterprise. *Harvard Business Review on Managing the Value Chain.* Boston, MA, Harvard Business School Press: 221–250.

Zack, M.H. (1999) 'Developing a Knowledge Strategy.' *California Management Review.* **41** (3): 125–145.

ZweigWhite (2001) *Information technology & E-business Survey of A/E/P & Environmental Consulting Firms.* Massachusetts, USA, ZweigWhite.

Part 2
Relationship-based Procurement Attitudes and Behaviours

Chapter 7
Developing Cross-team Relationships

Derek Walker and Keith Hampson

Choices of procurement method were presented in the second chapter. It was noted that other than when using the traditional procurement method, the adopted choice should not affect the capacity of a relationship based strategy (partnering or alliancing) to achieve the project goals through collaborative and cooperative teamwork. In this chapter we present the essential features of partnering or alliancing and explore how it may be successfully applied.

7.1 Introduction

Figure 7.1 illustrates the foundation for partnering that is consistent with project alliancing (Bennett and Jayes 1995, Chapter 1 page 5). It is based essentially upon team spirit. The interesting aspect of this simple model is that all three essential elements require trust, commitment, honesty, integrity, good communication skills and technologies between parties. This reinforces the points initially made in the Introduction to this book. The first feature, *mutual objectives*, naturally leads to discussion on trust and commitment. The second feature, *problem resolution*, leads to discussion of trust and commitment through trust based upon performance certainty – given the expectation that problem resolution will be undertaken in a

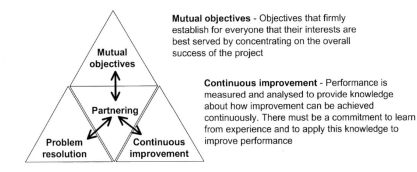

Mutual objectives - Objectives that firmly establish for everyone that their interests are best served by concentrating on the overall success of the project

Continuous improvement - Performance is measured and analysed to provide knowledge about how improvement can be achieved continuously. There must be a commitment to learn from experience and to apply this knowledge to improve performance

Problem Resolution - Resolve problems with an escalation strategy to solve them at the lowest organisational level possible

Figure 7.1: Three essential features of partnering
Source: Adapted with permission from Bennett, J. and Jayes, S. (copyright © 1995)

non-adversarial manner. It also leads to issues of the application of power and nature of management style and leadership. The last of these three essential features, *continuous improvement*, is linked to the way in which organisations learn. It also leads to issues of trust and commitment because it involves teams and individuals feeling safe in measuring performance to learn from their experience. This requires openness and honesty rather than recording idealised or false performance records in order to hide mistakes or to attempt to extract unwarranted credit.

Partnering and alliancing to a similar extent is founded upon team spirit and the honesty associated with notions of trust, commitment, and the application of power and influence. Excellent and effective communication is essential for successful relationship building.

7.2   The role of power and influence – building mutual objectives

Procurement arrangements were summarised and illustrated in Figure 2.2 in terms of client influence over design and procurement. At the high end of the scale, the client or client representative has the major input and influence upon the working relationship. Clearly, the contractor takes the majority of cost risk in the traditional and cost reimbursable procurement approaches. Also in these approaches, the contractor cannot make design decisions nor has substantial power to effect or affect design decisions. These remain the prerogative of the client and/or client representative. On the other hand in BOOT-type projects and design and construction, it is the contractor who holds most influence over design detail and even the design strategy. It is true that the client would have established a design and functional brief that in large part determines the overall outcome, but the contractor has the power to shape this form. For example, cladding type for a building may remain open subject to performance measures and the frame type may be steel, cast-in-place concrete, prefabricated or a hybrid system. This power dimension can also shape the nature of the working relationship between teams and determine how conflicting opinions about the relative merits of design solutions are tolerated and treated. Conflict and diversity are linked to power distance, which has been shown to be linked to culture (Hofstede 1991).

Ragins defines power as 'the influence of one person over others, stemming from an individual characteristic, an interpersonal relationship, a person's position in an organisation, or membership of a societal group'. She points out that groups with power are interested in maintaining their influence and resources, and may do so by supporting policies, practices, and prescription that exclude other groups from power. She also argues that society and those in the organisation with power substantially shape power relationships among groups in organisations (Ragins 1995, p. 96).

In some national cultures that are collectivist in nature, there are strong distinctions between members of in-groups and out-groups (Hofstede 1991). Powerful individuals are in a position to shape culture through influencing values, assumptions and ideologies. Building shared mutual goals is a leadership exercise using power and influence constructively to convince others that they share project objectives that coincides with their own individual interests. Leadership under these conditions requires considerable energy and intellectual pursuit of argument to build consensus and align interests with those of project outcomes. The implication of this view is that the organisations leaders' mindset can determine, to a large extent, the nature of power behaviours exhibited in teams, projects or within an organisation. Thus choice of procurement arrangement will not necessarily guarantee good or poor cross-team relationships. In undertaking an investigation into team relationships stemming from alternative procurement systems it would be of value to observe how power is used and how it affects the project organisational culture(s).

Power has also been defined as an agent's (person wielding power) capacity to influence a target's (person subject to this power) attitude and behaviour. Authority is concerned with perceptions about the prerogatives, obligations and responsibilities associated with particular positions in the organisation. Influence is restricted by the target's willingness to do what is asked, if this conflicts with the target's moral code then the target will not accept the agent's right to compliance (Yukl 1998). The way in which authority affects behaviour is important in an organisation as it impacts upon the effectiveness of this action. Commitment results when the target has absorbed and accepted suggestions of the agent. If this acceptance is grudgingly given or is not wholly accepted then the result will be compliance. If the agent disagrees with the agent then overt/covert resistance will follow. Yukl describes three levels of reaction to authority.

- **Instrumental compliance**: the target is willing to do whatever the agent requests, but only for reward. Power used by the agent is fear/punishment or reward. If the agent loses power to reward or the value of the reward ceases to be attractive, compliance will cease;
- **Internalisation**: the target becomes committed to support the agent's proposals aligning goals/vision accordingly. Commitment is independent of rewards offered as values and beliefs are the driving forces; and
- **Identification**: target complies to curry favour of agent. Relationship and affiliation motivate this behaviour. If the agent becomes less attractive then commitment is withdrawn (Yukl 1998).

The implications of these reactions are highly pertinent to procurement strategies and team relationships. But what is the nature and derivation of power? Yukl defines three source groups of power and describes their characteristics (Yukl 1998).

Position power derived from statutory or organisational authority:

- Formal authority;
- Control over rewards;
- Control over punishments;
- Control over information; and
- Ecological (physical/social environment, technology and organisation) control.

Personal power derived from human relationship influences or traits:

- Expertise;
- Friendship/loyalty; and
- Charisma.

Political power is derived from formally vested or conveniently transient concurrence of objective and means to achieve these:

- Control over decision processes;
- Coalitions;
- Co-option; and
- Institutionalisation.

These are raw forces to be used or abused and can be manipulated in procuring construction facilities. The way in which they are deployed explains the dangers of brainwashing, the reality–rhetoric gap and the failure to maintain a consistent and empowering attitude to deal with talented knowledge workers with much to offer (Green 1994; Green and Lenard 1999; Green 1999b).

The following outlines seven forms of power (Greene and Elfrers 1999, p. 178):

(1) **Coercive** – based on fear. Failure to comply results in punishment (*position power*);
(2) **Connection** – based on 'connections' to networks or people with influential or important persons inside or outside organisations (*personal + political power*);
(3) **Reward** – based on ability to provide rewards through incentives to comply. Is expected that suggestions be followed (*position power*);
(4) **Legitimate** – based on organisational or hierarchical position (*position + political power*);
(5) **Referent** – based on personality traits such as being likeable, admired, etc. thus able to influence (*personal power*);
(6) **Information** – based on possession to or access to information perceived as valuable (*position, personal + political power*);
(7) **Expert** – based on expertise, skill and knowledge which through respect influences others (*personal power*).

The nature of power and influence, the sources of this power and the way in which it is used to contribute to or manipulate cooperative relationships underpin all procurement strategies and the relationships that develop from these. It is interesting that a number of books have appeared providing advice on the use of power to undermine the competitor and to win against a perceived enemy. The most famous of these have been written by Machiavelli and Sun-Tzu. A recent book on power and its use, which features ideas from Machiavelli, Sun-Tzu and others, relates to winning power and holding power for personal gain and not to achieve a goal that is shared by others (Greene and Elfrers 1999). Positional power, however, is the least effective of the three outlined in building commitment to shared objectives, win–win outcomes and constructive dialogue whether in resolving differences or building shared understanding.

Hersey and Blanchard (1982, p. 178) argue that application of **coercive** power for alliancing and partnering should provide only a backstop and dispute resolution systems should be designed to inhibit or minimise the exercise of this form of power. **Connective** power should be used insofar that it becomes a natural outcome of achievement of project objectives. **Reward** and incentive need to be structured to form support mechanism for commitment and shared effort and not merely as principal objectives to be striven for. **Legitimate** power should flow from sound organisational design and serve to clarify and ensure goal achievement. **Referent** and **information/expert** power are the two power sources most applicable to best promote a leadership style that facilitates success of expert teams working together to achieve mutual and shared goals.

As the number 5 to 7 power forms above are recognised as being most effective in building relationships with highly skilled knowledge workers it is worth considering the development and management of those power bases in more detail. Table 7.1 and Table 7.2 provide a useful summary for this purpose.

These indicate the manner in which power sources are used (or can be abused) and this helps to explain how a human relationship factor such as trust is affected by the power relationship. It can now be appreciated that causes of claims and disputes can be explained as being associated with relationship issues. Disputes often stem from individuals resisting an inappropriate (in their perception) use of

Table 7.1: Increasing and maintaining power for referent power

How to increase + maintain power	How to use power effectively
• Show acceptance and positive regard	Use personal appeal when necessary
• Act supportive and helpful	Indicate that a request is important to you
• Don't manipulate and exploit people for personal advantage	Don't ask for a personal favour that is excessive, given the relationship
• Defend someone's interest and back them up when appropriate	Provide an example of proper behaviour (role model)
• Keep promises	
• Make self-sacrifices to show concern	
• Use sincere form of ingratiating the target	

Source: Leadership in Organizations by Yukl, Garry (Copyright © 1998) reprinted by permission Pearson Education, Inc., Upper Saddle River, NJ (Yukl 1998, p. 197)

Table 7.2: Increasing and maintaining power for information/expert power

How to increase + maintain power	How to use power effectively
• Gain more relevant knowledge	• Explain the reason for a request or proposal
• Keep informed about technical matters	• Explain why a request is important
• Develop exclusive sources of information	
• Use symbols to verify expertise	• Provide evidence that a proposal will be successful
• Demonstrate competence by solving difficult problems	
• Don't make rash and careless statements	• Listen seriously to target's concerns
• Don't lie or misrepresent the facts	• Show respect for the target (don't be arrogant)
• Don't keep changing positions	• Act confident and decisive in a crisis

Source: Leadership in Organizations by Yukl, Garry, (Copyright © 1998) reprinted by permission Pearson Education, Inc., Upper Saddle River, NJ (Yukl 1998, p. 197)

influence by team leaders using one or more of the power types outlined above. Indeed many of project failure causes relating to communication failures can be viewed in this light as inappropriate power application responses (Kumaraswamy 1997; Kwok 1998). Withholding, delaying or obstructing information flow are examples of either an overt or covert act of resistance to information being demanded or expected. Delay/inertia is a useful, and often powerful, response adopted by those coerced to deliver perceived unfair demands from those applying positional power.

The whole issue of power use and communication has been explored by a number of management thinkers interested in the management of projects. One such view expressed is the scope in terms of positional interrelationship between team members, their peers, those they report to and those reporting to them (Lovell 1993). Figure 7.2 illustrates a project manager and important inter-relationship of influence and negotiation that takes place at the interface of power use, communication and commitment to agreed action.

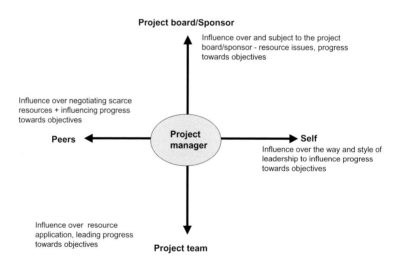

Figure 7.2: The project manager and 360° influence mapping
Source: Reprinted from the *International Journal of Project Management*, (vol. 11, no. 2, p. 94, copyright 1993) with permission from Elsevier Science

A project manager has several options available when trying to assert influence. These include presenting ideas in a rational and clearly communicated manner, challenging alternative ideas, and threatening to actually withhold crucial information. Likewise, other team members can apply the same tactics when dealing with the project manager. Thus communication and power/influence are closely linked.

Willingness to comply with authority and to assert authority forms a matrix of likely reaction quadrants (Lovell 1993).

(1) When acceptance of authority and assertiveness is high, **active consensus** takes place in a mature and productive manner.

(2)   When acceptance of authority is high but assertiveness is low, **passive loyalty** results. In such cases, project goals may appear to be mutually arrived at but are not. This can result in project goals being half-heartedly supported.

(3)   When acceptance of authority is low but assertiveness is high, **covert resistance** results. As in case 2 above, project goals may appear to be mutually arrived at but in reality, are not. In this situation, project goals may be covertly rejected and secretly undermined.

(4)   When acceptance of authority and assertiveness is low, **peer rivalry** takes place in an immature and unproductive manner. Project goals may be actively sabotaged as each party attempts to win at the expense of others or if not totally rejected then passively compromised through inertia.

This helps to explain the symptoms of what may be termed as 'team' or 'non-team' playing, 'playing the system', or 'playing political games' in which authority is subverted through the system of governance and organisational style. These sorts of power reactions take place at all levels, which may account for varying quality of relationships between teams from different firms who jointly contribute to the realisation of projects. The above also helps us understand the underlying mechanisms taking place during negotiations and mutual adjustment when making agreements and commitments. This is particularly true when establishing the relative importance of issues to negotiating parties, appreciating the needs and pressures of others and establishing and maintaining trust. The two dimensions under scrutiny are the degree to which each team attempts to satisfy its own concerns and the attempt to satisfy the other party's concerns.

Useful categorisation of peers can be based upon degree of trust and agreement with **high trust/low agreement** leading to opponents and **low trust/low agreement** leading to adversaries. The significance of this to partnering issues such as continuous improvement and problem resolution is that opponents can make a positive contribution through effective argument and casting perceived problems in a number of different lights. This can lead to a better understanding, which may produce a better decision or outcome with **high trust/ high agreement** – representing allies in quadrant 1. Opponents can be constructive whereas adversaries are generally destructive, as their aim is to thwart the intended outcome. This is where much of the trust required in relationships breaks down and where the quality of communication deteriorates.

Lovell offers useful techniques for persuasion including the use of:

- **assertiveness** – using power of logic, facts or opinion;
- **reward/punishment** – using pressure and persuasion to control others;
- **common visioning** – identifying a shared or common vision for the outcome;
- **participation and trust** – involving others in the decision making and problem-solving process to gain commitment (Lovell 1993, p. 76).

Additionally, interrelated factors determining appropriate selection of influencing tactics for a particular influencing attempt require consideration of:

- consistency with prevailing social norms and role expectations about use of tactics (that is, the societal view of power pointed out by Ragins (1995);

- the influencing agent possessing the appropriate power base for use of the particular tactic;
- appropriateness for the objective sought;
- level of resistance encountered or anticipated; and
- the cost of using the tactic in terms relation to benefit (Yukl 1998).

Yukl (1998) offers tangible help in Table 7.3 providing tactics to gain commitment from supervisors (UP), colleagues (LATERAL) and those supervised (DOWN).

Table 7.3: Use of influencing tactics

Influence tactic	Directional use of tactic	Sequencing of results	Used alone or in combinations	Likelihood of commitment
Rational persuasion	More UP than DOWN or LATERAL	Used more for initial request	Used frequently both ways	Moderate
Inspirational appeals	More DOWN than UP or LATERAL	No difference	Used most with other tactics	High
Consultation	More DOWN and LATERAL than UP	No difference	Used most with other tactics	High
Ingratiation	More DOWN and LATERAL than UP	Used more for initial request	Used most with other tactics	Low to moderate
Personal appeal	More LATERAL than DOWN or UP	Used more for initial request	No difference	Low to moderate
Exchange	More DOWN and LATERAL than UP	Used most for immediate follow-up	No difference	Low to moderate
Coalition tactic	More LATERAL and UP than DOWN	Used most for delayed follow-up	No difference	Low
Legitimising tactic	More DOWN and LATERAL than UP	Used most for immediate follow-up	Used most with other tactics	Low
Pressure	More DOWN and LATERAL than UP	Used most for delayed follow-up	No difference	Low

Source: adapted from *Leadership in Organizations* by Yukl, Garry, (copyright © 1998) reprinted by permission Pearson Education, Inc., Upper Saddle River, NJ (Yukl 1998, p. 229)

These tactics are listed as follows:

(1) **Rational persuasion** – the agent uses logical argument and factual evidence;
(2) **Inspirational appeals** – the agent attempts to arouse target's enthusiasm by appealing to target's value system or by increasing target's self-confidence;
(3) **Consultation** – the agent seeks target participation in planning strategy or change or to modify an existing proposal;
(4) **Ingratiation** – the agent uses praise, flattery, friendly or helpful behaviour;
(5) **Personal appeals** – the agent appeals to target's feelings of loyalty/friendship;
(6) **Exchange** – the agent offers an exchange of favours, indicates willingness to reciprocate, promises gainsharing;
(7) **Coalition tactics** – the agent seeks the aid of others to persuade the target with the support of others;

(8) **Legitimising tactics** – the agent claims authority or right for target support;
(9) **Pressure** – the agent uses demands, threats, frequent checking or making persistent reminders.

These important issues of power and influence have led management analysts to consider the concept and impact of empowerment. This is recognition of the indispensable nature of the employee and the potential contribution that he/she can make. **Empowerment** has been defined as the 'vesting of decision making or approval to employees where, traditionally, such authority was managerial prerogative' (Hammuda and Dulalaimi 1997). Trust and empowerment are closely and powerfully linked to effective teamwork. An alternative term that is being used these days for empowerment is 'enabling' which has a broader meaning for providing resources to enable people to achieve goals and objectives.

Newcombe draws interesting conclusions for implications of power in the construction industry. He notes that clients should realise that the criteria used to select consultants, contractors and the form of contract may be less important than the approaches to power structures. The fragmentation and friction engendered under the traditional system often fails to produce the expected result. Building expert/information power is an important aspect of relationship interactions for consultants, construction managers and project managers. Newcombe (1996) argues that

> If the empowerment approach is adopted then skill in building networks of contacts such as designers, trade contractors, clients, suppliers and stakeholders with an interest in a project will be necessary. Sharing power with other people requires cool nerve and judgement. Working in an empowered organisation will be very frightening for people used to the traditional system.

So we see that building team spirit through developing mutually agreed goals and objectives is a task easier preached than practised. It requires a degree of leadership and management style that is rarely found. This is because often those with the power and authority vested by the client to deliver projects have limited capacity or willingness to create the required team empowerment, or have the leadership qualities and psychological awareness to generate the necessary levels of trust and commitment. Often, project managers or the procurement method chosen does not allow the creativity and problem-solving ability of highly skilled teams to flourish. We can get teams together to develop and present a charter of mutually agreed objectives but it takes trust and commitment as well as an extraordinary level of group and individual leadership to sustain the rhetoric. We also need staff (as opposed to managers) willing to 'grasp the nettle'.

7.3   Performance improvement through innovation, continuous improvement and knowledge management

Another dimension of essential features of partnering or alliancing is a commitment to continuous improvement and organisational learning, that is, knowledge management and its use to enable innovation. **Organisational learning** codifies experience gained by individuals and teams in a form that adds value to an organisation. Learning is an asset, comprising intellectual property, which can be re-used to add competitive advantage (Love *et al.* 1999). It is also an effective means by which innovation can be introduced to organisations. Innovation is often generated from team members' personal experiences brought with them from one temporary organisation, or teams within them, to another (Walker and Lloyd-Walker 1999).

For once-off construction clients the issue of organisational learning may not be of immediate importance – except in an indirect way that general learning should add to general productivity and subsequent performance improvement. For clients with a more focused interest in continuous improvement and benefit gains from lessons learned, the diffusion of knowledge and organisational learning is a critical issue. Procurement methods have a direct impact upon the diffusion of organisational learning. In the study on construction time performance (CTP) commissioned by the Construction Industry Institute Australia (CIIA) in 1996, it was discovered that only on the process engineering projects had there been any attempt to formalise learning from a post project evaluation (Walker and Sidwell 1996). The construction industry, both for general building and civil engineering, seems to view structured harvesting of lessons learned as being of low priority and indeed an unnecessary expense (Walker and Sidwell 1996; Walker and Lloyd-Walker 1999). This represents a source of waste that should concern all parties in the construction procurement stream of activities.

Solutions to this particular waste problem have been identified as encouragement and development of effective innovation and developing a more effective means of harvesting lessons learned. Lenard argues, from the basis of results of a recent research project involving 17 case studies (Lenard 1999), that innovation observed from his case studies is based upon:

- the client's recognition of the need for innovation;
- contractual incentives to encourage innovation;
- creation of a symbiotic learning environment; and
- open communication at all levels.

Newcombe suggests that in the same way that UK safety regulations now require that construction companies should appoint a Safety Supervisor, that companies should also appoint a Learning Facilitator. This person would be required to debrief project participants to help convert individual tacit knowledge into a learning or knowledge base that underpins team and organisational learning (Newcombe 1999). This idea supports the idea proposed by Walker and Lloyd-Walker, that procurement systems should provide for project debriefing to enable clients and project teams to promote continuous improvement through creating a similar process to that suggested by Newcombe (Walker and Lloyd-Walker 1999).

Developing a learning environment focus presents challenges. A continuum of trust operates with a gradation of individuals', teams' and organisations' belief that their competitors will not take unfair advantage of lessons learned. From the individual's trust point of view, this means that they not lose their competitive advantage and be 'used up' and discarded (Green 1999a). This is a particularly sensitive issue when developing expert systems that routinise and electronically replace experts. From a firm's point of view, there is an issue of trust when passing on lessons learned, which are rightly considered a critical competitive asset. From the point of view of a temporary organisation or team groups spanning several firms on a project, there may be an issue of trust about opinion-related knowledge regarding performance of individuals or teams engaged in the project. In a sense the issue of organisational knowledge and trust may be about competitive advantage and objectivity. It is interesting to observe the reactions of companies unfamiliar with knowledge sharing through entering a partnering/alliancing arrangement and how they respond to clients with a knowledge management focus. Barlow *et al.* (1998) provide some keen insights from a series of partnering case studies in the UK involving over 40 companies. They observed some clients viewed a prime reason for partnering as being to improve innovation and organisational learning as one of their survival techniques – the motivation for others was to gain competitive advantage. Smaller suppliers and subcontractors were reported to often perceive participating in many meetings to discuss options of performing and coordinating their work with others as an imposition of time, energy and resources. After adapting to this unfamiliar environment they did feel, however, that they had gained innovative techniques and new ideas by sharing information which was worthwhile.

Clearly, there needs to be a shift in the construction industry's general culture to move towards a climate of sharing knowledge to enhance the entire industry's capacity to take advantage of lessons learned from each project. Partnering and strategic alliances with their ethos of shared problem solving and open communication may facilitate this cultural change process. Uher (1999) argues that the construction industry needs a new breed of proactive leader with vision and courage to challenge existing dogmas and approaches to managing the project delivery process.

This chapter concerns itself with exploring how cross-team relationships can be improved. We suggest that this is related to the choice of procurement form because the traditional approach fails to draw the construction team into the design process at preliminary and early development stages, thus valuable opportunities for knowledge sharing, more open relationship building and development of trust are missed. The literature provides interesting findings comparing the performance of traditional and non-traditional construction projects. In one research study of 69 projects, the authors concluded that the chosen procurement approach may assist in optimising but not directly influencing project performance (Naoum and Mustapha 1995). They note that the traditional procurement approach may be suitable for simple projects but that non-traditional complicated procurement methods may be better suited where projects become complex. This may be due to complex construction methods being required to implement the design brief or that complexity is introduced through the nature of building services or complicated and/or potentially conflicting client objectives. Others have argued that non-traditional procurement methods are better suited for dealing with project uncertainty for complex situations where the client is represented by a committee, often with competing objectives (Smith and Wilkins 1996).

One approach that brings the construction team into potentially closer contact with the design team is to overlap the design and construction process of the project through phasing the project. The project is developed as a series of overlapped design then constructed work packages. This is generally known as **fast-tracking**. A number of researchers have identified fast-tracking as a link between construction time performance and procurement method. This link is, at yet, inconclusive (Dulaimi and Dalziel 1995; Lam and Chan 1995; Walker 1995; Hashim 1996). However, the general consensus supports the view that non-traditional procurement facilitates greater construction management team involvement, which if the team responds favourably, increases chances of project success. Traditional procurement forms, while not excluding potential project success, present barriers to successful communication and inter-team relationship building as well as providing inadequate time for sound construction project planning. This project planning phase when undertaken collaboratively with cross-team contribution assists individual, team and organisational learning as it exposes participants to complex problems that need to be solved by viewing options from multiple points of view. Thus, opportunities for both innovation diffusion and knowledge transfer are increased. Co-location of design and production teams has also been identified as a stimulus to learning as there are better opportunities for cross-disciplinary problem solving as well as more face-to-face interaction and possible socialising that builds trust and commitment (Womack *et al.* 1990; Ragins 1995; Luck and Newcombe 1996; Barlow *et al.* 1998; Sobek *et al.* 1998).

A series of 64 case studies on construction time performance (CTP) of Australian construction projects concluded that traditionally procured projects are less likely to achieve as good construction time performance as non-traditionally procured projects (Walker 1997). There were neither novation nor reimbursable projects in the Walker 1997 study, however, a third of the cases included were construction management, project management and D+C projects. CTP is just one dimension of project success but it is closely linked to cost performance and does provide a good indicator of overall project performance. One interesting finding from the analysis indicated that the non-traditional procurement method, when compared to traditional project procurement methods, performed at a 21 per cent better CTP level. Further evidence provided reasons for this through investigation of data on some 100+ variables measured. It was also concluded, taking into account the literature, that to achieve good CTP the following progression applies:

- project team members (the client's representative, the design team, the construction management team) bring with them expertise and knowledge for potential project success, that is, expertise assets;
- these assets create a latent capacity for project success;
- when this capacity is effectively activated it results in good team performance; and
- this performance translates into project success, particularly good construction time performance.

This model for success, illustrated in Figure 7.3, is likely to be independent of any particular procurement method, but research has consistently pointed towards non-traditional procurement as providing an environment where this model can be most effectively applied. This research (Walker 1997) clearly indicated, from a construction time performance point of view, that there is a close link between team capacities, their behaviours and project timeliness outcomes. This reinforces

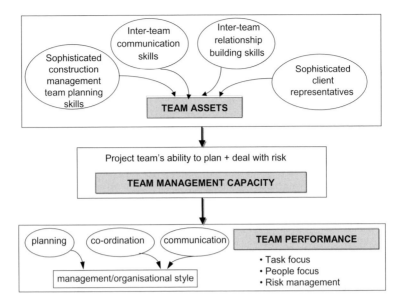

Figure 7.3: A model for construction time performance project success
Source: Walker (1997, p. 48)

the argument we propose that innovation and knowledge management foster improved project performance.

While Figure 7.3 indicates how knowledge is managed to achieve the level of team performance experienced, it does not indicate how different procurement options influence knowledge transfer. This is because knowledge management is largely independent of procurement type. The client, for example, could influence knowledge transfer on a traditional lump sum project or on design and construction or any other system. This can be achieved by simply stipulating a knowledge management and transfer mechanism in the contract conditions. If this were to be instigated then a small extra price would be paid by the client to ensure that the general market gains a competitive lift, which, in turn, could lead to lower costs due to improved industry productivity. Scuderi and Hampson support this idea by proposing a **knowledge management** strategy for a public works agency to improve industry effectiveness. They suggest enhanced contractor specific relationships and knowledge sharing from project to project will build a stronger local construction industry (Scuderi and Hampson 1999).

In attempting to create a more constructive environment in which trust, flexibility and openness to sharing ideas and intellectual capital is encouraged, we can look at the problem and attempt to minimise its impact. In analysing performance data there may often be conflicting interpretations of patterns and meaning. Conflict is not of itself bad. Much innovation arises out of conflicting interpretations of observed phenomena. In fact, economist Joseph Schumpeter, seminal author of innovation and technology literature, refers to much of this process as **creative destruction** (Schumpeter 1962). Conflict is often the result of different people having a different perspective on an issue. One person may see things one way and others may see the same thing quite differently. This is a positive and natural occurrence. Conflict requires clarification of the variance between the way one person sees something and the way another does. Senge calls this the process of dialogue, in fact he maintains that this process helps people understand issues more clearly 'in dialogue people become observers of their own thinking' (Senge 1990, p. 242). This is particularly important when considering plans or discussing possible impact of actions.

The fact that conflicts arise signifies either that healthy debate is taking place or negative contradictions are being proposed to protect entrenched (perhaps poorly thought through) positions. In discussing creative conflict Senge states that for this to happen:

- all participants must suspend judgement and assumptions (that is respect the other person's opinion and maintain an open mind that it may be valid);
- all participants to dialogue must consider each other as colleagues (rather than adversaries); and
- there must be a facilitator to hold the context of dialogue. This may be a system or workplace culture that allows the dialogue to flow towards better understanding of consequences hitherto unforeseen and to bring about creative win–win solutions (Senge 1990, p. 243).

Others agree with this view of conflict as being a positive attribute when used constructively. One interesting view of what is often called **the learning organisation** is the practitioner as inquirer (Argyris and Schön 1996, p. 35). This concept has practitioners consciously using self-dialogue and colleagues as sounding boards to question their understanding of any given situation. They see these as organisations that use conflict constructively to fully examine paradox and uncertainty in order to solve complex problems. Many problems leading to poor decision making result from errors caused by lack of available information, misinterpretation, and information being hidden when needed. There are also a lot of attitudinal issues leading to destructive conflict. If communication problems and pro-actively shared information could be openly laid on the table for discussion, without fear of defensive behaviour taking place, then much of what appears to be unknown could be discovered through constructive conflict.

Conflict can lead to **single loop learning** (in which rules are applied to modify behaviour and 'fix' problems) at the superficial level. Hierarchical power systems and structures prevalent in 'traditional' procurement systems encourage this. The rules and behaviours may lead conflict participants towards dispute resolution through claims. Possibilities exist to operate outside the rules or at least to use rules as a guide only in **double loop learning** (where insights become available that can lead to navigation around rules to the resolution of problems that make little sense). Senseless or retrograde rules act as barriers to effective team relationships. This is not about corruption or avoiding the framework of contracts and sound regulation or transparency of action. This is about seeing the issues in the context of a broader system of procuring a satisfactory project outcome within the project goals and vision.

Buildability exercises are a good example of double loop learning as is much the effort associated with quality management leading to continuous improvement. A particular design solution (the rule) may be resolved through a better and more workable solution.

This concept is taken further to challenge principles in **triple loop learning**. An example of this could be the principle of the building product being challenged through value analysis and other methodologies that may challenge the basic need for a feature, design solution or entire facility. Triple loop learning occurs when barriers are removed to liberate us so that we can explore wider possibilities that challenge system-wide assumptions, dogma, or deeply held belief systems. This may lead to invention rather than innovation (McKenna 1999). Innovation is generally about incremental improvement whereas invention is about a break-through where a paradigm shift occurs. Thus, conflict of opinion is inevitable and not a bad thing at all, in fact it indicates strength and not weakness in an organi-sation.

Newcombe argues that traditional and D+C procurement systems inhibit double loop learning and promote single loop learning. Significant changes to the design at tender are rarely challenged through the traditional procurement approach and the D+C approach frequently leads to standardisation at the expense of innovative design (Newcombe 1999).

The basic thrust of the argument presented thus far has maintained that the 'traditional' procurement system has forced a constraining effect on creativity. Further, project participants are restrained from sharing information in an open manner unhindered by fear of being exploited. There appears to be much confu-sion about which of the non-traditional procurement systems are most effective or even under what circumstances they should be adopted to counter this con-straining and restraining tendency. The common thread, however, is that a dif-ferent attitude is possible between participants on non-traditional procurement approaches. This attitude is the key to allowing cross-team learning to take place and explore other possibilities of meaning from different views of events. Love *et al.* (1999) explore this possibility for cross-project learning opportunities. This facilitates innovation and joint problem solving between the design team, client representative, construction team and supply chain teams. Attitude modifies conflict. Aggressive self-defence can result in one conflict outcome from one attitudinal pattern. Constructive dialogue, on the other hand, can elicit a win–win solution through negotiation where broader solution sets are entertained. Current interest in procurement systems that encourage interaction, genuine information and ideas sharing and joint problem solving is based upon a rejection of the power paradigm characterised by the 'traditional' procurement approach.

Green points out that the traditional leadership and working approach in the construction industry essentially follows the machine metaphor of 'the system' being like a well-oiled machine, however, as he correctly observes the construction industry is a 'people' industry and not a bureaucracy (Green 1994). People in the construction industry at most levels seem to want to 'get on with the job', the 'can-do' mentality is prevalent and has been shown so in at least one study of 45 projects in Australia which included measuring organisational style and organisational form (Walker 1994). That same 'can-do' attitude was reportedly introduced from the USA to the Broadgate project in London in the early 1990s (MacPherson 1991). Thus, the artificial them-and-us mentality is not supported in the industry as a preferred *modus operandi*. Rather it becomes the 'normal' system through default because the traditional procurement system sets teams (particularly the design and construction teams) in potential conflict through its win–lose propensity.

The partnership-style paradigm that is promoted by cooperative and colla-borative arrangements is the culture or political metaphor where systems are established to define vision and goals and to facilitate a shared and supportive milieu in which these joint goals can be realised. Thus, project teams go forward together to realise joint aims that satisfies numerous teams' agendas. The project culture or 'way things are done around here' and the political motivation to achieve aims through joint action is exemplified by the range of partnership and alliance arrangements that will be discussed in detail later. This attitude is pro-ductive but open to manipulation. Trompenaars reminds us that culture is a socially constructed reality. While often being unarticulated, it is shared by its members. This 'reality' is an historically determined phenomenon evidenced by values, rituals, heroes, symbols and practices (Trompenaars 1993). Green cautions against what he calls 'brain-washing' in mechanistic organisations. He uses busi-ness process reengineering (BPR) and lean construction as examples of a pre-vailing machine metaphor. He sees this as a danger for maintaining a 'command and control' mentality that is so typical of the 'traditional' procurement system, even when applied in the alternative procurement forms that seek to engage team spirit (Green 1998). We can see from Green's arguments that positive attitudes to collaborate to achieve project goals have a darker side and that this danger should not be dismissed. He sees cultural engineering as a powerful and potentially dangerous tool, especially when diversity is discouraged and opportunities for constructive conflict and debate is lessened.

In seeking to gain true benefits of diversity the danger of groupthink is ever present. One of the symptoms of groupthink is **cultural homogeneity**, where everyone thinks, acts and reacts in the same way. This has been exemplified as 'All team members must wear the same coloured shirts and play the same game in much the same way. Differences are either invisible or they are erased by cloning to the team's culture, putting on the team's shirt. If we follow the sports metaphor through, these sorts of teams are set in competitive relations with other teams, have exclusive and very clearly bounded memberships, and share their experiences only in accordance with strictly imposed rules' (Cope and Kalantzis 1997, p. 176).

For creativity and innovation to flourish, there needs to be a new paradigm of organisational form that encourages diversity for its advantages of seeing the same things in so many different ways. It is possible to design questioning processes to take place as part of a project design and design review process – using tools such as value analysis, constructability and business process re-engineering (Green and Lenard 1999). Greater opportunities are exposed and more options are canvassed in discussing problems and solutions. There is also a better chance of more fully exploring the consequences of planned actions where greater diversity exists. It is this theory that helps to explain why the 'traditional' procurement system has failed society and clients and why alternatives tend to fare better.

7.4   Problem resolution – acting like one big team

It is inevitable that different people will see problems in varying ways – that is the core of diversity. In previous sections in this chapter we have seen that diversity and challenging current methods lie at the core of innovation and organisational learning. We also saw that for trust and commitment to flourish in partnering or alliance relationships, an appropriate leadership style needs to be chosen that demonstrates sensitivity in the balance and type of power applied.

Encouraging diversity allows alternative and even radical reappraisal of methods, techniques, plans and even objectives to be considered. This requires an organisational structure and form that encourages constructive dialogue to occur. Teams and individuals must feel not only able to dispute issues but feel unimpeded in doing so. The aim is not to stifle opinion or debate but to arrive at a shared resolution of issues to the satisfaction of those involved while maintaining the momentum of the progress towards achieving project objectives. Two elements of this dialogue are required to be in place. First, teams and individuals must feel empowered to speak their mind so that all relevant facts, views and feeling are properly aired. This releases blockages in understanding and facilitates true communication, thus a basis for action can be established.

Second, a system must be in place that manages conflicting views and opinions to facilitate appropriate action. This is necessary because 'When problems are not resolved quickly, peoples' positions harden, details get forgotten and further problems accumulate so that what may have begun as a simple issue grows into a major dispute' (Bennett and Jayes 1995, chapter 1, p. 7). Problems must be resolved at the lowest possible level within an organisation as quickly as possible otherwise unresolved issues bog down progress.

Loosemore cautions us on the dangers of not balancing power and responsibility and notes that when problems arise, power struggles tend to develop between different interest groups (who seek to off-load responsibility for these problems) (Loosemore 1999). His case study on a traditionally procured project typifies the process of crisis generation and (mis)management prevalent in an environment where dispute resolution procedures are poorly developed. Often, threats of using more formal dispute resolution procedures such as resorting to posturing about legal rights and claims only entrenches positions of mistrust. Such actions only result in half-hearted or very low commitment to project goals being accompanied by high commitment to personal goals. Sometimes threats invite acts of retribution. Clearly, the most sensible solution to problem resolution is to keep disputes as 'cool' as possible, i.e. take the 'heat' out of the debate. Also, there is a need to resolve disputes at their source by establishing an escalation procedure where only those problems that cannot be resolved at one level are passed up the management authority chain for resolution.

A project partnering/alliancing charter is an agreement between parties to commit to a series of principles (Bennett and Jayes 1995). One important element of this principle is to establish a mechanism for resolving disputes at the lowest organisational level. Like risk, the principle is that those individuals with the means to deal with problems should do so and not pass this responsibility on to others. Often, even though a risk or problem is not 'fairly' or even 'technically' owned by an individual or team (entity), it is in everyone's interest that this entity resolves the issue quickly if it can do so. There is a large element of trust involved in taking this course of action that the 'swings and roundabouts' effect will ensure that the party consenting to this action will not be disadvantaged in the longer term. This is all part of the process of mutual adjustment that takes place in most negotiations. It may feature in both traditional and non-traditional procurement practices where good interpersonal relations are present.

Problem and dispute resolution procedures adopted in partnering/alliancing provide for the types of problem to be defined and reasonable timeframes for resolution stipulated. The reason for escalating a dispute may be hardening of diverse positions or may simply be a result of the party not being authorised to commit required resources to resolve the dispute. In cases where a dispute is escalated unnecessarily, the person escalating the dispute may well lose face in doing so. This provides a self-regulating mechanism for ensuring that problems are indeed resolved at the lowest level possible. It is normal for more than 90 per cent of problems to be solved at the lowest level (Bennett and Jayes 1995, chapter 3, p. 41). A flow-chart model problem resolution is illustrated and explained in Figure 7.4. The advantages of having an agreed dispute resolution procedure in place can be summarised as follows:

- problems are solved and disputes resolved quickly;
- solutions are generally found amicably through mutual adjustment;
- trust is generally reinforced;
- commitment is maintained and strengthened through joint problem solving;
- people generally feel empowered by the experience of resolving problems using their own discretion and authority;
- it builds self-confidence; and
- it allows diversity of opinion to be legitimised and also legitimises opposing points of views being accepted.

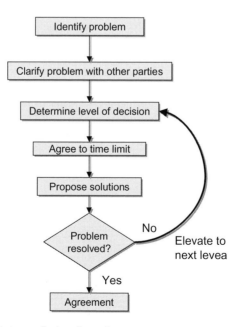

Figure 7.4: Problem resolution flow chart
Source: Adapted with permission from Bennett, J. and Jayes, S. (copyright © 1995) *Trusting the Team*. Reading, UK, Centre for Strategic Studies in Construction, The University of Reading

Throughout this section of this chapter we have stressed that trust and commitment underpins the three essential elements of partnering and alliancing (mutual objectives, problem resolution, and continuous improvement). We have also shown that balancing power and responsibility is crucial in the development of trust and commitment especially when such relationships have been formed out of a dependence upon a party's resources, knowledge or goodwill (Doz and Hamel 1998). So what exactly do we mean by the terms **trust** and **commitment** and how are these qualities created and maintained?

7.5   Trust, commitment and relationship building

Partnering and alliancing are based upon a need for trust to generate commitment and constructive dialogue. **Trust** is part of an outcome from negotiation. The everyday interrelationship experiences of project team members are shaped by the way in which they negotiate. This can include communication exchanges, decisions about activities or plans for design and/or construction and a host of other issues that contribute to teams acting like partnerships rather than bands of individuals working on the same project.

The notion of trust is complex. It has many layers of meaning. 'Simply put, trust means confidence – confidence that others' actions are consistent with their words, that those people with whom you work are concerned about your welfare and interest apart from what you can do for them...' (Rogers 1995). At one level, reasoned expectations will be fulfilled. You can trust that a certain thing will or will not happen under a given set of conditions. This raises interesting notions of the difference between scepticism and cynicism and how predictability is related to past experiences and future expectations based upon that experience. Trust at its naïve level is almost synonymous with hope. In most relationships, personal and business, trust is as much about something happening as not happening. If we do 'the right thing' and we trust people we expect that rewards (either tangible or intangible) will follow. Similarly if we are dealing with somebody we perceive as 'shifty'; we expect or trust that they will take advantage. This may result in pre-emptive action or precautions being taken. Trust is bound up with past experience both directly with the person(s) concerned and indirectly, through projected or anticipated experiences, thus trust is an intensely emotional and human phenomenon. Figure 7.5 illustrates a model of the range of influences that can affect our perception of trust (Whiteley *et al.* 1998).

Commitment is the physical and mental manifestation of the concept of trust. It is the proof of trust. It is the willingness to reciprocate energy invested through trust in the process of transformation of this energy into tangible results. Thus a 'trusting' supervisor may back off from detailed specification and control of how tasks may be performed. Commitment, means that another party will take this trust on board and 'live up to' the spirit of the bargain by probably committing more personal pride and obligation to 'do the right thing' than would otherwise be the case. Loyalty occurs when trust and commitment are tested. It can be viewed as the bankable capital of goodwill to reciprocate trust in times of adversity. One demonstration of an act of loyalty is to sacrifice something in the short term to maintain a long-term relationship intact and functioning for mutual advantage.

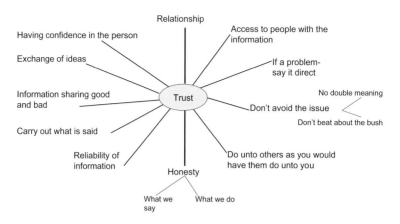

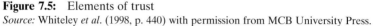

Figure 7.5:   Elements of trust
Source: Whiteley *et al.* (1998, p. 440) with permission from MCB University Press.

This clearly illustrates the linkages between trust, communication, commitment and management style. Figure 7.5 also illustrates how much work is involved in gaining and maintaining trust. Clearly, trust does not come pre-packaged in a particular procurement system though characteristics of some systems establish an environment in which trust is more likely to flourish than others.

Trust is a murky swamp of emotions. It is full of history, transfer of emotions from 'similar situations' and eternal hope. This phenomenon is largely ethereal in nature. Given that trust is largely about perceptions and those perceptions are based upon experience, it is interesting to investigate how person-to-person and organisation-to-organisation negotiation shapes these perceptions. Negotiation is a daily part of life – from resolving workloads, organising time-off, to exchanging information in a meeting where give-and-take is part of the process.

Considering which procurement systems to choose on the basis of engendering trust is problematic. The real issue is not which procurement system is best, but rather which management style and system best facilitates conditions that engender trust. A number of interesting conclusions emerged from a US study of 132 people of the effect of anger and compassion upon negotiation performance (Alfred *et al.* 1997). One of these was that **venting** (that is demonstrating the emotion in a physical or vocal form) is more likely to increase than to decrease anger in the person venting. Venting becomes a rehearsal for the real thing and if this is not exposed in debate then the venting remains as a festering sore – thus creating a situation where anger becomes tangible and is worse than defusing the situation as it emerged. This is significant in that in negotiation and day-to-day relationships, it appears better to clarify and resolve conflict as it arises or at least quickly, rather than let it fester. This reinforces the importance of having a dispute resolution system as described and illustrated in Figure 7.4 because so much of the dispute resolution system depends upon parties negotiating in good faith (i.e. ethically).

The implication for dispute and conflict resolution is to recognise the legitimacy of people holding different positions and to clarify them as quickly as possible to allow mutual understanding to diffuse dispute situations. Findings from the research of Alfred *et al.* (1997) on negotiating are summarised as:

- compassion serves to increase helping behaviour and decrease punishment behaviour;
- negative emotional regard will diminish negotiators' willingness to work with each other in the future;
- conversely, positive emotional regard will enhance the negotiators' desire to work with each other in the future;
- the root of anger lies in the judgement made by one group about why another group behaves in a manner harmful to them; and
- people tend to over-attribute another person's behaviour and motives, thus misconceptions can easily develop.

Much of the relationship building and maintenance process critically depends upon the way parties to a relationship perceived each other – Figure 7.5 illustrates this clearly. So why should we concern ourselves with ethics and negotiation when considering procurement issues? There are at least three standards for evaluating strategies and tactics for negotiation: ethics, prudence and practicality. **Ethics** refers to standards of personal or group morality. **Prudence** refers to judgements made upon effectiveness. **Practicality** refers to issues of cost-effectiveness, timeliness and other issues of a more concrete nature with respect to achieving goals. Trust and perception can be highly biased. Generally, we tend to perceive others in absolute terms (right or wrong), but rationalise our behaviour in relative terms (Lewicki *et al.* 1994, p. 386). Ethical negotiation besets people on a day-to-day basis. They make their judgements on themselves and others on this basis. Lewicki's research has shown that at least five influences can encourage negotiators to suspend their own sense of ethical standards. These include:

(1)   acting as an agent for someone else and responding to their pressures at any price acceptable to those they represent rather than adhere to their own standards;
(2)   viewing the whole process as a game;
(3)   being part of a culture that tolerates or encourages bending or breaking the rules to achieve success;
(4)   being so loyal to a group that you can convince yourself that what you do is acceptable in order to be rewarded for loyalty (in either financial or other terms); and
(5)   following orders and not considering the ethics of actions taken.

It is easy to dismiss project negotiation tactics as being part of a special process. However, we continually negotiate what we do, what we take responsibility and commitment for, and how we agreed to do things. We need to recognise that our actions and decisions affect many people in project team(s). The way we conduct ourselves will reflect the way others are prepared to deal with us and vice versa.

7.6   Trust and relationship maintenance

The issue of breaking trust is also crucial in any relationship as it sows the seeds of confidence destruction which often leads directly and, more importantly, indirectly into disputes and counterclaims. The term **psychological contract** emerged from the 1960s to 1980s as a means of describing the emotional capital invested in relationships. It mostly applied to employee and employer contracts – formal or informal. Psychological contracts embrace specific perceived obligations – as distinct from general expectations – that form general beliefs held about a relationship. Expectations emanate from a wide variety of sources such as past experiences, observation, rumours, reports or stories, and a range of other intangible impression generated things. Psychological contracts entail beliefs about what one party believes that he/she is entitled to because of a perception of what was offered verbally and in writing. 'Although psychological contracts produce some expectations, not all expectations emanate from perceived promises, and expectations can exist in the absence of perceived promises or contracts' (Robinson 1996).

The interesting aspect of Robinson's research was that it focused upon the impact of an employee's belief that a breach of trust had occurred and the effect that this had upon the employer/employee relationship (Robinson 1996). In many ways this is typical of general relationships between people working together in teams . She maintained that behaviour and attitudes could be explained by (in the case she cites) an employee's belief that a breach had occurred regardless of the reality of any breach. It was the belief of a breach of trust that was important in explaining behaviour and attitudes. Her paper raises useful insights into the way trust and attitudes and behaviour are intertwined, and how we might learn from this work in construction procurement strategies. Figure 7.6 illustrates the

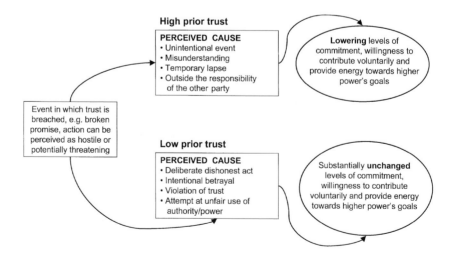

Figure 7.6:   Relationship between experiences of breached trust and commitment or goodwill
Source: Robinson (1996)

relationship between breached trust experiences and how continued commitment is affected based upon a longitudinal study of 125 newly hired managers in the USA over a 30-month period.

Robinson empirically tested a group of 125 Master of Business Administration (MBA) students in the USA who were working and studying part time. The study measured the psychological contract in terms of employee perceived actions and resulting employee willingness to contribute to attaining organisational goals over a 30-month period. Some of her findings, illustrated in Figure 7.6 and Figure 7.7, are useful for understanding how people interact when breaches of trust are perceived to have occurred in a supervisor/worker power situation where one party has the authority to direct the action of another.

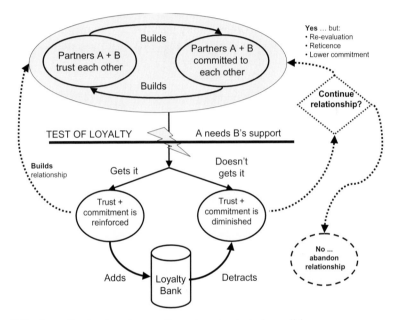

Figure 7.7:   Loyalty, trust and commitment under tested conditions

The implication of this work is that certain construction procurement systems may require greater levels of prior trust than others. The traditional procurement system with its emphasis on a fixed sum for a fixed stage of design encourages builders to 'create' disputes to compensate for low profits (NBCC 1989). In procurement options where the manager of the project is acting as a consultant, the client and other team members expect a 'professional attitude' such as sharing information. Kwok measured trust and commitment as variables in his study of construction alliances in public sector building works in Queensland (Kwok 1998). Robinson measured trust over time by administering three surveys to identify how commitment, loyalty and willingness to contribute changed in the light of experiencing breaches of trust (Robinson 1996). This approach can be of use to study how trust, commitment and loyalty operates within various procurement systems, particularly alliances.

Figure 7.7 illustrates what happens when loyalty is tested. One possible outcome illustrated is relationship reinforcement. Another is the relationship being called into question or diminished. Procurement systems bring with them varying expectation of commitment and levels of trust. The client/consultant or designer/contractor relationship is similar in many ways to an employer/employee relationship in mutual vulnerability where expectations exist about likely behaviour which affect openness, commitment and willingness to go beyond the expected. This illustration dealt with team members interacting with other individuals or teams. Again its substance is quite independent of procurement form, which explains why trust and commitment can occur or be absent in any procurement form. Earlier, we touched on how leaders can use or abuse power and in this chapter we discussed how trust and commitment are affected by power use. All this, of course, is also influenced by the organisational culture which may at one extreme be 'gung-ho' or at the other so conscious of meeting all stakeholders' aspirations that it becomes totally paralysed. Organisational culture is affected by its members and by the quality of the leadership and management style of those in positions of influence and authority (Yukl 1998; Bass 1990; Bass and Avolio 1994; Avolio 1996; Bennis and Nanus 1997).

7.7   Teams and leaders – building a foundation for partnerships

The way a leader behaves is governed by the characteristics of leaders and followers. These encompass the entire project teams in the facility supply chain as well as the design and management facilitation groups. We saw how trust and commitment are linked through empowerment and how this can liberate creative energies in team members. In an empowerment approach, the dominant team members relinquish authority to the follower because they probably have knowledge power or aspects of personal power. We also saw how different forms of power can be used to better liberate team members' energy and how loyalty, trust and commitment are affected by breaches of trust. This brings us to the issue of how management style and organisational culture affect the individual and their perceptions of what to expect from team mates, co-workers and allied project teams. The literature suggests that **organisational culture** releases energy, enabling performance excellence, in a hierarchical manner (Bass 1985, 1990; Bass and Avolio 1994; Avolio 1996; Bennis and Nanus 1997).

Avolio (Avolio 1996) categorised four distinct organisation hierarchies:

- **Level 1 – Adrift**. Few symbols/images provide any sense of community and/or cohesion. Members have difficulties identifying a core set of values or ideals. Decisions are not tied to precedents so members do not take responsibility for their actions.
- **Level 2 – Transactional (corrective)**. There is a heavy concentration on eliminating mistakes. People are generally recognised for what they do wrong versus what they do right. There is little evidence of collegial support or any significant willingness to be supportive of others. The general environment is characterised by risk avoidance where new ideas are hoarded by oneself.
- **Level 3 – Transactional (constructive)**. In the extreme cases everything gets done through negotiation and contracts – if contracts are fulfilled then

appropriate rewards are provided to employees. At the higher end, organisation's members provide each other with support/recognition, and they are willing to learn new skills/applications if adequately compensated for their efforts. Transactions not always so basic, and can be of longer duration. Consistent honouring of agreements eventually builds the basis of trust amongst organisation's members.

- **Level 4 – Transformational leadership**. The organisation's members more easily identify core values. There is evidence of a collective focus on building learning potential and performance. People feel comfortable questioning each other and realise that they are all working toward a central purpose and/or mission. Symbols and images signify important organisation values. Members trust each other to do what's right/moral.

The first two levels are characterised by group behaviours, whereas levels 3 and 4 more closely resemble teams at work rather than groups at work. Level 4 is seen to have the greatest proximity to partnership and alliance relationships though many project organisations which purport to use partnering arrangements is more closely characterised by level 3.

Figure 7.8 illustrates team development in terms of five management styles (Avolio 1996, p. 12). We can see the history of various management innovations in this progression. It is interesting to note that an alliance or partnering arrangement operating at the *laissez faire* level will probably have a poorer project performance than a 'traditional' system operating at a transformational level exhibiting use of the four Is.

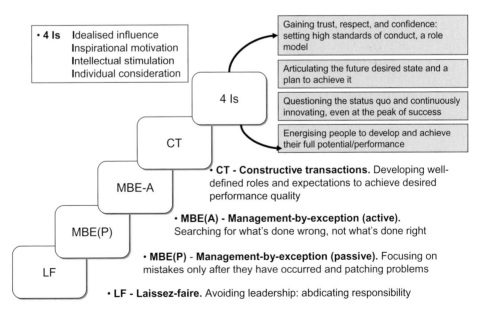

Figure 7.8: Leadership in developing teams
Source: Avolio, B. (copyright © 1996) What's All the Karping About Down Under? *Leadership Research and Practice*. W. Parry, K.W. South Melbourne, Pitman Publishing

In well-functioning teams where task leadership is assigned by competence and willingness to take responsibility for outcomes, followers will exhibit favourable characteristics of leaders as they take charge of tasks. Much of what is understood as intelligence is complex and multifaceted. In the supply chain and at the assembly stage of project delivery, numerous intelligent people, sometimes with few formal qualifications, assume a leadership role.

Numerous leadership academics have researched the nature of intelligence and intelligent leadership. Mant (Mant 1997) introduced the notion that intellectual 'firepower' is only one of several aspects to be marshalled to produce intelligent leadership. He shows that the most intelligent people may be carrying emotional baggage that inhibits them from acting intelligently and making sound judgements. The issue of judgement, wisdom and qualification (knowledge) provides interesting insights into why teams of highly intelligent people are capable of making disastrous decisions. One secret of teams working effectively and gaining and maintaining each other's trust and commitment is their level of emotional intelligence.

The literature suggests that emotional intelligence skills required can be learned. This is fortunate, as many team members require these skills to function optimally. They can be facilitated through good team coaching and leadership. Many of these contribute to an atmosphere where trust, commitment and loyalty can flourish. Goleman (1998, p. 95) identifies five facets of emotional intelligence. The first, self-awareness, is demonstrated by high levels of self-confidence in a self-depreciating rather than overt manner where the person understands such aspects of themselves as their moods, emotions and drives and the impact that this may have on others they interact with. The second, self-regulation, is demonstrated by integrity, comfort with ambiguity and openness to change. The third is intrinsic motivation with optimism and strong commitment to achieve excellence stemming from a desire to do a good job rather than satisfy quotas or performance targets – this is often evident from a passion for the importance of the desired outcome. The fourth, empathy, is expressed in such qualities as being able to nurture talent, be sensitive to different cultures and having a service mentality. The fifth, social skill, is often seen as an ability to persuade rather than cajole or bully through building solutions rather than imposing will.

Mant discusses useful insights into the nature of intelligence and how this contributes to sound leadership. He cites the example of Bob Clifford from Tasmania, a world leading designer and manufacturer of catamarans. Clifford performed very poorly at school but designs his catamarans through a keen sense of intuition drawing upon technical and scientific expertise from other team members when needed. Mant likens much of the kinds of intelligence evidenced by high performers as similar to that of artists – feeling, intuitive and emotional aspects of intelligence. This is much like the human body when intelligence is downloaded from the brain to the spinal column and limbs. When parking a car in the dark in a confined space we take on the car's body as part of our own and we park by 'feel'. Musicians also have this quality of intelligence when they 'feel' the music. Sculptors shape the forms they create by almost coaxing the image out of the materials they use. In this way intelligent team leaders (and followers acting in a leadership role while in charge of a task), use intuition and emotional intelligence to get the best from the resources they employ (Mant 1997).

The main thrust of this book has made it clear that the traditional lump sum procurement approach tends to detract from optimising a team's contribution to the project due to structural impediments of the procurement system. However, this still does not fully explain why some projects using the traditional system are successful and others using alternative procurement systems less successful. Part of the answer has been explained in the way trust, commitment and loyalty are harnessed by sound leadership and appropriate use of power and authority to best facilitate the highest levels of team contribution to project goals. Part lies in the way teams operate. A team is more than simply a collection of people. A team is a coherent entity focused on common goals. In the more advanced forms of organisational leadership described earlier, self-management and self-direction was offered as the way forward.

The following summarises characteristics of **self-managed teams** (Wageman 1997):

- Self-managed teams take responsibility for themselves;
- They organise their work, monitor performance and alter strategies and day-to-day actions to meet goals set by the team;
- They need a mature management style to set free their creative energies; and
- They can:
  o enhance company performance;
  o enhance organisational learning; and
  o enhance employees' commitment/participation.

Wageman argues from research that she undertook into organisations such as Xerox, that there are at least two basic influences on team success. First, how teams are designed and supported. Second, how the leader behaves in day-to-day interactions. Key issues she discovered from her research indicates how important high quality coaching is in sending cues to team members that they are responsible for their own performance. She also stresses that timely feedback as well as helping teams develops problem-solving strategies in team members.

In terms of design features, the following were identified as needed (Wageman 1997, p. 53):

(1) Clear, engaging direction;
(2) Task interdependence;
(3) Authority to manage the work;
(4) Performance goals;
(5) Skill diversity of team members;
(6) Demographic diversity of team members;
(7) Team size (to be feasible for self-management and coaching);
(8) Length of time the team has had a stable relationship (to develop coherency and team spirit);
(9) Group rewards;
(10) Information resources (sufficient to make decisions as a team and take both personal and group responsibility for performance);
(11) Availability of training (sufficient to effectively operate both individually and as a team); and

(12) Basic material resources (sufficient to effectively operate both individually and as a team).

Points 5 and 6 are important in a wider context. Diversity adds strength to teams as it offers more ways of visualising problems and solutions. Teams can be held in a mono-cultural straitjacket that inhibits wider consideration of the context of issues (Cope and Kalantzis 1997). This, of course, substantially depends upon the leadership style being matched to follower readiness in which the leader's support behaviour is matched to followers' necessary skills and knowledge and also their confidence and commitment to successfully undertaking the task (Hersey *et al.* 1996, p. 331). The quality of the leader's coaching provides an enabling mechanism to raise or lower the level of team performance. Table 7.4 illustrates this.

Table 7.4: Leadership coaching advice

Positive coaching behaviours	Potential negative influences
1. Providing reinforcers and other clues that the group is responsible for managing itself	1. Signalling that individuals (or the leaders/ managers) were responsible for the team's work
2. Appropriate problem-solving consultation	2. Intervening in the task
3. Dealing with interpersonal problems in the team through team-process consultation	3. Identifying the team's problems
4. Attending team meetings	4. Over-riding group decisions
5. Providing organisational-related data	

Source: Adapted from Wageman (1997, p. 53), reprinted from *Organizational Dynamics*, vol. 26, p. 53, 1997 with permission from Elsevier Science.

According to a number of influential management thinkers, the close of the twentieth century was accompanied by disquiet about the way in which the traditional corporate model serves society (Senge 1990; Hames 1994; Elkington 1997; Hames and Callanan 1997; Handy 1997; Gibsen 1998; Limerick *et al.* 1998). Amongst this criticism is the comment that

> People at all levels in organisations realised that traditional conceptions of capitalist organisation were frequently inefficient, unresponsive, cumbersome and demotivating. ... There is no shortage of evidence to indicate that people in their organisations responded by taking initiative in working together to provide new products and services, without waiting for command from above (Limerick *et al.* 1998, p. 1–2).

There are interesting aspects about this quotation. First, that it could have applied equally well to communism, indeed any totalitarian system and second, that it applies to the relationship between teams and leaders. The notion of flexibility as a team virtue is not new (Kanter 1989, 1995). There are calls for not so much more leadership but more flexible leadership (Limerick *et al.* 1998, p. 227). There has also been calls for 'leaders' to facilitate and allow the collective team's abilities to accomplish more than would otherwise be the case – to unleash the power of teamwork through greater autonomy, self-management and self-direction (Katzenbach 1997).

This brief reflection upon the literature strongly supports the idea that many teams are ready for the kind of leadership style that liberates them to get on with the job. A recent benchmarking study sponsored by the Construction Industry Institute Australia reinforces these conclusions (Testi *et al.* 1995). Figure 7.9 illustrates the way in which teams can be mobilised to achieve superior project outcomes. It is evident that there are approaches that enhance team management facilitating actions leading to success and there are approaches that do not.

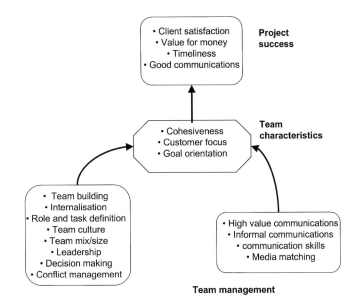

Figure 7.9:   Winning teams
Source: Adapted with permission from Testi, *et al.* (1995, p. vi)

Teams bring with them their own unique blend of problem solving, technical and interpersonal skills with varying strengths of complimentariness. Members bring with them attitudes and behaviour patterns based on past and current experiences. They accept varying degrees of accountability and responsibility depending upon their personalities, the workplace environment and the actions of the leader. The commitment to success that teams exhibit varies with their goal alignment, trust and loyalty to one another and/or the organisation (these points have been discussed in more detail earlier). Their performance is a function of this dynamic.

Teams that are substantially unfocused on common objectives merely resemble a working group or committee sharing ideas and perspectives. The **pseudo-team** has not focused upon collective purpose and is not trying hard to achieve any goals or objectives. The **potential team** has that desire to be focused but lacks clear direction or is inhibited by organisational or other performance barriers. **Real teams** are committed and focused towards common purpose but also display a working approach that demonstrates mutual accountability and commitment. High performing teams go beyond this capacity – they transcend their own and team goals to make the organisational goals their first priority (Katzenbach and

Smith 1993). However, teams fail to achieve goals even when this commitment is present. Often this failure can be explained by organisational barriers and hurdles placed by the nature and characteristics of the prevailing corporate culture (Hames 1994; Hames and Callanan 1997; Handy 1997; Gibsen 1998; Limerick *et al.* 1998).

Table 7.5 provides a variety of common dysfunctions leading to poor team performance (Robbins and Finlay 1997). Sometimes team dynamics poison team performance. Each team member will have a mutual propensity for acceptance by, and attraction towards, other team members. High attraction and acceptance leads to psychological membership with internalisation and commitment being fully realised. High attraction and low acceptance leads to preferential membership where commitment is somewhat conditional or tentative, as is the case with marginal membership having high acceptance but low attraction. The least committed state is low acceptance and low attraction, which leads to alienated membership of a team. Thus reasons for teams not working often lie with internal causation from team and individual dynamics as well as the characteristics of the organisation and/or its leadership style (McKenna 1999).

It can now be better appreciated the impact that the workplace environment has upon team performance – from a physical as well as psychological point of view. It is pointless for an unprincipled manipulative leader to cajole team members using **charismatic** or **transformational**[1] leadership styles when the team members see that their efforts are being ruthlessly exploited. Similarly, a truly superior quality leader and organisation that 'does the right thing' with well-focused ethical goals cannot succeed with a team that is dysfunctional. It has been stressed in this section that diverse skills are required to deliver complex projects. It has also been stressed that those people with these skills will come from a diverse background and have different world-views. There is widespread agreement that tolerance of diversity, indeed welcoming diversity is essential for the successful project teams of the twenty-first century (Hames 1994; Cope and Kalantzis 1997; Limerick *et al.* 1998; McKenna 1999).

7.8   Chapter summary

It was highlighted at the start of this chapter that many researchers have argued that the form of procurement is largely irrelevant and that the real issue is how the procurement option enhances or inhibits team members to maximise their constructive input to achieve project goals.

Essential features of partnering/alliancing introduced some important dimensions of the complex environment – both physical and psychological – that affects teams and leaders. The extent to which the client/client representative exerts design influence led to issues of organisational learning and enabling innovation through design/construction team synergy. It was important to explore how power and influence, trust, commitment and loyalty, and trust and negotiation are linked

---

[1] For more detailed examination of transactional and transformational leadership literature see, Bass, B.M. (1985); Bass, B.M. (1990); Bass, B.M. and Avolio, B.J. (1994); *http//learning.mit.edu/res/kr/barker/barker-webers.htm*

Table 7.5: Why teams do not work

Problem	Symptom	Solution
Mismatched needs	People with private agendas working at cross purposes	Get hidden agendas out on the table by asking what people want, personally, from teaming
Confused goals or cluttered objectives	People don't know what to do or it doesn't make sense	Clarify the reasons the team exists: define its purpose and expected outcomes
Unresolved roles	Team members are uncertain what their job is	Inform team members what is expected of them
Bad decision making	Teams may be making the right decisions, but in the wrong way	Choose a decision-making approach appropriate to each decision
Bad policies, stupid procedures	Team is at the mercy of an employee 'handbook from hell'	Throw away the book and start making sense
Personality conflicts	Team members do not get along	Learn what team members expect and want from one another, what they prefer, how they differ, start valuing and using differences
Bad leadership	Leadership is tentative, inconsistent, or stupid	The leader must learn to serve the team and keep its vision alive or leave leadership to someone else
Bleary vision	Leadership has 'foisted a bill of goods' on the team	Get a better vision or go away
Anti-team culture	The organisation is not really committed to the idea of teams	Team for the right reasons or don't team at all; never force people into a team
Insufficient feedback and information	Performance is not being measured; team members are groping in the dark	Create a system of free flow of useful information to and from all team members
Ill-conceived reward systems	People are being rewarded for the wrong things	Design rewards that make teams feel safe doing their job; reward teaming as well as individual behaviours
Lack of team trust	The team is not a team because members are unable to commit to it	Stop being untrustworthy, or disband or reform the team

Source: with permission from Robbins and Finlay (1997, p. 14).

to and advance our fundamental understanding about the reasons why a particular management approach may be appropriate. The management style and approach – mainly concerned with attitudes, the process of facilitating team spirit and the dynamics of team relationships – is more important than the procurement choice itself. The appropriate choice of project delivery system is about not only choosing the best delivery method, but also choosing the most appropriate management style and approach.

In the next chapter the nature and characteristics of partnering and alliancing will be discussed from a perspective of how it best provides the project delivery mechanism with a more inclusive management style can be deployed which better engenders trust and commitment.

7.9 References

Alfred, K.G., Mallozzo, F.M. and Raia, C.P. (1997). 'The Influence of Anger and Compassion on Negotiation Performance.' *Organizational Behaviour and Human Decision Processes*. **70** (3): 175–187.

Argyris, C. and Schön, D. (1996) *Organizational Learning II: Theory, method, and practice*. Reading, MA, Addison-Wesley.

Avolio, B. (1996) What's All the Karping About Down Under? *Leadership Research and Practice*, ed Parry, K.W. South Melbourne, Pitman Publishing: 3–15.

Barlow, J., Jashapara, A. and Cohen, M. (1998) 'Organisational Learning and Inter-firm "Partnering" in the UK Construction Industry.' *The Learning Industry Organization Journal*. **5** (2): 86–98.

Bass, B.M. (1985) *Leadership and Performance Beyond Expectations*, New York, Free Press.

Bass, B.M. (1990) *Bass & Stogdill's Handbook of Leadership: Theory, Research, and Managerial Applications*. New York, Free Press.

Bass, B.M. and Avolio, B.J. (1994) *Improving Organisational Effectiveness Through Transformational Leadership*. London, Sage.

Bennett, J. and Jayes, S. (1995) *Trusting the Team*. Reading, UK, Centre for Strategic Studies in Construction, The University of Reading.

Bennis, W. and Nanus, B. (1997) *Leaders' Strategies for Taking Charge*. New York, Harper Business.

Cope, W. and Kalantzis, M. (1997) *Productive Diversity – A New, Australian Model for Work and Management*. Annandale, NSW, Australia, Pluto Press Australia Limited.

Doz, Y.L. and Hamel, G. (1998) *Alliance Advantage – The Art of Creating Value Through Partnering*. Boston, MA, Harvard Business School Press.

Dulaimi, M.F. and Dalziel, R.C. (1995) *Construction Management as a Procurement Method a New Direction for Asian Contractors*. CIB W92 Symposium East meets West, Hong Kong, University of Hong Kong.

Elkington, J. (1997) *Cannibals with Forks*. London, Capstone Publishing.

Gibsen, R. (1998) *Rethinking the future*. London, Nicholas Brealey: 276.

Goleman, D. (1998) 'What Makes a Leader?' *Harvard Business Review*. **76** (6): 92–102.

Green, S. (1999a) 'Partnering: The Propaganda of Corporatism.' *Journal of Construction Procurement*. **5** (2): 177–186.

Green, S. and Lenard, D. (1999) Organising the Project Procurement Process. In: *Procurement Systems: A Guide to Best Practice in Construction*, eds Rowlinson S. and McDermott, P. London, E&FN Spon: 57–82.

Green, S.D. (1994) *Sociological Paradigms and Building Procurement*. East meets West – CIBW92 Symposium, Hong Kong University, University of Hong Kong.

Green, S.D. (1998) 'The Technocratic Totalitarianism of Construction Process Improvement: A critical perspective.' *Engineering Construction and Architectural Management*, **5** (4): 376–386.

Green, S.D. (1999b) Partnering: the Propaganda of Corporatism? In: *Profitable Partnering in Construction Procurement*, ed Ogulana, S.O. London, E&FN Spon: 735.

Greene, R. and Elfrers, J. (1999) *Power: the 48 Laws*. London, Profile Books.

Hames, D.H. (1994) *The Management Myth*. Sydney, Business & Professional Publishing.

Hames, R.D. and Callanan, G. (1997) *Burying the 20th Century – New Paths for New Futures*. Warriewood, NSW, Business & Professional Publishing.

Hammuda, I. and Dulalaimi, M. (1997) 'The Theory and Application of Empowerment in Construction: A Comparative Study of the Different Approaches to Empowerment in Construction, Service and Manufacturing Industries.' *International Journal of Project Management*. **15** (5): 289–296.

Handy, C. (1997) *The Hungry Spirit*. London, Random House.

Hashim, M. (1996) *The Effect of Procurement Systems on Performance of Construction Projects in Malaysia*. CIB W92 Symposium, North meets South, Durban, Natal, RSA.

Hersey, P. and Blanchard, K. (1982) *Management of Organisational Behaviour*. New York, Prentice-Hall International.

Hersey, P., Blanchard, K. and Johnson, D.E. (1996) *Management of Organizational Behaviour*. London, Prentice-Hall International.

Hofstede, G. (1991) *Culture and Organizations: Software of the Mind*. New York, McGraw Hill.

Kanter, R.M. (1989) *When Giants Learn to Dance: Mastering the Challanges of Strategy, Management and Careers in the 1990s*. London, Simon & Schuster.

Kanter, R.M. (1995) *World-Class: Thriving Locally in the Global Economy*. New York, Simon & Schuster.

Katzenbach, J.R. (1997) 'The Myth of Top Team Management.' *Harvard Business Review*. **75** (6): 82–91.

Katzenbach, J.R. and Smith, D.K. (1993) *The Wisdom of Teams – Creating the High-Performance Organization*. Boston, Harvard Business School Press.

Kumaraswamy, M.M. (1997) 'Conflicts, Claims and Disputes in Construction.' *Engineering, Construction and Architectural Management*. **4** (2): 95–112.

Kwok, T. (1998) *Strategic Alliances in Construction: A Study of Contracting Relationships and Competitive Advantage in Public Sector Building Works*. PhD, Faculty of the Built Environment. Brisbane, Queensland University of Technology.

Lam, P.T.I. and Chan, A.P.C. (1995) *Construction Management as a Procurement Method: A New Direction for Asian Contractors*. CIB W92 Symposium, East meets West, Hong Kong, University of Hong Kong.

Lenard, D. (1999) 'Future Challenges in Construction Management: Creating a Symbiotic Learning Environment.' *Journal of Construction Procurement*. **5** (2): 197–210.

Lewicki, R.J., Litterere, J.A., Minton, J.W. and Saunders, D.M. (1994) *Negotiation*. Sydney, Irwin.

Limerick, D., Cunninton, B. and Crowther, F. (1998) *Managing the New Organisation: Collaboration and Sustainability in the Postcorporate World*. Warriewood, NSW, Business & Professional Publishing.

Loosemore, M. (1999) 'Responsibility, Power and Construction Conflict.' *Construction Management and Economics*. **17** (6): 699–709.

Love, P.E.D., Li, H. and Hampson, K.D. (1999) *Cooperative Strategic Learning Alliances in Construction*. Proceedings of the CIB W55 and W65 Symposium 1999, Cape Town, South Africa, CIB.

Lovell, R.J. (1993) 'Power and the Project Manager.' *International Journal of Project Management*. **11** (2): 73–78.

Luck, R.A.C. and Newcombe, R. (1996) *The Case for Integration of the Project Participants' Activities within a Construction Project Environment*. Glasgow, Scotland, E&FN Spon.

MacPherson, I.G. (1991). Broadgate: Phases 5, 9 and 10 a Construction Manager's Point of View. *Fast Build*, ed Stacey, D. London, Thomas Telford.

Mant, A. (1997) *Intelligent Leadership*. Sydney, Allen and Unwin.

McKenna, R. (1999) *New Management*. Sydney, Irwin, NSW, McGraw Hill.

Naoum, S.G. and Mustapha, F.H. (1995) 'Relationship Between the Building Team,

Procurement Methods and Project Performance.' *Journal of Construction Procurement*. **1** (1): 55–63.

NBCC (1989) *Strategies for the Reduction of Claims and Disputes in the Construction Industry – No Dispute*. Canberra, National Building and Construction Council.

Newcombe, R. (1996) 'Empowering the Construction Project Team.' *International Journal of Project Management*. **14** (2): 75–80.

Newcombe, R. (1999) 'Procurement as a Learning Process.' *Journal of Construction Procurement*. **5** (2): 211–220.

Ragins, B.R. (1995) Diversity, Power, and Mentorship in Organizations: A cultural, structural and behavioural perspective. *Diversity in Organizations – New perspectives for a changing workplace*, eds Chemers, M.M., Oskamp, S. and Costanzo, M.A. London, Sage: 282.

Robbins, H. and Finlay, M. (1997) *Why Teams Don't Work – What Went Wrong and How to Make it Right*. London, Orion Publishing Group Ltd.

Robinson, S.L. (1996) 'Trust and Breach of the Psychological Contract.' *Administrative Science Quarterly*. **41** (4): 574–599.

Rogers, R.W. (1995) 'The Psychological Contract of Trust – Part 1.' *Executive Development*, **8** (1): 15–19.

Schumpeter, J. (1962) *The Theory of Economic Development*. Cambridge, MA, Harvard University Press.

Scuderi, P. and Hampson, K.D. (1999) *Knowledge Management in Government Through an Information Management Strategy*. 2nd International Conference on Construction Process Reengineering, Sydney, Australia.

Senge, P.M. (1990) *The Fifth Discipline – The Art and Practice of the Learning Organization*. Sydney, Australia, Random House.

Smith, A. and Wilkins, B. (1996) 'Team Relationships and Related Critical Factors in the Successful Procurement of Health Care Facilities.' *Journal of Construction Procurement*. **2** (1): 30–40.

Sobek, D.K., Liker, J.K. and Ward, A.C. (1998) 'Another Look at How Toyota Integrates Product Development.' *Harvard Business Review*. **76** (4): 36–49.

Testi, J., Sidwell, A.C. and Lenard, D.J. (1995) *Benchmarking Engineering and Construction – Winning Teams*, Adelaide, CIIA and University of South Australia.

Trompenaars, F. (1993) *Riding the Waves of Culture: Understanding Cultural Diversity in Business*. London, Economics Books.

Uher, T. (1999) 'Partnering Performance in Australia.' *Journal of Construction Procurement*. **5** (2): 163–176.

Wageman, R. (1997) 'Critical Success Factors for Creating Superb Self-Managing Teams.' *Organizational Dynamics*. **26** (1): 49–60.

Walker, D.H.T. (1994) *An Investigation Into Factors that Determine Building Construction Time Performance*. PhD, Department of Building and Construction Economics. Melbourne, RMIT University.

Walker, D.H.T. (1995) 'An Investigation Into Construction Time Performance.' *Construction Management and Economics*, **13** (3): 265–274.

Walker, D.H.T. (1997) 'Construction Time Performance and Traditional Versus Non-traditional Procurement Systems.' *Journal of Construction Procurement*. **3** (1): 42–55.

Walker, D.H.T. and Lloyd-Walker, B.M. (1999) Organisational Learning as a Vehicle for Improved Building Procurement. *Procurement Systems: A guide to best practice in construction*, eds Rowlinson, S. and McDermott, P. London, E&FN Spon, London, UK. **1**: 119–137.

Walker, D.H.T. and Sidwell, A.C. (1996) *Benchmarking Engineering and Construction: A Manual For Benchmarking Construction Time Performance*. Adelaide, Australia, Construction Industry Institute Australia.

Whiteley, A., McCabe, M. and Lawson, S. (1998) 'Trust and Communication Development Needs – An Australian Waterfront Study.' *Journal of Management Development.* **17** (6): 432–446.

Womack, J.P., Jones, D.T. and Roos, D. (1990) *The Machine that Changed the World – The Story of Lean Production.* New York, Harper Collins.

Yukl, G. (1988) *Leadership in Organisations.* Sydney, Prentice-Hall.

Chapter 8

Developing a Quality Culture – Project Alliancing Versus Business-as-Usual

Michael Keniger and Derek Walker

We argue in the introduction to this book that a quality culture is required to be adopted in its fullest sense if satisfactory project outcomes are to be delivered. This chapter addresses this vital issue. We view a quality culture from a broader perspective than just the delivery of a quality assured product or service. We step back and first ask the question who is being served by a project? Then we ask, how can this best occur and provide positive outcomes for most if not all project stakeholders? The immediate response to the first question for many projects would be the 'the paying customer', the client. However, the literature suggests sophisticated project initiators should and often do consult with a wide range of project stakeholders (Morris 1994; Briner, *et al.* 1997; Cleland 1999). It is generally at the briefing project stage that stakeholder interests are canvassed and addressed. These are strategic issues requiring careful needs analysis to be undertaken to: map stakeholders; determine their interests, needs and aspirations; develop a range of feasible solutions for testing; and gain commitment to proceed with a project strategy to fulfil the need, usually through delivering a physical product and/or service. A range of briefing and value analysis tools and techniques may be employed at this stage and these are adequately discussed elsewhere, see for example (Barrett and Stanley 1999; Smith and Jackson 2000). Further, when undertaking a product/service development project to introduce an innovation or to expand potential market reach, a high degree of customer focus and consultation is clearly evident from the literature (Bensaou and Earl 1998; Magretta 1998; Malone and Laubacher 1998; Prahlad and Ramaswamy 2000; Chase and Dasu 2001; Seybold 2001).

In Chapter 2 we saw that there are a wide range of project delivery choices. We believe that it is beyond the scope of this book to discuss in depth the briefing process beyond providing a short introduction to link the briefing process with development of quality culture when initiating a project.

8.1 Project briefing and development of a quality culture

Methodologies designed for producing a quality outcome when preparing a project brief are well advanced for construction and other project types. PRINCE, for example, is an IT-specific project management methodology initiated by the Central Computer and Telecommunications Agency (CCTA) in the UK. PRINCE-2 was developed from PRINCE with wider applicability to a range of project management applications including building and construction engineering. PRINCE2 identifies a start-up the project (SU) process and initiates the project (IP) process. It follows standard project management processes identified in the literature by authoritative authors of project management texts such as Morris 1994 and by experienced institutional bodies with extensive experience in developing detailed project initiation practices. The Royal Institute of British Architects Plan of Work, first published in 1967 and regularly updated since is one such example. These relate specifically to the initial stages of a project.

The project start-up process involves a project initiator obtaining a mandate to commence developing a strategy to meet a specific need (Bentley 1997, p. 19). This may range from a need for an improved service delivery (health, education, commercial, etc.) to the erection of a physical icon of national significance such as the National Museum of Australia. Often, a physical building may not be required because the identified need may be met by alternative means. For example, a virtual museum may have been one solution in which experiences of artefacts, information and access to learning about national treasures may have been developed. This would have been a purely information and communication technologies (ICT) based project solution. A more conventional design solution would be to design and build a museum, which would in all probability have ICT displays and exhibits accessible from within the museum and/or from the internet. A working party would be formed to investigate strategic options and to present a case for proceeding along a range of project delivery paths to establish why one particular path should be adopted rather than any other. This is the strategic needs analysis phase referred to earlier (Smith and Jackson 2000).

In a best practice guide for the traditional procurement approach, seven phases are identified for the briefing process. Phase 1 involves project initiation with a global brief being developed. In phase 2 a feasibility study is undertaken with a project feasibility report being developed. In the project definition phase (3), a basic brief is developed. This leads to the outline plan phase (4) where the feasibility of the basic brief is tested and an outline plan developed. If this is accepted then it forms a brief for the provisional plan phase (5). This plan is further developed and again tested and if successful moves towards phase 6, the definitive plan in which a definitive brief is developed. This, when further tested for feasibility, moves to a seventh phase, the final specification (Barrett and Stanley 1999, p. 2). The focus is generally on developing a solution that meets fitness for purpose (quality) at a feasible cost. In construction industry sectors influenced by the UK approach, quantity surveyors generally undertake feasibility testing in terms of cost while in USA influenced construction industry sectors cost-engineers are used. Under either approach the process is generally rigorous with well-defined procedures and a long history of professional expertise available to assist project initiators. There are no shortages of guides and texts that may help when following the traditional approach to project brief development, see, for example, Salisbury (1990, Chapter 3).

The traditional procurement approach, as we saw in Chapter 2, is often considered slow in realising construction projects. It has also been criticised for inhibiting the possibility for the construction methods experts to provide advice on constructability or buildability (Francis and Sidwell 1996; Sidwell and Mehrtens 1996; Griffiths and Sidwell 1997). Similarly, the traditional briefing approach has also been criticised for deficiencies in practice in the engagement of consultants and the ability of the client and design consultant to develop an effective brief that meets the needs of a wider community of interest in project outcomes. Human communication factors and lack of commitment by project initiation parties to explore and develop options adequately has been cited (Barrett and Stanley 1999, pp. 6–7). After analysing five case studies in depth, Barrett and Stanley developed two major solution strategies (empowering the client and managing the project dynamics) and three supporting strategies (appropriate user involvement, appropriate visualisation techniques and appropriate team building) (Barrett and Stanley 1999, p. 18).

The interesting aspect of the work of Barrett and Stanley (1999) is that the five identified solutions appear to be better addressed using a relationship-based approach to project procurement. Also, they are pertinent to developing a quality culture rather than a mechanistic tick-the-box mentality to producing work of an acceptable quality. They illustrate how their five solutions may be implemented by using a matrix to describe a contingency-based brief strategy. This scales support for people along one axis against support for technical knowledge along another axis. They also integrate a third organisational maturity dimension for the project initiator (Barrett and Stanley 1999, p. 29).

Barrett and Stanley's idea is adapted from Hersey and Blanchard's model of four leadership styles (Hersey *et al.* 1996). If we take their model but look at it with a traditional to relationship-based continuum perspective, we get interesting insights into how a relationship-based approach to briefing may create a quality culture that can be described from two extreme ends of that spectrum.

The 'instructing **S1**' style is the least relationship-based solution building option though it potentially needs high levels of skill in briefing ability and willingness that are unlikely to be evident. The project initiator knows exactly what is technically needed and simply instructs that a particular building be designed and built with precise project outcomes in mind. The project initiator will, however, either have underdeveloped briefing skills or have little willingness to be engaged in a briefing process beyond the barest minimum involvement. This can be effective for simple projects where the needs, technologies and construction approaches are easily understood. Many traditional procurement projects may fall into this category. A quality culture requirement should be part of the brief.

The 'cooperating in design brief development **S2**' style may be appropriate when the project initiator and the design team share the technical knowledge of the end product and project objectives so that working together presents real possibilities for improving the project outcome. In this style the project initiator would also have both some degree of ability to brief effectively and some willingness to do so expressed by at least occasionally making a commitment to participate actively in the briefing process. This style is appropriate for a project that is complex and requires a high level of solution building and there would be real potential for developing a relationship-based procurement approach. These types of project (often characterised by complexity in technology, scope, scale or rapid delivery) lend themselves to alliancing arrangements. A quality culture can naturally evolve through the design team and project initiator working together in joint problem solving and learning activities. If the design team or project initiator does not intervene with a facilitation mechanism, then serious problems in developing a satisfactory quality project outcome may well emerge.

The 'collaborative **S3**' style is one where the project initiator needs a higher degree of guidance through people involvement in unearthing their tacit knowledge of detailed technical matters to develop a design brief effectively. Project initiators would have high skills in working with various stakeholders and/or design specialists to build design solutions through the briefing process. The project initiator would have good skills and willingness to participate fully in the briefing process. These types of project may also benefit from an alliancing arrangement. A quality culture can also naturally evolve through the design team and project initiator working together in joint problem-solving activity and should be formally factored into the relationship to compensate for lower levels of technical knowledge support given by the project initiator.

The 'delegating **S4**' style is relevant for simple projects where the needs, technologies and construction approaches are easily understood. The project initiator would have high levels of skill and willingness to brief the design team and may well have standard procedures (including TQM requirements) but because of the nature of the project being technically simple and performance standards well understood, it will only a be necessary to provide a 'watching brief' in order to ensure that all parties are competent.

The Barrett and Stanley (1999, p. 29) model indicates that the briefing process is mainly about gaining joint understanding of how project objectives/requirements may be realised. The project initiator is primarily responsible for clearly communicating project objectives and providing guidance in terms of technical matters (usually through other experts). This will generally involve two types of difficulty. First, it is unlikely that the project initiator can be fully effective in explicitly diffusing technical knowledge about the desired project outcome to those being briefed. A certain amount of tacit knowledge, that which is internalised and difficult to explain or communicate (Nonaka and Takeuchi 1995; Davenport and Prusak 2000), will need to be demonstrated or shared through an experience of socialisation such as joint problem solving/sharing perspectives (Nonaka *et al.* 2001) so that aspects of a brief's relevance, context and importance are clearly understood. Only then can this tacit to explicit knowledge conversion cycle be complemented by combining explicit knowledge with existing documentation, procedures, and shared institutional knowledge (Nonaka and Takeuchi 1995, p. 52).

Much of the guidance given in texts about briefing is concerned with checklists and more easily identified building elements, for example Salisbury (1990), however Barrett and Stanley (1999, p. 86) clearly argue that visualisation is a fundamental part of the briefing process. They argue that visualisation techniques should be employed to increase the potential for shared understanding, adequately resourced and be used effectively. Scaled models and prototypes, while generally effective are time consuming to make, expensive, and not easy to modify and change rapidly. Fortunately computer-based visualisation and virtual reality (VR) technologies are well advanced with many applications readily available such as fly-through or walk-through 3D-CAD packages, simulation tools and animation software that can help people experience virtual worlds (Retik and Hay 1994; Fisher *et al.* 1997). The advent of knowledge management tools, where these are effectively used, also makes the process of accessing explicit knowledge less daunting. Knowledge mapping is one such tool that numerous companies are now developing. These are guides rather than repositories (though they may link information bases) and allow users to track down lessons learned from past experiences, link to particular people with specific expertise (such as British Petroleum's Virtual Teamwork), or learn how some particular process works (Davenport and Prusak 2000, p. 72).

In taking part in the briefing process it is clear that a quality culture lies beyond adherence to or mastery of any specific quality standard. Quality standards such as ISO9000 or ISO14000 are more concerned with compliance rather than commitment. Commitment is present with these standards to the extent that they seek to ensure that plans are made that assure a quality process is planned for, implemented and checked and verified (compliance). However, none of these standards assure that relationships between project teams are of a sufficient quality to ensure that shared mental models match expectations of the facility to be built or that a high level of cooperation will prevail to overcome problems and to ensure satisfaction with the realised project. We perceive that a quality culture encompasses the following:

- Systemising quality and ensuring its integration into all aspects of design and production to deliver fitness for purpose;
- Driving out waste through simplification of processes, minimising or eliminating unnecessary redundancy (except where required for safety and security) and enhancing environmental sustainability of projects through concentrating upon whole-of-life issues;
- Developing the quality of trust and commitment in people-to-people relationships to minimise wasteful expenditure of creative energy in defensive routines and negative conflict resolution techniques;
- Intelligently using information and communication technologies to facilitate effective understanding between project teams of their varying perspectives and to allow relevant lessons learned to be internalised to promote innovation and continuous improvement;
- Coordinating the aspirations, needs and requirements of a broader base of project stakeholders than the paying client or the project teams so that a win–win environment is more widely achieved;
- Designing a project brief, project team selection process, rewards and penalty system, project monitoring and influencing system that effectively addresses the above points.

It can be seen from the last point that a key inhibitor to achieving a quality culture is failure to design a framework in which it can flourish. When a procurement system is used that encourages conflict, distrust, win–lose behaviours and lowest cost rather than best value aspirations it should be not surprising that a quality culture as described above fails to take root.

The Barrett and Stanley (1999, p. 29) model provided a useful framework to link the project initiator's knowledge of technical matters and the likely quality and degree of briefing interaction with people involved in designing solutions. For simple projects, technical knowledge to develop effective design solutions requires low levels of people involvement because technical details are either well understood or, at least, well documented. Most projects, however, are far from simple and so it is necessary to involve people who can generate and build solutions more closely. They include not only the 'client' or 'design team' but also other stakeholders such as those in the project supply chain, users and others affected by the project outcome.

Developing and building a brief that provides 'project understanding' can be supported by empowering stakeholders by providing an organisational framework that gives them a voice and means of expression, providing tools to visualise outcomes and developing a project team culture that recognises the need for an integrated quality management system spanning fitness-for-use, stakeholder safety, environmental sustainability and relationships between stakeholders that promote win–win approaches not only to project team members but also for the project – a best-for-project approach.

The next section in this chapter describes how a case study project using a relationship-based procurement system (project alliancing) was able to achieve many of the quality culture elements described above.

8.2 The National Museum of Australia – a case study[1] in developing a quality culture to match stakeholder needs

The general motivation for choosing a project alliancing approach is explained in Chapter 4 section 4.1.4 and section 4.2 provides a detailed description of the selection process. In this section we trace the history of how the National Museum of Australia project brief was developed and how this led to a set of quality measures and a monitoring and evaluation system that facilitated a quality culture.

8.2.1 History of the selection of the project team

Although first mooted in the early seventies the National Museum of Australia was not established until 1980. For a variety of reasons a suitable building for the Museum did not receive sufficient political support to enable it become a reality until the mid-1990s. This extensive delay was partly because a history of major government projects experiencing extensive cost and time over-runs helped to dissuade successive governments from proceeding with the project. Having made this decision, the further decision was made to complete the Museum in time for the centenary of federation celebrations, which severely constrained the programme for design, procurement and construction. Table 8.1 illustrates the pre-construction history of the project.

The necessity to deliver a high quality, public building was understood from the start of the project. Similarly, the need to hold costs within the limits of the

---

[1] We would like to acknowledge that this section is substantially based upon the *Interim Report of the Independent Quality Panel for the Acton Peninsula Project* at handover on 11 March 2001. Elements of that report are included within this text.

Table 8.1: The timeline to establish the project alliance for the Acton Peninsula Project

INCEPTION	**1980**		National Museum of Australia (NMA) established under *National Museum of Australia Act 1980*
	1996	**16 August**	The Minister for Communications and the Arts, Senator Richard Alston appoints Advisory Committee on New Facilities for NMA and the Australian Institute of Aboriginal and Torres Strait Islander Studies (AIATSIS)
		December	Advisory Committee – gives recommendations to Senator Richard Alston – Acton Peninsula site is the preferred site
DESIGN COMPETITION	**1997**	**6 June**	Competition launched for National Museum of Australia by Prime Minister of Australia John Howard and the Minister for Communications and the Arts, Senator Richard Alston
		2 July	Michael Keniger and John Davidson AM appointed as advisers to the design competition
		31 July	Stage One competition entries received (extended from 10 July 1997)
		8 August	Technical assessment of Stage One entries
		11 August	5 short-listed entries asked to provide more information by 15 August
		19 August	The 5 short-listed entries announced by the Minister for Communications and the Arts, Senator Richard Alston
		22 August	Stage Two began with issue of vol. 1 of brief
		27 August	First briefing session
		9 September	Stage Two issue of vol. 2 of brief
		15–16 September	Second briefing session
		10 October	Stage Two entries received
		13–25 October	Stage Two entries on display – by appointment only
		23–25 October	Stage Two, design presentation
		29 October	**The Winners announced by the Minister for Communications and the Arts, Senator Richard Alston**
		15 November	Competition concept design on display
		8–9 December	Public Works Committee meeting
SELECTION – BUILDING AND SERVICES CONTRACTORS	**1998**	16–18 May	Call for proposals in 5 newspapers and Telstra Transigo
		22 May	Industry briefing
		29 June	**Receive proposal**
		21–22 July	Nominate initial short list of proponents
		28–30 July	Interview process
		31 July, 3 Aug, 7 Aug	Short-listed proponents notified on separate dates
		4–5 Aug, 7–8 and Aug, 11-12	2-day workshop with each short-listed proponent
		13 August	**Preferred allliance team notified**
		17–18 August	Risk–reward structure determined with preferred alliance team
		21 August	Preferred alliance approved
		22 August	Alliance agreement executed
	1999	15 April	Business-as-usual (BAU) estimate established – to be known as target out-turn cost (TOC)

Source: Hampson *et al.* (2001, p. 11)

approved budget was paramount as was the achievement of the target opening date at the beginning of the Federation Centenary in January 2001 (later altered to March 2001). These requirements were to be met by a client agency, the Department of Communication, Information Technology and the Arts (DoCITA), that had little experience of major public capital works. It was decided that a design competition should proceed and initially procurement options were left open for the construction phase although the brief for the competition mentioned that the appointed architect would be expected to work with an innovative form of contract. The inexperience of the client group was revealed by the decision to launch the competition without the full ratification and endorsement of the Royal Australian Institute of Architects (RAIA). The opposition of that body brought adjustments to the competition conditions and advisers nominated by the RAIA were appointed to assist the selection panel. Following detailed scrutiny of the 76 entries, five were short-listed for the second stage of the competition with Ashton, Raggatt, McDougal in association with Robert Peck von Hartel and the landscape architects Richard Weller and Tom Sitta being declared the winners and so recommended to be commissioned for the design of the Acton Peninsular Project.

The outcome of the competition was subject to a penetrating examination by the Parliamentary Public Works Committee, which undertook a public investigation of all aspects of the project and of the competition. The inquiry ranged across the brief, the site selection, the budget and the anticipated cost of the selected scheme, the proposed form of contract, the expertise of the advisers and the nature of the winning scheme. In part, this review was necessary to reassure government that the lead agency, the Department of Communication, Information Technology and the Arts (DoCITA) could deliver such a major project of national significance on time and on budget. This caution was due to that department's lack of experience and expertise in managing complex capital works projects.

The outcome of the inquiry was to confirm the selected scheme as the scheme that should proceed, that the budget for the project was fixed, that the programme for construction was also fixed and that project alliancing could be adopted as the contract form. The inquiry also confirmed the need to sustain the integrity of the selected design throughout the construction process together with a further expectation that the quality of the scheme should be sustained throughout all stages. Essentially, the deliberations and recommendations of this enquiry gave strength to the creation of a culture of quality for the Acton Peninsula project.

The adoption of project alliancing as the contract form was a courageous decision for the DoCITA team, led by Dawn Casey, which had so little experience of construction. The collaborative arrangements between the principal parties underpinned by a 'no blame, no fault' agreement involved many risks. Alliancing had been used previously for major infrastructure projects but this was to be the first use of the procedure for a building construction contract. A more cautious appraisal would have reasoned that the Acton project was not the ideal vehicle for such an experiment because of its intricacy and the complexity of client needs that together amplified the risks. In hindsight, it is unlikely that the project would have been completed on time, on budget and to such a high standard under a more conventional form of contract. Virtually all of the alliance procedures had to be defined and generated for the specific needs of the project and a culture of collaboration had to be fostered amongst the members of the alliance and amongst the construction team on site. In particular, the procedures for the sustenance of design integrity and the achievement of built quality had to be determined and articulated in a robust and sustainable form.

The organisational structure of the project delivery team illustrates how the order and pattern of communication and responsibility was designed to facilitate a culture of quality. Earlier in this chapter we stressed the need, on complex projects in particular, for close stakeholder involvement in shaping a project brief and actively contributing to the decision-making process. We also stressed that as many stakeholder groups as is practicable that will influence project success should also be involved in effective policy and strategic decision making. Further, a quality culture will best be developed when a win–win system of rewards is in place to align team goals with project goals. This strategic approach to developing a quality culture should be reflected in the project governance arrangements. The National Museum of Australia delivered this strategy through its linkage of the major end-user stakeholders to the alliance leadership team and so through the project management team to the construction coordination committee. This relationship map is illustrated in Figure 8.1.

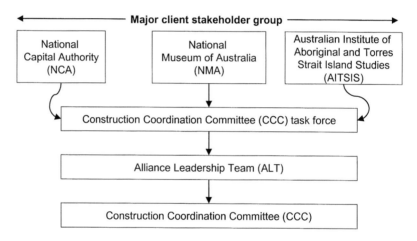

Figure 8.1: Management structure of the Acton Peninsula project alliance
Source: Hampson *et al.* (2001, p. 20)

Pre-agreed commercial outcomes for all alliance parties follow the success of the project as measured against project objectives. The inherent logic in the risk/reward structure was:

Outstanding performance = Outstanding return
Normal performance   = Normal return
Poor performance   = Poor return

The driver for using a project alliance on the project was to maximise the value for money within available funds rather than to reduce total cost, which is a primary goal of many projects. As there is a potential monetary reward in an alliance agreement for cost savings, a corresponding risk/reward measure was needed to ensure that costs were not minimised at the expense of the design integrity or quality. This was vital in securing the support of the project architects to enter into an alliance with the building and services contractors and provides a unique dimension to the delivery of this project.

Once the Public Works Committee had endorsed the employment of project alliancing the Construction Coordination Committee (CCC) invited proposals from building and services contractors to select the most capable team to contribute to the design and to execute the construction works within an alliance framework. Following an intensive process of submission, interview and a two-day workshop with each short-listed team, the preferred consortium was invited to attend a risk/reward workshop led by the alliance facilitators and legal advisors. Also in attendance were representatives of both the Commonwealth of Australia and Australian Capital Territory (ACT) governments and the successful project architects.

The purpose of the workshop focused on the terms and conditions of the alliance agreement, which included agreeing the profit and overhead percentages and risk/reward structure for all parties. It was agreed that the Acton Peninsula alliance agreement would have four risk/reward components: cost; time; design integrity; and quality. Quality would be interpreted broadly to encourage excellence in the following ten aspects of the project:

(1) Workmanship/finishes
(2) Environment/ESD
(3) Safety
(4) Lifecycle management
(5) User satisfaction
(6) Public relations
(7) Industry and public recognition
(8) IT leadership
(9) Indigenous employment opportunities
(10) Involvement of small/medium enterprises

Up to A$2 million would be payable to the non-Government members for outstanding quality, and a penalty of up to A$2 million payable for poor quality. There was no existing pool of money to reward outstanding quality. It was anticipated that savings would be achieved through innovation in the value management stage of the project, and that the Commonwealth would put such savings into a 'quality pool' to encourage the achievement of excellence. The sustenance of design integrity was agreed to be a crucial objective of the project and that its importance would be underscored by the removal of all profit should it not be achieved. The definition of design integrity and the monitoring of its achievement was undertaken by a panel chaired by Steve Ashton of Ashton, Raggatt, McDougal that included independent members appointed for their expertise in design evaluation.

The alliance agreement was signed within a week of the risk/reward workshop Its conditions required the alliance leadership team (ALT) to develop quality performance levels corresponding to 'business-as-usual', 'outstanding' and 'poor' together with the methodology for measurement. It was agreed that the ALT would be assisted by an independent panel to assist it in verifying and monitoring the performance levels and measurement methodology.

The scope of work increased to include the exhibition design, fabrication and installation and the exhibition designers, Anway & Company, were selected in September 1998 and invited to join the alliance. Accordingly, the risk/reward arrangements were renegotiated and the quality pool increased to A$3 million (commensurate with the increase in scope) whereas the penalty amount remained as A$2 million. A revised alliance agreement added the exhibition designers and further amended the risk/reward arrangements by creating a quality pool funded by the Government through the Department of Communication, Information Technology and the Arts (DoCITA).

8.2.2   Alliance leadership team and internal quality working group development of a quality culture

At its first meeting, held on 31 August 1998, the alliance leadership team (ALT) developed a charter for the alliance. The preamble reflected the quality aspirations for the project:

> *This Alliance will create an exceptional Australian cultural precinct on Acton Peninsula, Canberra. The project is the flagship for the Centenary of Federation and will be a source of pride for all Australians. The way we go about it will lead the way for construction projects in the future.*

At its second meeting the ALT established an internal quality working group to commence the task of defining quality measures and measurement methodology. The working group included members of the ALT and the project management team – the senior full-time project personnel. The primary role of the working group was to develop draft quality measures to provide a starting point for discussion with potential members of the independent quality panel. There were no known precedents available to the project team for defining and measuring quality on a complex building and landscape project such as the Acton Peninsula project. Advice and assistance on methodology was obtained from Sydney Water, based on its experience with Australia's first public sector alliance – the Northside Storage Tunnel project. But it was self-evident that many of the measures for a subterranean infrastructure project would be of little direct relevance to the museum.

It was left to the working group members to define what would be an outstanding result in their own fields of endeavour. This had the double benefit of focusing the team on quality issues from the outset of the project, and creating a sense of ownership of the measures and processes. The project team created its own vision of what it wanted to achieve through this process. This approach is characteristic of a quality culture rather than an ethos of quality control.

The working group proposed, and the ALT agreed, that the initial list of ten quality benchmarks be consolidated to six. By December 1998 a working draft of the quality measures was taking shape, and potential names for the independent quality panel were being discussed. In January 1999 the ALT endorsed a framework for the Independent quality panel that set out the proposed role and operation of the panel.

The draft quality measures were circulated for the ALT meeting on 18 February 1999, but not discussed. The February and March 1999 ALT meetings were dominated by intensive discussion and negotiation of the target out-turn cost (TOC) estimate for the building and external works. The TOC is the cost benchmark against which under-runs or over-runs are shared on pre-agreed percentages.

Reaching agreement on the target out-turn cost proved time consuming because initial estimates were higher than anticipated, and there were differences of estimates between estimators. While a saving of A$4.8 million was achieved through project engineering, this was more than offset by increased costs attributable to further knowledge of project complexity, omissions in earlier estimates, or market prices being higher than previously expected. The result was insufficient savings being generated for the alliance to create a quality pool. This demonstrates the difficulties involved in developing a feasible incentives scheme to encourage the achievement of excellence.

The ALT recognised that to proceed without a quality pool would create the wrong commercial drivers, as the only means of enhancing profit would be through cost reduction with no counter balancing quality incentive. Consequently, DoCITA agreed to fund the quality pool of A$3 million from its client contingency with the proviso that the first call on the quality pool was to fund the Commonwealth's share of any cost overruns. Under this arrangement the quality pool served a dual purpose – as a buffer for the Commonwealth's financial risk, and a positive incentive to encourage outstanding performance. DoCITA was both a 'client' and a project alliance partner on behalf of government, thus it needed to develop a transparent and independent quality management system. The joint effort exhibited by all parties in collaborating to develop an equitable and workable risk/reward system was further evidence of an emerging quality culture.

8.2.3   The role of the independent quality panel in developing a quality culture

In mid-April 1999 the ALT agreed that the draft quality measures were ready to be given to the independent quality panel (the panel) comprising:

- Professor Michael Keniger, Professor of Architecture, Head of School of Architecture and Planning and Head of the Department of Architecture at the University of Queensland. He had previously been involved in the project as the architectural adviser to the construction coordinating committee (CCC) during the design competition.
- Mr Leon Paroissien, former Director, Museum of Contemporary Art, Sydney, and editor of contemporary arts journals. He had previously been a member of the Evaluation Committee for the selection of the exhibition design team.
- Mr Ron Black was appointed in July 1999 as the third independent member of the panel. Mr Black is a management consultant and a former General Manager of ACT Capital Works and a senior executive in the National Capital Development Commission.

The panel was empowered to recommend secondment of additional specialist advice, as considered appropriate. The purpose of the panel was to provide the

ALT with independent expertise and accountability for the measurement of quality. Specifically, the panel:

- Provided independent verification of the benchmarks, including any subsequent changes should these prove to be necessary;
- Provided independent verification of the procedures for measuring the benchmarks, including auditing arrangements;
- Provided a final report to the ALT recommending the final total quality score; and
- Provided ongoing advice throughout the project duration on issues relating to quality and to assist the project team in obtaining the highest possible quality score.

The panel provided a recommendation, together with its reasons, of the final total quality score. The ALT was responsible for the final determination, although given the independent expertise of the panel; the ALT needed clear, demonstrable and defensible grounds to amend the final total quality score. The alliance's probity auditor provided oversight of the process as and when required.

The panel's first task was to comment on the draft quality measures developed by an internal alliance working group. These were revised to incorporate the panel's comments and then resubmitted to, and endorsed by, the ALT in June 1999. The panel held its first formal meeting on 25 June 1999 to discuss its proposed method of operation. From this point on, the panel became the forum for further refinement of quality measures and processes and the internal alliance working group ceased to operate.

Project management team members and other key project personnel (architects and designers) attended the panel's meetings and scoring sessions. This further reinforced a quality culture through joint problem solving and shared agreement of what quality meant and how it could be achieved. This culture was further amplified by meeting with the key site staff to discuss quality measures and their implementation. This helped to reduce any sense that the quality panel was operating merely as an inspectorate.

Over the first six months of the panel's operation the systems and processes for measurement were developed for which there was virtually no precedent. The process for measuring quality is discussed further in section 5.2.4 together with the quality measures and the scoring process developed and used by the panel. The panel visited comparable public buildings in Canberra and museum exhibitions in Sydney and the new Museum of Melbourne to establish what would constitute 'poor', business-as-usual' and 'outstanding' results. This enabled the panel to determine and articulate quality standards and benchmarks whilst defining, for the project management team, critical success factors and issues requiring close attention.

As progress with construction and fit-out advanced, on-site scoring commenced. The panel also provided advice by identifying areas where it thought it would be possible to increase the quality score by addressing the issues it bought to the attention of the project management team. Examples of key issues raised by the panel during on site inspections, and the alliance response are provided in Table 8.2.

Table 8.2: Examples of quality issues raised by the quality panel during on-site inspections and the alliance response

Issue raised/needing attention	Outcome
Administration area – steelwork in the 'knot' area assessed at below BAU.	Lessons learnt led to substantial improvements in subsequent areas
Use of plasterboard – need for protection of corners	Protective strips used
Temporary exhibitions gallery – visible joints in plasterboard	Rectified by use of textured paint and lighting solutions
Resolution of the window treatments and effect on visitor experience. The high lux levels are a concern for conservation of objects, but external views should not be lost	Windows treated with Helio screen to reduce lux levels but still maintain external views
GFA – Low light fittings on bulk-head at GFA upper may be touched by visitors. Heat emanating from the bulbs was a concern	Replaced with recessed fittings
Circa will need well-trained operators and documented operating procedures to ensure safety of operations. Crowd management and the management of queueing will be critical	Manuals and training provided
Legibility of text – some graphic labels are difficult to read. Low light levels in some areas exacerbated the problem	Graphic labels reproduced with increased font size. Lighting refocused and/or light levels increased where possible within conservation constraints
Signage – inadequate guidance for orientation and wayfinding	Study commissioned to redo external signage and internal wayfinding
GOAD – trip hazards are an OH&S concern	Rectified
Hoffman Room – moving parts are a safety issue	Moving parts secured
GOAD – difficult for visitors to find the exit	Directional banner on flagpole installed. Monitoring need for permanent architectural solution

This list is not intended to be exhaustive, but to illustrate the positive impact of the process on the final project outcome. It also indicates how the broader elements of aesthetic design, functionality and safety were viewed and addressed. This demonstrates a quality culture linking design and construction improvement through an effective feedback mechanism.

As the panel explicitly noted, the quality measurement benchmarks were developed concurrently with the early phase of the project. As the project progressed, the measures and processes were refined and amended in the light of experience of their operation. For example, the ALT agreed to a request from the Chair of the Museum Council to delete the measure of the cultural balance of the exhibitions, because exhibition content was more properly the responsibility of the Museum Council, not the alliance. This ability to refine and adjust the measures further indicated the presence of a quality culture.

The ALT also agreed to recast the public relations measure. On reflection, it was considered that promoting the museum to a national audience and promoting tourism are the responsibility of the museum and the relevant tourism authorities, and well beyond the scope and resources of the alliance. The fifth quality measure was renamed 'Public and Industry Recognition'. It retained an incentive for the alliance to actively participate in activities that promoted the site (through open days, tours, etc.), and to promote the project and project alliancing within the construction industry, but not to perform the core PR function of the museum. A new component was added to the first quality measure (see page 213) (quality of built finishes) in October 1999 addressing the management of defects during the defects liability period. The process of measurement was progressively developed and refined with use, however, the integrity of what these measures aimed to achieve remained unchanged with the ALT and the independent quality panel discussing, clarifying and interpreting measures from time to time. Given the lack of precedent and pioneering nature of some of the measures, the quality measures and processes stood up remarkably well over the life of the project with relatively few (and largely minor) amendments.

The alliance recognised that the ambition to achieve excellence could only be achieved with the mutual support and commitment of all the subcontractors and employees on site. Therefore the alliance entered into an innovative project agreement with the ACT Building Trades Group of Unions to encourage and reward the achievement of excellence in the areas of: workmanship and finishes; safety; environment; workplace relations; and programme. The performance measures for workmanship and finishes were tailored to each trade and reflected in the remuneration of the employees.

The maximum available allowance was A\$1.75 per hour for a score of 100 per cent. A score of 75 per cent, for example, would result in a payment of A\$1.31 per hour (i.e. A\$1.75 × 0.75). Performance was assessed and payment made quarterly in arrears. The project agreement was established and managed as a separate system from the alliance quality measures with the independent quality panel not being involved, although many of the measures were consistent. The project agreement extended a key principle of the alliance agreement, that rewards are linked to project results, down to individual employees. The project agreement engendered a positive workplace culture. With over 1.7 million hours worked on the project, there was no time lost due to site-related industrial issues.

A performance audit by the Australian National Audit Office (Auditor-General of the Australian National Audit Office 2000) reviewed the quality management of the project and concluded that:

> *The ANAO supports the establishment of a system to assess/measure the quality of the project and considers that the quality measures and incentives, which are quite innovative for the construction industry generally, have the potential to achieve a sound quality result.*

8.2.4 The quality measurement (QM) process adopted on the National Museum of Australia project

Quality has proved to be a property that is hard to define and even harder to measure – especially in the construction industry. Most evaluations of quality confine themselves to the definition and tracking of processes, which are intended to ensure that quality, is achieved. This is essentially the basis of quality assurance (QA), which most accredited consultants, contractors and subcontractors hold in some form or another. The 'quality' sought for the Acton Peninsula project is possibly best defined as 'the relative nature or standard of something, the degree of excellence possessed by a thing' (OED). This definition supersedes both conventional QA expectations and the conventional contract expectations of achieving good practice as determined by compliance with the specification and with Australian Standards and statutory codes.

The sense of quality sought was that which is appropriate to the flagship project of Australia's celebration of the Centenary of Federation and to be appreciated as such by the public at large. Essentially, the completed project was intended to be of such a level of excellence as appropriate for the National Museum Australia, a leading public institution, and the process of construction was also to embody a similar quality. Various strategies were employed to encourage the achievement of quality as summarised in the quality framework and as endorsed by the ALT:

The monetary incentives for quality apply primarily to the workmanship of the buildings and presentation of the exhibitions. Quality will however also be measured in relation to aspects of the construction phase of the project.

Quality on the Acton Peninsula project was not an absolute measure viewed in isolation from the project circumstances. For example, more expensive building materials were not automatically assumed to lead to a higher quality score than less expensive materials. The philosophy of quality on the project considered the appropriateness of decisions within the broader cost, time, design integrity and quality objectives of the project as determined and agreed by the members of the alliance. This again demonstrates a proactive and project focused quality culture.

The quality measurement (QM) process established for the Acton Peninsula project required the clients and their alliance partners to explicitly articulate their quality objectives for the project. This exemplified a true quality culture fully supported by project leaders and the adoption of these objectives helped to develop quality measures that allowed those at the workface to visualise and attain the desired standards.

In order for the approach adopted by the alliance to be actuated it was necessary for quality measures and processes to be developed in relation to the key factors involved in a major construction project that included:

- Inputs to the process (typically, efficiency measures).
- The process itself.
- Outputs (i.e. quality of the finished project).
- Outcomes, (i.e. the impact of the project and the extent to which it achieves its objectives).

The quality measures developed for the Acton Peninsula project were necessarily broad and divided into six categories:

QM#1 – Buildings, landscape and site works.
QM#2 – Exhibitions.
QM#3 – Environment.
QM#4 – Indigenous employment opportunities.
QM#5 – Public and industry recognition.
QM#6 – Safety.

It is now relatively common for public bodies to specify objectives beyond those of time, cost and quality of product, such as employment of local labour, environmental protection, or knowledge transfer. 'Public and industry recognition' and 'indigenous employment opportunities', however, are innovative objectives, which extend the responsibilities of a major government project.

In order to be able to establish a transparent and objective process of measurement a mix of process, output and outcome was used as the basis of evaluation. Input/efficiency measures were not included as the cost risk/reward structure already provided a strong financial incentive for alliance participants to act efficiently.

Effectively, QM#1 – QM#2 were primarily concerned with outputs (e.g. quality of workmanship) and outcomes (e.g. visitor satisfaction). QM#3 – QM#6 were primarily concerned with the processes of construction. Further, QM#1 – QM#2 were measures of the primary project objectives – derived from the core business of the end-users – whereas, QM#3 – QM#6 tended to be secondary objectives derived from broader Commonwealth policy or industry or community expectations.

As noted earlier, the alliance agreement required the ALT to develop performance standards corresponding to 'business-as-usual (BAU)', 'outstanding' and 'poor'. A quality reward would only be paid for performance that exceeded the BAU performance of the alliance participants. To establish the BAU it was necessary to establish benchmarks that would provide an agreed reference point for all parties to the process. These were developed from:

- historical performance within organisations;
- continuous improvement during the project;
- comparisons with other projects or organisations; and/or
- industry-wide benchmarks.

The initial benchmarks were established by the alliance's internal quality working group that referred to the knowledge and experience of conventional standards of performance drawn from their own organisations; by comparison with other projects; and by referring to industry-wide statistics as appropriate.

These fledgling standards had to be accompanied by processes to encourage the achievement of quality and to facilitate its measurement through all levels of the project structure. The detailed structure of the quality measurement system had to be developed in its entirety within a very short timeframe. Effectively, much of the innovative and unprecedented system had to be developed over the early months of the contract and whilst the quality culture itself was being fostered on site.

In respect of the first point, the system that was developed covered to an extent, design detailing, as part of the workmanship of buildings, and covered both the design and construction of exhibitions.

The principal elements of the QM system included:

- an assessment group for each QM;
- evaluation tools and techniques appropriate to each QM;
- working papers that clarified and expanded the framework document;

- benchmarking against relevant buildings and building elements and exhibitions;
- score sheets; and
- quality records.

Each of these elements represented many aspects of the overall project and were complex in their internal relationships. To assist with measurement they were divided into components and subcomponents, which could be more closely defined. Table 8.3 sets out the principal structure of the six areas of measurement and their internal structure. It was recognised that some of the components were more fundamental in contributing to the meeting of the project's objectives either because of their nature their scale or their eventual public presence. As financial incentives were involved in the achievement of excellence a series of weightings

Table 8.3: Quality measure weightings

QM#	Measures	Components		Weightings
1	Buildings and Siteworks			30
		1.1 Quality of built finishes	18	
		1.2 Non-conformances	6	
		1.3 Defects	6	
2	Exhibitions			30
		2.1 Design quality	9	
		2.2 Use of content	9	
		2.3 Integration of technology	3	
		2.4 Accessibility	3	
		2.5 Visitor experience	6	
3	Environment			10
		3.1 Environmental management	1	
		3.2 Waste management	2	
		3.3 Water quality	2	
		3.4 Air quality	1	
		3.5 Energy efficiency/LCC	3	
		3.6 Ecologically sustainable development	1	
4	Indigenous employment opportunities			10
		4.1 Enhancing opportunities in construction period	1.4	
		4.2 Enhancing opportunities beyond construction period	1.4	
		4.3 Training	2.9	
		4.4 Employment	2.9	
		4.5 Supportive workplace	1.4	
5	Public and industry recognition			10
		5.1 Promoting the site	2	
		5.2 Industry recognition of alliancing	4	
		5.3 Stakeholder image	4	
6	Safety			10
		6.1 Management processes	1.7	
		6.2 Safety outcome	5	
		6.3 Individual intention	3.3	
	TOTAL		100	100

were developed to reflect the contribution of each component to the overall quality score.

The subdivisions and their respective weightings could be varied to suit any individual project and the quality panel came to the eventual view that the weightings of the two core measures, QM#1 and QM#2, should have been greater given their impact on the public presence of the project.

A 21-point measurement scale from −10 (poor) to +10 (outstanding) was adopted, with 0 representing business-as-usual. As the scale was to be used for widely differing elements and activities it was accompanied by brief descriptors to aid with interpretation as given in Table 8.4.

Table 8.4: Quality management measurement scale, the 'tape measure'	
Score	**Description**
+10	Outstanding in all respects
+8	Outstanding in almost all respects, particularly the key components. Very good in the remainder
+6	Very good in most respects. Balance are either outstanding or good
+4	Varies between good and very good
+2	Varies between BAU and good
0	Business as Usual (BAU) – commonly achieved or typical for large government projects. Meets or exceeds specifications
−2	Just acceptable. Meets specifications in almost all areas, with the balance either just below or BAU
−4	Just fails. Meets specifications in many areas, with balance being just below specification
−6	Significant failures. Fails in many areas, with balance being either side
−8	Not acceptable, or in limited instances accepted at a significantly reduced valuation
−10	Poor – unsatisfactory, unsafe or unsound. Fails significantly to meet the objectives. Needs removal or replacement

The 21-point scale was referred to as the 'tape measure' and its use entailed the detailed discussion of a range of variables to match both the nature of the element being measured and circumstance in terms of time, budget, or degree of difficulty.

Again, in retrospect, the panel considered this scale to be too finely grained. In practice, coarser grained divisions were found to be more useful that related to the 5-descriptor bands above and below BAU.

Those involved in the measurement process included:

- the project management team which assisted with the generation and collation of statistical information, such as safety statistics and employment data;
- external technical experts – where specialist expertise was required to measure or validate results, such as measuring water quality; and
- the independent quality panel – particularly for qualitative measures where judgment was required, such as quality of built finishes and exhibition design.

Assessment groups were established for each measure, typically comprising a member or members of the panel and relevant senior personnel from the project management team. It was originally intended that the assessment groups would develop detailed procedures to ensure effective and consistent scoring and would be responsible for the initial scoring of performance against the quality measures, with verification given by the independent quality panel. In practice, the on-site scoring was typically done by the entire quality panel combined with relevant personnel from the assessment groups for QM#1 and QM#2 together. This approach helped to embed responsibility for quality throughout the project team with the joint undertaking of review and reflection and the joint determination of any required action.

It was anticipated that quality measurement would be undertaken:

(1)   Regularly throughout the course of the project with progressive scores contributing to a final score (e.g. water quality, safety statistics).
(2)   At completion of project, but with 'progressive tracking of performance' (e.g. buildings).
(3)   At the end of the twelve-month period (e.g. visitor satisfaction).

The construction process components required progressive monitoring over the life of the project. For example, measurement of water quality was undertaken monthly. Other components (such as quality of built finishes), which focus on outputs/outcomes, could not be finally scored until that component was completed. As considerable financial benefits were influenced by the outcomes of QM#1 and QM#2 the ALT sought progressive feedback – via trend scores – to help it guide the project towards the meeting of the quality goals.

Progressive tracking of the achievement of quality on-site commenced as soon as practicable. Assessments and indicative scoring started on certain areas such as the administration area, Facades and Main Hall knot steelwork in February 2000. This led to the development of scoring spreadsheets and the comprehensive and detailed scoring of building and site works began in September 2000. The compressed timeframe for the project, and for exhibitions in particular, enabled the panel to provide initial advice and guidance on the basis of drawings, prototypes and other materials as they became available. In turn, this reinforced the quality culture of the project and further extended the role of the quality panel in being proactive in engendering quality in its many forms.

Now that the project is complete, the quality system and results obtained seems logical, straightforward and effective. This was not the team's perception in the early stages of the project when the development of the measurement system presented a major challenge. The development of the quality system entailed considerable resources from each of the members of the ALT and this became difficult to sustain at some points in the construction process, which caused the quality panel some concern. Ideally, the key elements of the system should have been in place prior to construction commencing but this was not possible and many aspects required development and refinement as the project progressed. The advantages of this 'learning through action' was that there was a broader involvement in the development of the quality system which helped to secure its ownership by all the key members of the construction team.

In practice, it was difficult to undertake complete scoring, especially in the later stages with the overlap of the many construction, finishing and fit-out trades. The major spaces tended to be full of extensive scaffolding and this together with the protective coverings and the provisional lighting hampered thorough assessment and the preparation of trend scores. This could well be less of an issue for a building of less complex form. As it was some of the components could not be measured until after the completion of the defects liability period. The ALT agreed that an amount equivalent to 50 per cent of the anticipated quality pool bonus payment be retained until this time to allow for any undiscovered issues and problems.

The selection of alliancing as the project delivery strategy was driven by the need to provide an appropriate balance between cost, time and quality objectives. The project had to achieve quality levels expected of a national cultural institution, and also to be delivered within a tight timeframe and budget. The alliance agreement provided strong incentives to reduce the cost of the project. Equally, there were strong incentives to complete on time. The quality pool and the quality measurement system were established so that the incentive structure did not drive 'cost cutting' or 'shortcuts' to save time at the expense of quality. The use of alliancing together with the development of a comprehensive quality management system were both pioneering aspects of the project and for the construction industry.

The quality management system became an important management tool that enabled the project quality objectives to be achieved and for their achievement to be monitored. The system of progressive scoring and reporting provided an effective early warning system, and offered valuable and timely input to decision making by the project management team and the ALT. Panel members jointly spanned diverse specialised fields, but formed a united, positive and complementary presence on the project. The panel consistently provided expert, considered and independent advice. The QM system as it took shape provided an excellent foundation for achieving quality and it is one of the key reasons why the Commonwealth and the ALT had confidence in the quality measurement results.

An important part of the quality panel's role was to identify where scores could be improved, within sufficient time to enable the project team to rectify any shortcomings. Although advice on solutions was often offered, it was not the panel's role to identify or evaluate the solutions – which was left to the project management team. Examples of issues identified early by the panel included problems with the plated steelwork and plasterboard to walls and ceilings in galleries, particularly the Temporary Gallery. In each case the project management team assessed the advice received and took appropriate action to address the issues. This improved the project outcome and enabled the alliance to increase the quality score. Again, this process of considered evaluation and reflective action demonstrates clear evidence of a quality culture that evolved and strengthened during the course of the project.

By providing proactive advice the panel came to be seen as a positive force on the project. This is distinct from the traditional superintendent or inspection roles, which are more likely to be reactive – drawing attention to problems after they occur.

Whilst there was often vigorous debate, meetings of the quality panel and the project management team were characterised by a spirit of cooperation and goodwill between the parties. This was no easy achievement, as the project team was under constant pressure from project deadlines and could easily have reacted in an uninterested or negative way to the panel's reviews and advice. Instead, the panel's feedback was eagerly sought. To some extent this reflected the positive financial incentives offered by the quality pool. It also recognised the personal commitment of the project team personnel in achieving the high benchmarks that they had helped to set.

The QM framework provided for quality assessments and scores to be calculated were set at the following points:

- An interim score at opening on 11 March 2001.
- A final score at 11 March 2002.

The weightings for the six measures and the score for the project as a whole were assessed as illustrated in Table 8.5 at the interim period:

Table 8.5: Interim quality score for the National Museum of Australia project

No.	QM	Weighting %
1.	Buildings	30
2.	Exhibitions	30
3.	Environment	10
4.	Indigenous employment	10
5.	Public and industry recognition	10
6.	Safety	10
Score for total project (out of +10 maximum)		8.31

There were only a small proportion of components and subcomponents where the quality scores were below the high level generally achieved across the project. The quality panel considered that these scores were soundly based and provided a fair assessment, which properly reflects the outcomes, achieved at the time of handover.

The framework document provides for a 12-month post-completion assessment period, similar to the defects liability period (DLP) of traditional contracts. The factors included for monitoring, however, are more extensive than for a standard DLP, e.g. ease of operation and maintenance, speed of defect rectification, visitor satisfaction, etc. Whilst undertaking the review task in this period is minor in comparison with the project as a whole, it had its complexities. Principally, the cohesive construction team had been disbanded and the momentum established during the construction had been replaced by a continual need to identify and rectify defects whilst the Museum was in operation. One factor is the interface of the ALT with the new arrangement of end-users.

Although the alliance agreement remains in place during this time, new operating relationships have to be formed with the Museum, the Institute, and the National Capital Authority (NCA). Further, the QM system has to be maintained during this period despite the reduced level of contact and resources, as quality data needs to be collected and analysed for reports to the ALT and project management team at regular intervals. A tenant request control system proved effective in dividing responsibilities for defects between tenants and alliance. Monitoring of this system over the 12-month period was included in the QM process.

8.2.5 Lessons learned for future adoption of quality measurement process on similar projects

This was the first building project in Australia to use a QM system of this type. This QM system made a significant contribution to the success of the project and also provided valuable information on the design and use of QM systems generally. To assist in the consideration of QM systems for future projects, key elements for an effective system, based on the experience of the project, are provided below.

8.2.5.1 *Incentives*

The QM system should be underpinned by an incentive system, such as the risk/reward mechanism in a project alliance. There should be associated benefits that reward the general workforce for producing high quality results.

8.2.5.2 *A management tool*

An appropriately designed QM system should provide an effective management tool for the project partners, by progressively monitoring performance and so informing decision making and assisting with the planning of the best use of resources.

8.2.5.3 *Key elements of a QM tool*

The quality measurement framework should:

- Divide the project objectives into constituent performance indicators (quality measures) that may need to be further broken down into sub-measures as appropriate to the size and complexity of the project.
- Specify agreed methods and standards of measuring outstanding, business-as-usual (BAU) and poor quality work. These methods will generally include both quantitative (e.g. ratio of lost time injuries rate to hours worked) and qualitative (e.g. benchmarked against the standard of workmanship of a component of an existing building).
- Involve all key stakeholders in the generation of the quality measures and in the definition of performance standards.
- Be informed by independent assessors that direct the system and offer considered assessment and advice (i.e. the quality panel).
- Cater for measuring a range of performance indicators.
- Use external experts and assessment groups for specialist measures (e.g. for environmental measures).

- Define roles and responsibilities, including the appointment of 'champions' for each quality measure.
- Use a scoring scale that is straightforward and easy for all to understand and employ in a consistent way.
- Have appropriate reporting arrangements to provide feedback to enable quality to be improved in action by identifying measures linked to key performance indicators.

8.2.5.4 *Management of QM system*

Adequate resources must be allocated to develop and manage the QM system. The project leadership must communicate the quality goals and requirements to project staff. Once 'outstanding' quality has been defined, the project management team should develop action plans to strive for its achievement. A 'champion' should be nominated for each quality measure, whose job is to implement the action plans. Responsibility should also be clearly assigned to collate the collection of information and coordinate performance measurements, progress and final reports.

8.2.5.5 *Probity considerations*

An independent person or panel, such as the quality panel, skilled in the relevant fields, should be appointed to oversee the system and provide advice and guidance to assist in achieving the best possible project outcomes, and provide independent assessments. This person or panel should be proactive in their guidance and advice.

8.2.5.6 *Ownership*

The QM system should be developed with input from all the key stakeholders and communicated widely to the workforce.

8.2.6 Detailed descriptions of quality measures used on the National Museum of Australia project

The assessment criteria of each of the quality measures used on the National Museum of Australia project are presented below.

QM#1

QM#1 consisted of three components:

1.1   Quality of built finishes
1.2   Non-conformances
1.3   Defects

QM1.1 – Quality of built finishes
The primary process for this measurement was visual inspection and collective judgment by the independent quality panel and the assessment group. The

buildings and site works were divided into subcomponents – for example, the Main Hall, the Administration area – and assessed separately. The total score for the component was calculated by assigning weightings to each sub-component.

Trend scores were given during the construction process. An interim score was assessed as at opening, with the final score to be assessed after 12 months.

QM1.2 – Non-conformances
Non-conformances are taken to include incomplete and/or substandard work and items contrary to specification if they adversely affect the quality of the facility, or cause increased operating costs. They are also taken to mean non-conformance with any statutory requirements, which would preclude obtaining final authority approvals, for example in relation to public spaces, fire and safety etc. The final score for this component was assessed by the project management team at opening, from the 'Issues list' generated as a result of inspections by the technical adviser to the Commonwealth, and inspections by the architects and exhibition designers. All authority approvals were in place at opening.

QM1.3 – Defects
The project management team assessed an interim score as at one week after opening. Judgment was used in consideration of the circumstances involved, for example, where long purchasing or shipping lead times precluded defects rectification within one week.

The alliance established a formal process with the end-users to log and track defects and other tenant issues. A 'tenant request control sheet' was maintained to track the date requests are issued, the response, and date closed. Post-completion meetings are held regularly to discuss outstanding issues. The final score will be determined after 12 months on the basis of the information produced by this tracking system.

QM#2

QM#2 consisted of five components:

2.1   Design quality
2.2   Use of content
2.3   Integration of technology
2.4   Accessibility
2.5   Visitor experience

QM2.1 – Design quality
The primary process for measurement was visual inspection and collective judgement by the independent quality panel and the assessment group. The exhibitions were divided into subcomponents, e.g. the North Permanent Gallery, the Gallery of First Australians, and assessed separately. The total score for the component was calculated by assigning weightings to each subcomponent.

Trend scores were given during the fitout process. An interim score was assessed for this report as at opening, with the final score to be assessed after 12 months.

QM2.2 – Use of content and QM2.3 – Integration of technology
The process for measurement was the same as for QM2.1.

QM2.4 – Accessibility
The process for measurement was the same as for QM2.1, but supplemented by feedback from the Museum after the soft opening, and after two months of operation. It was intended that a focus group would be used to provide feedback, however, this did not eventuate.

QM2.5 – Visitor experience
Visual inspection and judgment of the independent quality panel and assessment group was used to give an interim score. In addition, visitor surveys conducted by the National Museum will be reviewed to ascertain visitors' satisfaction ratings. The final score will be given after 12 months.

QM#3

QM#3 consisted of six components:

3.1   Environmental management
3.2   Waste management
3.3   Water quality
3.4   Air quality
3.5   Energy efficiency/life cycle costs
3.6   Ecologically sustainable development

If the project was successfully prosecuted for a breach of any environmental legislation, then the maximum score achievable for the measure was zero.

QM3.1 – Environmental management
Quarterly reviews were undertaken of the environment measures since January 2000. Actions taken by the alliance and outcomes achieved were described in a final report by a specialist consultant, National Environmental Consulting Services (NECS).

QM3.2 – Waste management
The waste management contractor monitored the percentages of waste going to landfill and recycling. The final score was derived from the report prepared by environmental consultants NECS.

QM3.3 - Water quality
Water quality was monitored monthly by the environmental consultant NECS. The final score was derived from the report by NECS.

QM3.4 – Air quality
A complaints register was maintained by the alliance. The final score was derived from the report by NECS.

QM3.5 – Energy efficiency/life cycle costs
The benchmark for this component was developed after many of the design decisions had already been taken. Measurement of this component was difficult as the process for measurement was not adequately defined. The score was based on the final report prepared by NECS.

QM3.6 – Ecologically sustainable development
The benchmark for this component was developed after many of the design decisions had already been taken. Measurement of this component was difficult, as the process of measurement was not adequately defined. The Environmental Code for Best Practice developed for the Sydney Olympic Games was used as a guide.

QM#4

QM#4 consisted of five components:

4.1 Enhancing opportunities in the construction period
4.2 Enhancing opportunities beyond the construction period
4.3 Training
4.4 Employment
4.5 Supportive workplace

QM4.1 – Enhancing opportunities in the construction period
The indigenous employment quality measure was prepared in consultation with the Department of Employment, Workplace Relations and Small Business (DEWRSB), which is the Commonwealth Department having carriage for indigenous employment issues.
    The alliance developed an action plan with the Construction Industry Training and Employment Association (CITEA). CITEA maintained statistics that were the basis for scoring and prepared a final report and score for the alliance.

QM4.2 – Enhancing opportunities beyond the construction period, QM4.3 – Training, QM4.4 – Employment and QM4.5 – Supportive workplace
The process for measurement these was the same as for QM4.1.

QM#5

QM#5 consisted of three components:

5.1 Promoting the site
5.2 Industry recognition of alliancing
5.3 Stakeholder image

QM5.1 – Promoting the site
The assessment group undertook measurement of QM#5. Throughout the project reports were provided to quality panel meetings about activities undertaken such as site visits and presentations. A compilation of media and industry coverage was provided with the meeting papers. In addition the quality panel was invited to attend events, such as open days and seminars.

QM5.2 – Industry recognition of alliancing and QM5.3 – Stakeholder image
The process for measurement was the same as for QM5.1.

QM#6

QM#6 consisted of three components:

6.1   Management processes
6.2   Safety outcome
6.3   Individual intention

Note: if there had been a death or injury that prevented the injured person returning to work, then the maximum possible score for the measure was zero.

QM6.1 – Management processes
The enhanced safety plan was reviewed by alliance auditors Australia Pacific Projects, as part of their performance audits of the project. APP found that the safety plan is consistent with a best practice model and reflects a high level of commitment by the alliance. The project management team assessed the final score.

QM6.2 – Safety outcome
The target lost time injury rate of 1:35,000 worked was derived from Bovis Lend Lease injury reports across all sites in the Canberra region in the previous two years. The score for this component was based on statistics collated and retained by the project management team.

QM6.3 – Individual intention
The alliance performance manager scored the site safety meter weekly. It is a measure of conformance to the safety procedures adopted on the project.

8.3   Chapter summary

This chapter began with an introduction discussing the concept of initiating a briefing process that facilitates and encourages a quality culture. The briefing process was singled out as an important focus from the many phases of project development because it sets the scene for a quality outcome.

We presented a briefing model based upon a contingency approach recognising the project initiator's: technical knowledge of project needs and requirements, working behaviour towards various project team members to help build design solutions and maturity in terms of ability and willingness to engage in the briefing process. This model helped to differentiate between less complex projects where an instructing or delegating briefing style may be employed and more complex projects where a cooperating or collaborating briefing style may be experienced. The project briefing experience may help to explain how the relationship between the project initiator and design and construction teams affects mutual under-standing about what is wanted, needed and required. This understanding forms a core component of any quality culture because it establishes a framework that supports exchanges of ideas and perspectives, which lead to solutions to problems. Sophisticated and experienced project initiators are generally aware of the value of fully contributing to the project briefing process as a well-informed project participant. This relationship can best promote a quality culture by extending the positive interaction throughout the construction phase.

The essence of a project culture was highlighted as six key points that have much to do with relationship skills and qualities that build upon solid technical skills. Gaining mutual understanding of project goals, stakeholder aspirations and project constraints and opportunities for any project were underscored as vital ingredients positively influencing project success. We argue that a relationship-based procurement system such a project partnering or alliancing provides an appropriate procurement solution for more complex projects because it promotes deep levels of team interaction, mutual learning and exchanges of rich and varied sources of knowledge. This is all derived from close interaction characterising relationship-based procurement systems that support a quality culture.

This chapter has stressed, as does the entire book, the advantages of linking project performance with organisational learning and behaviours in which perspectives and worldviews can be more freely exchanged. The second section of this chapter relating to the National Museum of Australia project as a case study, provides ample evidence of what a successful quality culture may look like. Figure 8.2 illustrates how a quality culture may develop from constituent concerns and foci.

The first two levels in Figure 8.2 focus upon a quality **output** such as a product and/or service. The base element of any quality culture is a sound QA system, which, typically, is solely focused upon outputs such as the product itself and systems and processes involved with production. The focus on occupational health and safety and environmental management systems develops as a more sophisti-

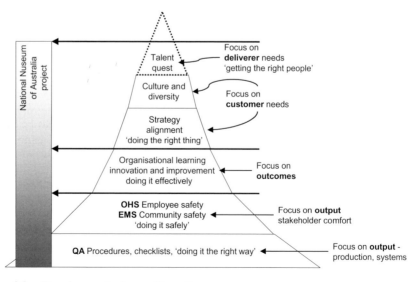

Figure 8.2: Development of a quality culture

cated appreciation of the security and comfort needs of the wider range of project participants and stakeholders emerges together with the need to answer statutory obligations and community expectations.

A focus on effectiveness emerges as a quality culture further develops. Innovation, incremental and breakthrough improvement and project whole-of-life concerns become important **outcomes** sought by more sophisticated project initiators. Learning from experience (lessons learned) and development of a learning organisation become important tools for improved effectiveness of project teams. These compatible aims provide a strong focus on project outcomes.

The next level of a quality culture provides a sharp focus on customer needs. Project delivery teams may concentrate on 'doing the right thing' by aligning strategic organisational and project aims and finding common ground with other project participants. This is where project partnering, alliancing and other forms of relationship-based procurement systems enter the quality culture. This is reinforced by a sharpening focus upon diversity and the nature of organisational cultures working on projects. A realisation develops that strategic alignment and customer focus do not equate to homogeneity and uniformity. Diversity of opinions and culture that informs creative abrasion which, in turn, sparks creativity, is encouraged. This renews customer focus because project stakeholders and clients are diverse and need multiple points of interaction to best express their requirements and aspirations as well as to co-develop innovative solutions.

The construction industry shares, with all other knowledge-based industries, a challenge to attract and maintain the best talent that can provide the necessary ingredients of intelligence and flexibility to realise a quality culture. The requirements for a quality culture indicated in Figure 5.3 include technologies and systems that support quality together with the behavioural characteristics of organisations and individuals that allow a quality culture to exist and flourish. The dotted triangle in Figure 8.2 indicates that the next phase of a quality culture quest is to provide an environment that not only attracts the best talent but also allows them to realise their creative urges by focusing upon delivering project success in a more fulfilling way. We argue that such talented individuals will not only meet and exceed project client/customer/user needs but also seek to satisfy their own intrinsic motivational needs to 'do their best possible'. Chapter 10 will pursue this theme more fully.

Figure 8.2 indicates a hierarchy of needs with the quest for talent as the final point yet it underpins each level as a necessary ingredient. We believe that it only becomes fully apparent as the upper levels of the hierarchy of a quality culture are reached that special people and special approaches need to be introduced. It also becomes evident that information and communication technologies (ICT) are supportive of a quality culture. They support the use of quality systems and procedures yet have proven to be unable to replace informed judgement. The promise of expert systems and robotics in the 1980s has not been realised nor supplanted human intuition. The current internet and web-based agents and technologies as well as knowledge management software support systems only serve to provide intelligent support and rapid search and processing capacity. Indeed one part of the talent quest is to find and exploit the capability of people to use ICT. The other part is to attract and retain creative people who can develop systems and methodologies and novel solutions to the problem of delivering high quality outcomes as well as outputs. The evidence presented in section 8.2 of this chapter indicates that the QM system developed through the cooperative and collaborative process on the National Museum of Australia project was innovative, rigorous and intelligent.

The analysis of the Acton Peninsula project suggests that it appears to have benefited from a high-level quality culture as indicated in Figure 8.2 above. Evidence presented in Chapter 10 from research undertaken on that project indicates that while the project teams were encouraged to learn from each other and to take advantage of training and education courses, these were generally undertaken in their own time. The higher levels of culture indicated in the partially shaded rectangle represent best current practice. We anticipate that in future, attention will be focused on the development of quality systems that incorporate measures of the quality of human relationships and job satisfaction together with reflective feedback of how the organisational framework and management system drives or inhibits commitment to project success. This is discussed further in Chapter 10, which provides some insights into how these properties of future quality systems may develop.

8.4   References

Auditor-General of the Australian National Audit Office (2000) *Construction of the National Museum of Australia and the Australian Institute of Aboriginal and Torres Strait Islander Studies, Audit Report*. Canberra, Australia, Australian National Audit Office.

Barrett, P. and Stanley, C. (1999) *Better Construction Briefing*. Oxford, UK, Blackwell Science Ltd.

Bensaou, M. and Earl, M. (1998) 'The Right Mind-Set for Managing Information Technology.' *Harvard Business Review*. **76** (5): 119–128.

Bentley, C. (1997) *PRINCE-2 A Practical Handbook*. Oxford, UK, Butterworth-Heinemann.

Briner, W., Hastings, C. and Geddes, M. (1997) *Project Leadership*. Aldershot, UK, Gower.

Chase, R.B. and Dasu, S. (2001) 'Want to Perfect Your Company's Service? Use Behavioural Science.' *Harvard Business Review*. **79** (6): 79–84.

Cleland, D.I. (1999) *Project Management Strategic Design and Implementation*. Singapore, McGraw Hill.

Davenport, T.H. and Prusak, L. (2000) *Working Knowledge – How Organizations Manage What They Know*. Boston, MA, Harvard Business School Press.

Fisher, N., Barlow, R., Garnett, N., Finch, E. and Newcombe, R. (1997) *Project Modelling in Construction ... Seeing is believing*. London, Thomas Telford.

Francis, V.E. and Sidwell, A.C. (1996) *The Development of Constructability Principles for the Australian Construction Industry*. Adelaide, Construction Industry Institute Australia.

Griffiths, A. and Sidwell, A.C. (1997) 'Development of Constructability Concepts, Principles and Practices.' *Engineering, Construction and Architectural Management*, **4** (4): 295–310.

Hersey, P., Blanchard, K. and Johnson, D.E. (1996) *Management of Organizational Behaviour*. London, Prentice-Hall International.

Magretta, J. (1998) 'Fast, Global, and Enrepreneural: Supply Chain Management, Hong Kong Style.' *Harvard Business Review*. **76** (5): 103–114.

Malone, T.W. and Laubacher, R.J. (1998) 'The Dawn of the E-Lance Economy.' *Harvard Business Review*. **76** (5): 145–152.

Morris, P.W.G. (1994) *The Management of Projects: A New Model*. London, Thomas Telford.

Nonaka, I. and Takeuchi, H. (1995) *The Knowledge-Creating Company*. Oxford, Oxford University Press.

Nonaka, I., Toyama, R. and Konno, N. (2001) SECI, *Ba* and Leadership: A Unified Model of Dynamic Knowledge Creation. *Managing Industrial Knowledge – creation, transfer and utilization*, eds Nonaka, I. and Teece, D. London, Sage: 13–43.

Peters, R.J., Walker, D.H.T., Tucker, S., Mohamed, S., Ambrose, M., Johnston, D. and Hampson, K.D. (2001) *Case Study of the Acton Peninsula Development*, Government Research Report. Canberra, Department of Industry, Science and Resources, Commonwealth of Australia Government: 515.

Prahlad, C.K. and Ramaswamy, V. (2000) 'Co-opting Customer Competence.' *Harvard Business Review*, **78** (1): 79–87.

Retik, A. and Hay, R. (1994) *Visual Simulation Using VR*. 10th ARCOM Conference, Loughborough, UK, Association of Researchers in Construction, ARCOM, The University of Salford, UK.

Salisbury, F. (1990) *The Architect's Handbook for Client Briefing*. London, Butterworth & Co.

Seybold, P.B. (2001) 'Get Inside the Lives of Your Customers.' *Harvard Business Review*. **79** (5): 80–89.

Sidwell, A.C. and Mehrtens, V.M. (1996) *Case Studies in Constructability Implementation*. Adelaide, Construction Industry Institute Australia.

Smith, J. and Jackson, N. (2000) 'Strategic Needs Analysis: Its Role in Brief Development.' *Facilities*. **18** (13/14): 502–512.

Chapter 9

Developing an Innovative Culture Through Relationship-based Procurement Systems

Derek Walker, Keith Hampson and Stephen Ashton

We investigate in this chapter the phenomena of innovation and how a relationship-based procurement approach may be harnessed as a driver for innovation. At the beginning of Chapter 7 we presented a model showing three essential features of partnering, one of these was continuous improvement, generally accomplished through innovation in product, process, and/or service delivery approach.

This chapter links discussion presented in Chapters 3, 6, 8 and 10 (relating to theories of the learning organisation, innovation and promoting a team culture that seeks innovation and invention) with examples of innovation observed on a major successful building alliancing project constructed in Australia. We also explain in this chapter how innovation is a vital ingredient of a relationship-based procurement approach. We follow this and explore how successful innovation is dependent upon a clear strategy through using an example of a major and successful innovation implemented on the National Museum of Australia project. We then summarise this chapter and provide some implications for the future in the construction industry.

9.1 Introduction

Innovation is part of a change strategy. It is a realisation that a current state must be changed in order for organisations to survive or prosper. The purpose of innovating is to achieve competitive advantage or more prosaically to survive.

In section 4.2 we discussed how performance improvement could be achieved through attention to innovation, continuous improvement and knowledge management. In Figure 7.3 we illustrated a model with the project team's management capacity to plan for and deal with risk as a central driver of success in construction time performance. Imai promotes the idea of *kaizen* or continuous improvement as a series of small steps in reflection upon practice and using this feedback mechanism as a driver for making small improvement changes – essentially it is a process of calibration (Imai 1986, p. 25). He shows this to be a people-oriented and highly collective activity using both tacit and explicit sources of worker knowledge. He recommends it as being appropriate for slow-growth economies with mature systems and processes that need adjustment or minor improvement. This approach was highly recommended in the 1980s when many Western economies were experiencing relatively low growth rates of only several percentage point increases each year. Further, this was before the widespread introduction of an e-economy and its major and sudden impact upon competition and globalisation. Kaizen's value lay in its complete focus on a culture that supported continuous improvement.

Later in the 1980s a more radical movement was gathering pace – business process re-engineering (BPR) where rapid change was seen to be generating a crisis that only revolutionary re-thinking processes and approaches would suffice (Hammer and Champy 1983). This movement, which led to widespread downsizing of organisations and loss of corporate memory, has been severely criticised (Hilmer and Donaldson 1996; Hames and Callanan 1997) even by those who first promoted BPR (Hammer 1996). Much of that criticism has centred upon the belief that while customer focus is both necessary and vital for organisational sustainability, it is undermined by a fearful workplace culture (Hilmer and Donaldson 1996), by loss of corporate memory, skills and knowledge through displaced workers (Prusak 1997; Davenport 2001) and by loss of the critical role of middle managers as being the brokers and facilitators of innovation triggered by internal shocks from top-management or external business turbulence (Nonaka and Takeuchi 1995, p. 130). While the impact of BPR and corporate responses to it continued to occupy the minds of many management theorists, the whole issue of strategic planning was called into question by one of its most ardent proponents (Mintzberg 1994).

Strategic planning had been seen as a corporate level holistic planning tool that enabled organisations to develop a capacity to meet gaps in competence and to re-invent itself in response to demand for new products, services or processes. Again, that was the mindset of the 1970s, 1980s and to a lesser extent the early 1990s, but turbulence continued to undermine many organisations' ability to forecast the future accurately and the call was made for developing scenarios instead of preparing forecasts (Mintzberg 1994, p. 248).

Strategy, however, has more recently been seen as important in preparing organisations to innovate and be flexible in the e-business age (Porter 2001). The need for innovation and breakthrough invention to be established during these times of uncertainty and volatility has been established, for example:

- to stay ahead of the competition. See, for example, Von Hipple *et al.* (1999) and Hargadon and Sutton (2000);
- to form alliances with customers to understand their needs better and how new uses can be made for innovative ideas. See, for example, MacMillan and McGrath (1997) and Kim and Mauborgne (1999); and
- to work better with the supply chain to develop process innovation (Womack *et al.* 1990; Womack and Jones 2000; Hammer 2001).

Thus, we now see enthusiasm emerging for a collaborative approach to solving problems jointly through innovation that requires diversity of perspective and creative application. This then sets the scene for the next section of this chapter.

9.2   The nature of innovation and how it can be fostered

We will use as a definition of innovation 'an idea, practice, or object that is perceived as new by an individual or other unit of adoption.' (Rogers 1995). Innovation obviously involves a perceived need to change from an 'old' state to a 'new' one. The purpose of innovation is Darwinian. It is about survival and growth – about ecological (market) niches being filled by the exuberance of a life force. Innovation is, therefore, a decision-making process to enact change in technology, process, services rendered or other management approaches.

9.2.1   The decision to innovate

Rogers divides organisational innovation into three type of decisions: optional innovation, collective innovation and authority innovation decisions.

(1)   *Optional innovation* decisions allow individuals within an organisation to decide whether to adopt an innovation regardless of the actions of other members of the organisation. This is a devolved and empowering bottom-up innovation decision-making approach.
(2)   Achieving consensus between members of an organisation makes *collective innovation* decisions. This is a bottom-up and lateral form of innovation decision-making approach.
(3)   *Authority innovation* decisions are made by small groups within an organisation but are applied across all members of an organisation. These groups may have power, status or technical expertise. This is primarily a top-down decision-making approach.
(4)   A fourth type can also be identified as *contingent innovation*. It is common for the different innovation decision processes to be combined sequentially and is contingent upon emerging circumstances. This may be a multi-directional as well as multi-phased decision-making approach.

Two processes of *innovation initiation* and *implementation* may be further broken down into five sub-processes. These are *agenda setting, matching, redefining/ restructuring, clarifying, and routinising*. Agenda setting and matching occur during the initiation phase whilst the remaining three sub-processes form part of the implementation phase of an innovation.

(1)   The *agenda setting* stage is the first in the process of innovation initiation and, possibly, eventual implementation. During this phase the problems of an organisation are identified and the need for an innovation to solve that problem is identified. Occasionally an innovation may be identified that will improve an existing process or solve an existing problem that may not have been previously been identified. This is particularly true of innovations involving the use of information communication technology.
(2)   *Matching* is the subsequent process of associating a particular innovation with an identified organisational problem. Once these two stages have occurred it is time to move onto the implementation phase.

(3) The first sub-process of the implementation stage is *redefining/restructuring*. During this process both the innovation and the organisation structure will be modified to suit the needs of the organisation.

(4) *Clarifying* then occurs during which the relationship between an innovation and an organisation is defined more clearly.

(5) *Routinising* occurs when the innovation becomes a core process within an organisation and it is no longer identified as an innovation.

9.2.2   The process of undertaking innovation

It is difficult to agree on the boundary between *kaizen* and breakthrough change. However, regardless of the characteristics of the proposed change the implementation of the change is of critical importance. Thus the way in which innovation (or breakthrough) is implemented is of vital importance to productivity improvement. It is important that impediments or hurdles to effectively diffusing innovation should be recognised and obviated.

Figure 9.1 illustrates the hurdles that are encountered when attempting to innovate. The first hurdle indicated is the decision-making process of what to innovate and how. This is followed by the hurdle of the time taken to implement effectively the innovation change. Following hurdles may be encountered, sometimes but not often out of the sequence indicated. The purpose, rationale and manner in which any change may take place require critical communication skills that present challenges within most organisations. Powerful social and/or political systemic barriers may also be encountered and all need to be taken into consideration and planned for. Each of these hurdles may potentially dilute enthusiasm and commitment to innovation. For a more detailed discussion on barriers and drivers of change in general, readers should refer to Chapter 10.

In a recent study the following statement about innovation in the Australian building and construction industry was made (PricewaterhouseCoopers 2001, p.1).

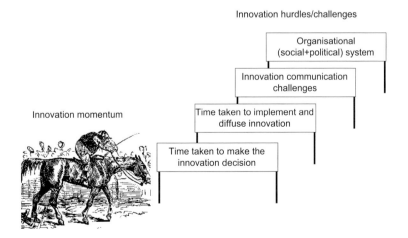

Figure 9.1:   Innovation challenges

Innovation was driven internally by senior management attitudes, marketing, information technology departments, and staff and employees. Joint ventures and collaborative efforts supported and facilitated the innovative process. These were evidenced by:

- Senior management teams devoting time to investigating the future and to understand the needs of the marketplace, the resources at their disposal (including suppliers) and the competitive business environment;
- Working environments that encourage creative solutions;
- Strong support for contractor relationships that encourage innovation (alliance contracting) and reduce risk;
- Strong support for joint ventures (JVs) and collaborative efforts to develop and commercialise innovative solutions; and
- Good project management for the identification, development and commercialisation of innovation.

Clearly, innovation is a complex issue encompassing organisational management and workplace culture. Delivering projects is a strategy for enacting change (Turner 1999) and so this chapter will concentrate upon construction project management as a means to demonstrate innovation as a change management process.

Tatum carried out a number of studies into innovation implementation, predominantly dealing with innovations in construction technology; i.e. manufacturing process and product innovations rather than management process innovations. He identifies five aspects of the innovation process in construction firms (Tatum 1989, p. 605). These are briefly described below supported by the work of others, and include:

- Reorganising forces and opportunities for innovation;
- The need to create a climate for innovation;
- The provision of new technologies;
- The need to experiment and refine innovations within the construction firm; and
- The implementation of the innovation.

9.2.2.1   *Reorganising forces and opportunities for innovation*

Tatum considered this driver to be the dynamic nature of construction markets and competition. He observes a 'market pull' element to increasing innovation and, complementary to this, a competitive rivalry that sees competitive advantage through innovation. He also perceives an entrepreneurial approach that recognises innovation opportunities for all activities regardless of whether or not that activity is performing well or not. The regulatory nature of the construction industry is observed to be a possible force for innovation. This is interesting given that other researchers have considered the regulatory nature of construction to be an innovation inhibitor, however, OHS and EMS requirements have forced at least one Australian contractor to approach QM in an innovative way and to use this requirement as a driver for flexibility and innovation (Shen and Walker 2001; Walker and Shen 2002). The development of new technology is also considered to be a generator of innovation; the 'technology push' process identified in other manufacturing industries. Finally the strategic focus of the organisation is seen to be a factor in generating innovation (Nam and Tatum 1989, 1992).

9.2.2.2 The need to create a climate for innovation

Several preconditions for innovation including the need for: vision; commitment for improvement; committing resources to allow autonomy; and toleration of failure of an innovation. The need to develop capabilities is clearly stated. These capabilities include not only the major technical discipline but also related disciplines. It also means that construction firms need to identify a technological gatekeeper and develop the entrepreneurial ability to carry forward innovations. Several authoritative academics support this argument (Leonard-Barton 1995; Lenard and Bowen-James 1996; Amabile 1998).

9.2.2.3 The provision of new technologies

Firms need to be able to adopt new approaches from external sources, modify and adapt available technologies, incrementally improve existing technologies and to develop new technologies. It is interesting to compare this with the innovation adoption process described by (Rogers 1995). Technological influences on product innovation in manufacturing industries such as automotive and aerospace are quite different from the construction industry. Technology push is less evident in traditional procurement processes because designers (such as architects) can be conservative and remote from the production of the constructed facilities. They tend not to directly see the impact that new innovative approaches may have upon production performance. Indeed in the commonly litigious construction environment often they quite sensibly fear potential risk of contractors or clients suing them for experimenting with untried or inadequately tested new technologies. This has been evident for decades (Nam and Tatum 1989) and has been more recently highlighted (PricewaterhouseCoopers 2001, p. 50)

9.2.2.4 The need to experiment and refine innovations within the construction firm

It is worth noting again that Tatum's research is based upon construction technology innovations, however there is no shortage of research in other areas supporting this proposition (Leonard and Rayport 1997; Hansen, *et al.* 2000; Hargadon and Sutton 2000; Thomke 2001). A recent study of innovation in the Australian construction industry measured innovativeness, 239 senior management of 1400 potential identified respondents were surveyed, using a web-based questionnaire followed up using e-mail and telephone to ensure a high response rate. High innovators (the top quartile of the sample using a measure of innovativeness) experimented through joint ventures, trying out previously unseen technology and processes, and undertaking incremental improvement exercises (PricewaterhouseCoopers 2001). In the general knowledge management literature, the role of experimentation was indicated as particularly evident (Nonaka and Takeuchi 1995; Nonaka *et al.* 2001).

9.2.2.5 *The implementation of the innovation*

This is necessary to be considered at both the project and organisation level. Sub-process within projects include the provision of resources, gathering the support of planning and bidding groups within a firm, gaining experience in the use of the innovation and development of a competitive advantage from that innovation. Feedback is seen as being essential throughout (Van de Ven 1986; Leonard-Barton 1995; Nonaka and Takeuchi 1995; Kulkki and Kosonen 2001).

9.2.3 Achieving improvement through innovation

Tatum's work was later developed and expanded to provide one of the most important pieces of construction-related innovation research which deals with the ways in which construction companies can organise themselves best to increase their innovativeness (Harkola 1994; Mitropoulos 1996; Hampson and Tatum 1997). He found that innovative organisations tended to have a longer-term viewpoint and were prepared to accept development problems with an innovation as long as more enduring benefits remained apparent. Innovative organisations tended to take a broader view of risk than more conservative organisations. They tended to adopt a 'vertical integration' approach where possible in that they would consider the implications of an innovation over all phases of a project. The importance of planning was recognised in innovative organisations with flexibility being a priority. This is also supported by Australian studies on construction time performance (Walker and Sidwell 1996). An innovation culture was also fostered within innovative organisations. This manifested itself through persistent pursuit of increased productivity, a highly competitive approach and continual improvement of every aspect of the organisation's business.

In order to achieve high levels of innovation Tatum recommends that organisations adopt flexible group arrangements to try and emulate the 'skunkworks'[1] approach sometimes adopted in manufacturing industries (Tatum 1989). Such groupings should maintain a high and diverse level of technical capability and be provided with slack resources to allow for experimentation. It is vitally important to develop inter- and intra-organisational links. A diffusion network allows innovation to be adopted throughout an organisation and between organisations involved in a construction project and form what has been termed a 'community of practice' (Wenger and Snyder 2000). The role of individual staff is crucial in an innovative organisation in that they can challenge the status quo and respond to diverse viewpoints of problems, causes and solutions. Such a process has been called 'creative abrasion' (Leonard-Barton 1995). Additionally, a technological gatekeeper[2] is an essential element within the innovative construction organisa-

---

[1] This is an approach where a small team is provided with some resources and a mandate to take a good idea and develop it as far as they can. The team has high levels of autonomy and commitment to problem solving (see the article by Tulley (1998). 'How Cisco Mastered the Net.' *Fortune.* **4** (138): 207–209. for a good example of the operation of a skunkworks).

[2] A person who has the natural capacity to act as a conduit for ideas and knowledge in an organisation by searching out new knowledge from cross-departments or externally and bringing these to the attention of colleagues and co-workers. This person acts as mentor, and coach as well as catalyst for adaptation of ideas to new circumstances.

tion but incentive is an important lubricant that allows knowledge to flow (Kikawada and Holtshouse 2001). This individual or small group/community identifies external technology and has the perception to visualise potential applications in the firm and has the means to transmit this knowledge through a network of colleagues to leverage intellectual assets (Kikawada and Holtshouse 2001, p. 290).

Key differences exist between the project-based approach common in construction and the more routine product and process development that occurs in manufacturing industries for example. Demands of cost, time and quality necessary to meet client requirements on each individual project often limit opportunities for innovation. Also, the timeframe for construction work is project- rather than product-based so that many innovations are not introduced because they might disrupt authorised plans. However, when planning flexibility and an innovative approach are encouraged and present this has been shown to be significantly associated with construction time performance success (Walker and Sidwell 1996; Walker and Shen 2002).

There is widespread agreement amongst leading project management academics on the need for a project champion. This is a person who provides great insights into project success and failure and strongly supports the need for a project champion who has the experience and personality to drive projects forward (Morris 1994, p. 255; Cleland 1999, p. 222; Turner 1999, p. 51). The power of a champion can be enhanced by the collective power of workers who work on their own initiative. Empowerment has been recognised for many years as an important social issue in helping to create innovation. Yukl (1988, p. 258), for example, links the positive effects of empowerment with leadership in generating excellence through ensuring that employees are not dominated to follow the leadership of an organisation and thus restricted in experimenting with new ideas through sheer power of subjugation or lack of motivation. Caution needs to be applied to developing empowerment and as Burdett (1999, p. 10) warns that for those empowered to act autonomously when they 'lack the a rich understanding of the overall context, confusion, conflict and self-interested actions are the inevitable result'. Empowerment clearly requires adequate knowledge and access to broader perspectives of an organisation and/or project goals. Thus to be successful, empowerment requires a knowledge-sharing dimension resulting in understanding of goals and agendas between leaders and followers.

Figure 9.2 illustrates the organisational culture that needs to be in place to facilitate the creation and development of innovation. There is clearly a requirement for not only the technical competence to innovate, but also an organisational environment dimension. It can be appreciated from the above that the extent of innovation generated within an organisation and its subsequent beneficial application with benefits derived depends upon the interaction of three clusters of cultural factors. Cultural aspects of a relationship-based approach to procurement are more fully explored in Chapter 8 for developing a quality culture and Chapter 10 when considering human capital.

The organisational leadership driver cluster may be supported by adequate funding mechanisms to initiate and develop innovation (PricewaterhouseCoopers 2001, p. 57). Enabling information and communication technologies (ICT), as discussed extensively in Chapter 6 of this book, also supports this cluster. There

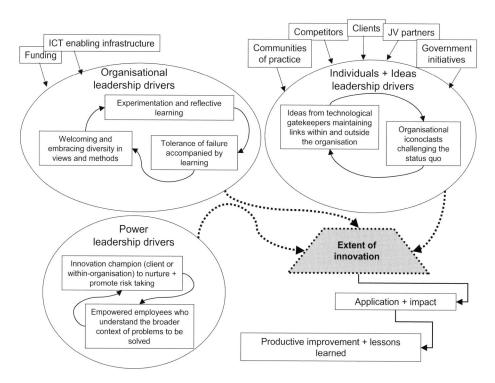

Figure 9.2: Leadership drivers and requirements for innovation

must be a supportive environment created within the organisation demanding leadership qualities that encourages dissent and diversity so that people are open to new ideas. This also requires a tolerance of failure as part of the experimentation process. Individuals must be open and sharing with ideas. There is also a need for those in an organisation to challenge the way 'things are done around here'. This requires leadership qualities of an organisation not readily found in the typical 'command and control' management style prevalent in many construction organisations. The need for power or energy to carry forward ideas into innovation is also highlighted in Figure 9.2 as a second cluster of factors affecting the extent of innovation. This power leadership driver is closely linked with supportive or inhibiting aspects of trust and commitment as outlined in detail in Chapter 10 of this book.

Workers who are the most likely generators of innovation because of their close proximity to processes and 'the work', need to be empowered to develop innovations, however, there also needs to be an innovation champion, with a high level of influence. Champions have been highlighted as a key project management driver for success by several of the project management academic leaders (Morris 1994; Cleland 1999; Turner 1999). The source of ideas that sparks the imagination of champions may derive from a wide range of external as well as internal sources. In Figure 9.2 external sources of ideas for innovation are indicated as emanating from competitors, clients, joint venture partners, and government initiatives (which often encourage involvement with academic institutions). An important internal source includes communities of practice (COPs). These groups gather periodically in real or virtual space to discuss technical 'work' issues and share solutions to problems and 'war stories' (Wenger and Snyder 2000). These groups may include quality circles where QA issues are discussed and debated that lead to innovation. Generally however, COPs are self-selecting rather than designated and form more wide-ranging groups than quality circles and their aim is principally dedicated to finding and disseminating technical solutions. There are numerous examples of these including successful COPs at Xerox where substantial cost savings through innovative practices have been developed (Davenport and Prusak 2000).

Drivers and the nature of innovation have been discussed above but productivity improvement through innovation is not assured without adequate mechanisms for diffusing both innovation and lessons learned. To gain a more profound understanding of innovation and its diffusion it is necessary to consider the nature of the innovation process more deeply. There are two distinct aspects of innovation that need consideration. These are the initiation of an innovation and the implementation and diffusion of innovation.

9.3   Innovation as a strategy

Burgelman and Rosenbloom (1989) proposed an evolutionary process framework for the formation of technology strategy. In summary, the idea is that technology strategy emerges from organisational capabilities, shaped by the **generative** forces of the firm's strategic behaviour, the evolution of the technological environment, and by the **integrative** mechanisms of the firm's organisational context and the environment of the industry in which it operates. Experience with a particular strategy is expected to have feedback effects on the set of technical capabilities. This evolutionary perspective is representative of a view of strategy-making as a social learning process in which strategy is inherently a function of the quantity and quality of organisational capabilities. This illustrates how merely wishing employees to be innovative is simply an insufficient response to the innovation challenge. A concerted approach is required in which organisations plan, coordinate, monitor and project-manage the process of innovation. Innovation, therefore, requires an effective strategy and project management approach to be achieved. Managing the strategic introduction of innovation is similar to managing projects; there are external influences and stakeholder issues that need to be addressed.

This model is useful since it combines the elements of strategic behaviour and organisational context (**internal mechanisms**) with technology formation and industry context (**external mechanisms**), each of which influence the evolution of a firm's technology strategy. The relationships between each of these forces shape technology strategy and represent dynamic forces in the firm's environment. This process model was further developed in an examination of technology strategy and competitive performance in construction at the level of the firm (Hampson 1993).

Hampson reviewed a range of determinants for technology innovation dimensions with the objective of selecting one set of parameters that ensured conceptual completeness, but avoided unnecessary duplication. The following dimensions were selected as starting points for the technology strategy portion of his research. The descriptions for each dimension indicate the source of concepts:

A. **Competitive positioning** – incorporates level of competence, product and process mix, competitive positioning, and the (Burgelman and Rosenbloom 1989) internal driving mechanism of strategic behaviour. Competitive positioning is the firm's relative emphasis and command of technology within the sector

B. **Sourcing of technology** – incorporates technology and the value chain, and sources of technological capability (Porter 1990). Sourcing is the acquisition of explicit (hardware) and implicit (knowledge) technologies.

C. **Scope of technology strategy** – incorporates technology portfolio and scope of technology strategy. The *scope* dimension identifies the core and peripheral technologies and measures diversity of the firm's technological approach.

D. **Depth of technology strategy** – incorporates emphasis on research and development and depth of technology strategy. Depth highlights relative level of research and development investment, and depth and specialisation of firm's technical capabilities.

E. **Organisational fit** – incorporates organisational policies and the internal driving mechanism of the organisational context from the (Burgelman and Rosenbloom 1989) model. Organisational fit assesses the reward system and communication structure of the firm.

Each of the five key dimensions can be further partitioned into attributes that provide a useful summary checklist of important issues or measures to consider. Table 9.1 summarises all technology strategy measures to be addressed within these dimensional areas.[3]

Competitive positioning requires strategic analysis to seek out a market or competency niche, which can be filled by the organisation. This is particularly relevant when working in a relationship-based procurement environment because the organisation is making itself attractive as a partner and seeking to develop a need for services or goods it can produce that fill a particular gap in a portfolio of required project competences and resources. The ability to be exusive and specialised not only generates potential business opportunities but can also provide the means to command profit margin premiums. Specialised contractors generally follow this strategy. They need not necessarily be glamorous or 'hi-tech', for example a demolition contractor may develop specialised skills in safe (OHS and EMS) ways of removing existing structures/facilities, or perhaps develop novel/special uses for disposed materials.

Sourcing of technologies may require strategies of acquisition, merger, JV or other cooperation options discussed in Chapter 3 (see Figure 3.2). The decision is seldom a simple one. The strength of desire for organisational learning has some impact. For example, in acquiring an innovative technology such as ProjectWeb

---

[3] Hampson (1993) details the rationale for selection and guidelines for use of each technology strategy measure.

Table 9.1: Technology strategy measures to be addressed

A   **Competitive positioning**
  (i)   Emphasis of technology in overall business strategy
  (ii)   Command of key technologies in sector
  (iii)   Command of unique technological position
  (iv)   Ability to be key technology leader in sector
  (v)   Monitoring of competitor technologies

B   **Sourcing of technology**
  (i)   Acquisition of explicit technology
  (ii)   Acquisition of implicit technology – head office management
  (iii)   Acquisition of implicit technology – site management
  (iv)   Emphasis on organisational learning
  (v)   Monitoring evolving technologies in sector

C   **Scope of technology strategy**
  (i)   Breadth of technological capabilities
  (ii)   Content focus of tech monitoring and development
  (iii)   Geographic focus of tech monitoring and development

D   **Depth of technology strategy**
  (i)   Emphasis on research and development
  (ii)   Depth of technical capabilities – head office management
  (iii)   Depth of technical capabilities – site management
  (iv)   Degree of specialist tasking

E   **Organisational fit**
  (i)   Reward systems – head office management
  (ii)   Reward systems – site management
  (iii)   Structuring of information flows – site to site
  (iv)   Structuring of information flows – site and head office

(discussed in Chapter 6) a strategy of training, development and enabling alliance partners to embrace this innovation had to be incorporated into the business plan.

The scope of technology strategy needs to be fully considered so that an innovation diffusion strategy is not stymied by a lack of capabilities to understand its potential contribution either at the outset of its implementation or during its formative phase. When introducing any innovation it is necessary to question its impact both from a short-term and more enduring perspective. Often, an innovation affords opportunities to reinvent service delivery and it can be used in quite unexpected ways. For example, using digital photography instead of standard progress photographs allows for not only a more effective way of recording progress but they can also be used as a feedstock for developing method statements, plans and for training and development within computer-enabled learning materials.

The depth of technology strategy begs questions relating to how innovation possibilities are to be generated. If an internal research and development strategy is envisaged then obviously there are staff selection and training/retraining issues to be addressed as well as other resource allocation questions to be answered. If outsourcing or joint development strategies are expected then skills and competencies need to be sourced to select appropriate partners and to manage transparently and fairly the process and to secure internal learning opportunities through the transaction.

Appropriate organisational fit is critical. Reward systems need to be carefully thought through and implemented. A substantial body of literature on organisational learning and knowledge management stresses the need for reward systems to match a knowledge sharing culture (Leonard-Barton 1995; Davenport *et al.* 1998; Hansen *et al.* 1998; Brown and Duguid 2000; Davenport and Prusak 2000; Davenport 2001; Kluge *et al.* 2001; Leadbeater 2001). The construction project management system is particularly vulnerable to lessons *not* being learned. Often teams are disbanded without adequate project debriefing and gathering lessons learned and if some semblance of this does occur, it often resides in inaccessible places to be of any practical use (Walker and Lloyd-Walker 1999).

In summarising this abbreviated section on strategy and innovation the following can be concluded. The external environment provides both incentives to innovate and challenges to an organisation's survival or prosperity that require a measured and thoughtful response. This response needs to be strategic because these external stimuli affect the long-term future of the organisation concerned. The strategy both affects and is affected by the internal environment of the organisation. There may well be a need for the organisation to reinvent itself, particularly if it seeks to move from a mass market with perhaps low profit margins typical of much of the construction industry into a niche player with specialised skills that command premium profit levels with competencies more difficult to replicate and/or compete against. The move towards BOOT-type schemes, alliances, or joint ventures require careful strategies to craft a team of specialists with unique offerings of value that best contribute to the project deliverable.

We have argued that a business strategy needs to be formulated for organisations to deliver innovation effectively. This strategy is affected by and affects the evolution of the technological environment, process or product innovation. It in turn affects and is affected by the industry context that provides the driver for innovation and change. The organisational context also is intimately affected by and affects both the business strategy and industry context. These four influences are linked and together shape the innovation strategy needed to be adopted. The complexity of this view is enhanced when issues of skills and competency development are linked to decision making on how best to leverage experience. To state this simply, there is no simple answer, other than by undertaking a comprehensive and thoughtful strategic response to forces and influences prevailing from within and external to the organisation.

One thing is certain, innovation is not an option but a requirement for survival and long-term prosperity. Innovation has been shown to deliver substantial benefits. In the Australian innovation study (PricewaterhouseCoopers 2001, p. 72) 'high innovators' in the sample surveyed when compared to 'low innovators' achieved substantially better results across a wide range of factors:

- recording three to four times the proportion of turnover from products and services developed in the last three years;
- recorded three to four times the cost savings as a result of process and organisational innovations in the last three years;
- created more jobs and capitalised on more new markets;
- were 50 per cent less likely to miss their project delivery deadlines; and
- were 50 per cent less likely to fail to achieve their stakeholders' objectives.

To achieve the above, effective strategy rather than good fortune is the key. In Chapter 6 we introduced a core ICT innovation, ProjectWeb that enabled the alliance team to effectively communicate and coordinate their activities. In the next section we outline the impact of another important innovation that lay at the core of the National Museum of Australia project delivery system – the three dimensional (3D) modelling design tool used on that project. We chose this innovation to demonstrate how this innovation affected those participating in the alliance and others critical to the project delivery. It was an innovation that required the kind of strategy illustrated in Table 9.1 and, similar to the ProjectWeb innovation discussed in Chapter 6, was an important determinant for the client in the selection of alliance partners who could deliver the necessary skills and experience to provide effective delivery of benefits from innovation. In this sense the following provides a strategically selected innovation case study.

9.4 Pushing the (3D) envelope in documentation – shop drawing and manufacturing procedures in the Australian construction industry on the Acton Peninsula project

The Acton Peninsula alliance (APA) was an innovative grouping of leading construction industry and government organisations, specifically formed to deliver the National Museum of Australia project for the Australian Commonwealth Government. The project comprised the National Museum of Australia and the Australian Institute of Aboriginal and Torres Strait Islander Studies, brought together on an 11-hectare site on the shores of Lake Burley Griffin in Canberra. The Commonwealth Government took a lead role in fostering new forms of industry groupings and delivery processes as part of its commitment to ongoing industry reform and innovation. The capacity to demonstrate effective delivery of innovation was a key selection criterion (see Chapter 4).

This section addresses a particular aspect of innovation that was significantly advanced by the APA – the development of more advanced integration of computer aided design, documentation and manufacturing than are normally found in the industry. These innovations were made possible by two main factors:

(1) By the research and development efforts of the organisations involved.
(2) By the way in which the project alliance structure allowed these organisations to operate and to interact with each other synergistically and free of the normal non-productive contractual behaviours commonly seen in the construction industry.

9.4.1   The traditional paradigm

For many years the construction industry has relied on a traditional paradigm for the way in which project information is created and communicated to the project participants. In its simplest form, ideas are created by designers attempting to devise real world solutions (normally three-dimensional objects) to clients' particular problems or needs. These imagined 3D objects (whether buildings, roads or industrial facilities) are then deconstructed into 2D drawings or representations of various aspects of the imagined object. These deconstructed representations are provided to the construction team, who are then expected to reconstruct the 3D imaginings of the designers as a real object. Further, for various contractual and practical reasons, it is common for the documentation produced by designers to be redrawn by fabricators to suit their own particular needs.

Computers were first introduced as technical drawing tools into Australian design offices in the early 1980s. Since that time their abilities and use have increased exponentially. Now it is almost universal that 2D drawings are provided in CAD formats, typically using software such as AutoCAD and MicroStation. Also 3D computer renderings of proposals as a presentation tool is in increasingly common use, usually based on software such as 3D Studio Max or Form Z. The majority of users of these techniques, however, appear to view them as replacements for the traditional techniques, i.e. 2D CAD instead of 2D manual drafting, and 3D renderings instead of hand-drawn perspectives and artist's impressions or physical models.

A smaller number of design professionals have begun to explore the potential of these computer tools beyond their common role simply as a more efficient mechanism for carrying out traditional tasks. This has led to developments such as:

- linking of 2D CAD plans to databases to extract attributes such as floor areas, quantities and other facility data;
- use of 3D software to create animations and fly throughs; and
- electronic transfer of drawing information between offices.

9.4.2   Paradigm shifts

In Melbourne, Australia, a small group of architectural practices, led by Ashton, Raggatt, McDougal (ARM), have pushed their use of 3D software well beyond the common role of the replacement of the perspective artist or the model maker. The software is being used as a design tool itself, becoming completely interactive with the design process to produce new ideas and possibilities which were beyond the reach of traditional manual techniques. Examples of this application include:

- Use of mathematical algorithms to make repeatable (i.e. economically manufacturable) but non-regular design ideas.
- Ability to use lighting software to design and verify lighting proposals before manufacture, both in terms of appearance and technical factors such as light levels.
- Testing of various material selections in terms of colours, reflectance and surface patterns before committing to purchase or manufacture.
- Construction of virtual reality mark-up language (VRML) models enabling real time journeys through and around buildings by clients.
- The production of scale models and prototypes directly from 3D computer aided design (CAD) files (e.g. stereo lithography).

These new ideas also push the design team beyond normal means of construction documentation, into areas such as:

- Supply of exact cross-sections of complex shapes directly to development of shop drawers;
- Accurate checking of coordination of structural framework and external linings of complex shapes prior to fabrication.
- Supply of accurate cutting patterns direct to fabricators.
- Supply of accurate 3D positioning information to site.

Further, in a recent meeting Professor John Gero of the University of Sydney stated the interesting observation that the CAD technology of today was developed over two decades ago as part of a series of computer science PhDs. It has taken several decades for this technology and innovation to reach the levels of maturity and acceptance within the construction industry of today. It seems that typically a 10- to 20-year gestation period is evident in bringing these technologies into common use. He also noted that the levels of innovation evident in computer games used by teenagers and young people today are the forebears of the simulation and virtual reality innovations that will be in widespread use in the construction industry in the decades ahead. While Bovis Lend Lease and Arup did not use these technologies on the National Museum of Australia project they are being discussed and considered as part of a research development into the use of these innovative technologies in the future. Bovis Lend Lease is a global construction developer, project manager and contractor and Arup is a global engineering consultancy.

9.4.3   3D CAD innovation on the National Museum of Australia project

A number of innovations in the use of computer capabilities that were used on this project have been realised – some believed this to be for the first time in the industry in Australia. ARM Senior Associate Paul Minifie and his team led the joint venture architects (Ashton, Raggatt, McDougal and Robert Peck von Hartel Trethowan) in this area by sourcing and adopting new software (originally from non-architectural applications) specifically for the National Museum of Australia project.

This allowed the architectural team to construct a full 3D surface model of the buildings, both internally and externally. The model communicated the 3D intentions of the designers directly and accurately, without the need to go through the 2D translation process and back again. Some examples of how these innovations have flowed through the design and construction process are briefly outlined below. This has only been possible because the team participants have made the commitment and have the ability to deal with this type of information exchange. Additionally, the substantial commitment made by the alliance to the implementation of ProjectWeb, the internet-based project communications tool developed by the Bovis Lend Lease project managers, as a project-wide means of information exchange, storage and retrieval was crucial to the orderly implementation of these innovations. ProjectWeb allowed the electronic exchange of information in many formats between the team members, in a fast and auditable way.

Commitment was sealed through the project alliance with two important mechanisms that demonstrate the importance of a relationship-based procurement approach. The 'soft' side of this was the trust and commitment developed through successful relationship building and maintenance described in Chapter 3 and Chapter 10 of this book but also elsewhere in Chapters 7 and 8. The 'hard' side of the relationship enabler was a 'no-litigation' clause that was part of the National Museum of Australia project alliance agreement. This latter provision allowed designers the confidence that best endeavours would be made to ensure that any problems or questions relating to dimensional integrity, etc. would be dealt with pragmatically rather than by the business-as-usual approach of immediately resorting to a strategy of collecting evidence of blame and using this information as a crude weapon to extract 'extras' for contract variations.

9.4.4   Design issues

The creation and maintenance of an accurate 3D surface model produced many benefits:

- The architects were able to produce a wide range of representations of exterior and interior appearances on an ongoing basis, both for inhouse design purposes and for presentations to stakeholders.
- Animations and VRML fly throughs were produced on a regular basis for various presentations, which is an easier task due to the ongoing availability of the 3D model. In other projects, the model is often produced simply for the purpose of providing a set of images, which is not particularly cost effective.
- Many design ideas were tested quickly and accurately using the model.
- Accurate information for complex shapes was given to the cost planner – surface areas for example.
- Accurate setouts for construction were generated quickly.
- The design team has also researched software which unfolds 3D surfaces to create flat cutting patterns or panel sizes – this was used for building accurate physical models of the more complex design areas, as well as communicating exact panel layouts to the façade contractors.

The 3D model was used to provide exact volumes and surface areas to the acoustic specialists Bassett Consulting Engineers. This allowed critical benefits of modelling and testing materials and conditions likely to prevail to minimise both time and energy in fine-tuning the interior design to meet acoustic requirements of the Museum.

This specialist lighting design consultant (Vision Lighting) made full use of the 3D model to construct accurate visual and technical tests of the proposed lighting design. The existence of the model improved the efficiency of their task. The ability to show the architects and the clients the exact effect of the design was also invaluable, and far superior to the use of so-called artists' impressions, which can be very misleading and are expensive to constantly reconstruct. This latter fact tends to force decisions to be made on the basis of very limited testing of options and a limited understanding by the client in particular of the final effect.

Anway and Company, the exhibition designers, welcomed the availability of the 3D model as a valuable tool to construct their own renderings and VRML model. The latter model allows the client to be walked around the design at the various stages of its development, and to gain a more complete understanding of the views and vistas at any point of the exhibition. To replicate this by manual sketches would have been extremely time consuming, and similar to the lighting design, would actually result in less testing of the ideas and decisions based on less information.

The landscape architects organisation, Landscape Architects Room 413, was also provided with the base 3D model of the whole site and buildings, and used this to coordinate the levels and contours of their design. They augmented the model with their own 3D work where necessary, which was then returned to the architects for inclusion in the base model. They used the combined model to provide renderings of the main central landscaped space, the Garden of Australian Dreams.

9.4.5   Engineering and manufacturing issues

The structural engineers (Arup) received the full 3D surface model from the architects as the starting point to define the area available for structural members. They in turn designed the structure using 3D structural analysis tools, and constructed their own 3D CAD model of all the main structural members. This electronic model once designed was given to the fabricator and sub-alliance member National Engineering. National Engineering appointed several steel detailing companies to provide the shop drawing services, again based on an accurate 3D CAD model. Cutting, scheduling, positional and detailing information is then produced from the final 3D model of the steelwork.

This process has eliminated a significant amount of steelwork detailing which would normally be carried out in the engineer's office. Typically these details are then redrawn by the fabricator and often redesigned to an extent in conjunction with the engineer. This more streamlined process was possible only because of three factors occurring simultaneously:

(1)   The innovative contractual arrangements in the project alliance.
(2)   The generation of the design information in a 3D format.
(3)   The engineer, fabricator and detailers having the willingness and capability to deal in this format of information.

The process also resulted in fewer errors and less redundant drawing and fabrication work. It resulted in close to zero coordination errors between steelwork and building finished surfaces, which would otherwise be a major issue for the complex interior and exterior shapes which occur in parts of the National Museum of Australia.

G James Pty Ltd, the sub-alliance facade partner in the project, dealt with all the glass and aluminium components with their sub-alliance partner National Engineering in fabricating the shaped steel sheet components. Their use of the 3D model in these areas was extensive. Benefits they experienced included:

- The exact shapes and sizes of all the aluminium panels were provided directly to G James as an unfolded surface model. This was sufficiently accurate to eliminate the need for panel shop drawings entirely. The unfolded model was also used to locate the different types of panels and all the colours.

- Similarly, exact dimensions of all aluminium glazing sections and glass sizes were worked through with G James, and adjusted in the model to ensure accuracy while accommodating the most economical and buildable sizes and spans to be used. This process was greatly facilitated by the alliance environment, and has involved the transfer of 3D technological skills from the architects to the subcontractors – a significant benefit to the industry.
- Strongly contoured precast concrete panels were used in sections of the facade. Full size prototypes were made from polystyrene using direct CAD CAM manufacturing techniques, again from a 3D model.

Typically, services engineers in Australia are not yet using 3D modelling to locate ductwork, pipework or to coordinate the locations of all services with the building envelope. A small number of these firms use 3D modelling to design thermal loads or to check the computational flow dynamics of a space. In this project the design team provided the engineers with relevant cross-sections from the 3D model for this purpose – with significant advantages gained in management and waste reduction over that normally anticipated and planned for.

9.4.6 The National Museum of Australia project experience of this innovation

The innovative approach used on the National Museum of Australia project allowed significant transfer of 3D information throughout the construction project progress. To the best of the designer's knowledge, the degree of integration throughout the team and in particular the use of the accurate surface modelling (as opposed to the more usual visualisation techniques using 3D Studio for example), and the surface unfolding are both pioneering advances in Australia.

This innovation represents a significant step in the Australian building and construction industry toward a possible information transfer standard, which sees the 3D model as the common data repository for the project. In future, it is anticipated that project information will be shared and manipulated by all participants enabling the ultimate 'design and construct' mechanism, where the full description of the internal and external surfaces is provided as the design intent document, with total construction flexibility in the spaces between, and full coordination of all services and structure done visually in 3D. In addition, client design sign-off could be based upon the 3D model.

9.5 Summary and implications for the future

We aimed in this chapter to indicate and explain how an innovative culture may be developed through a relationship-based procurement approach. We started by providing some conceptual insights into innovation, what the literature and what we mean by the concepts and how innovation may apply to a product or process. We avoided getting bogged down with discussion about product innovation and concentrated upon process innovation because we believe the content of this chapter is best served by concentrating on process innovation – our aim was to discuss innovation in the context of relationship-based procurement approaches, which is deeply concerned with management process.

We discussed the nature of innovation and how it can be fostered. Figure 9.2 provided a useful illustration of leadership drivers and requirements for innovation, which closely link with other aspects discussed throughout this book. The cultural and human factors were highlighted as pivotal for successful innovation generation and implementation and organisational strategy were underscored as necessary to deliver the appropriate facilitation process to achieve an innovative culture.

We provided as tangible evidence of successful innovation an example of the extension of 3D CAD technology used on the National Museum of Australia project through a significant part of the building construction supply chain. We explained how this innovation had a major and positive impact upon a number of project team members, some of whom were part of the project alliance and some who were not. This project is used throughout this book as an exemplar of a relationship-based procurement approach. Readers will no doubt find Chapter 6 of interest where another significant ICT innovation, ProjectWeb, is discussed and Chapter 8 where development of a quality culture was based in part upon innovative processes.

9.6   References

Amabile, T.M. (1998) 'How to Kill Creativity.' *Harvard Business Review*. **76** (5): 76–87.

Brown, J.S. and Duguid, P. (2000) 'Balancing Act: How to Capture Knowledge Without Killing It.' *Harvard Business Review*. **78** (3): 73–80.

Burdett, J.O. (1999) 'Leadership in change and the wisdom of a gentleman.' *Participation and Empowerment: An International Journal*, **7** (1): 5–14.

Burgelman, R.A. and Rosenbloom, R.S. (1989) Technology Strategy: An Evolutionary Process Perspective. *Research on Technological Innovation, Management and Policy*, eds Rosenbloom, R.S. and Burgelman, R.A. Greenwich, CT, JAI Press.

Cleland, D.I. (1999) *Project Management Strategic Design and Implementation*. Singapore, McGraw Hill.

Davenport, T.H. (2001) Knowledge Workers and the Future of Management. *The Future of Leadership – Today's Top Leadership Thinkers Speak to Tomorrow's Leaders*, eds Bennis, W., Spreitzer, G.M. and Cummings, T.G. San Francisco, Jossey-Bass: 41–58.

Davenport, T.H., Delong, D. and Beers, M. (1998) 'Successful Knowledge Management Projects.' *Sloan Management Review*. **39** (2): 43–57.

Davenport, T.H. and Prusak, L. (2000) *Working Knowledge – How Organizations Manage What They Know*. Boston, Harvard Business School Press.

Hames, R.D. and Callanan, G. (1997) *Burying the 20th Century – New Paths for New Futures*. Warriewood, NSW, Business & Professional Publishing.

Hammer, M. (1996) *Beyond Reengineering – How the Process-centred Organization is Changing our Work and our Lives*. New York, Harper Collins.

Hammer, M. (2001) 'The Superefficient Company.' *Harvard Business Review*. **79** (8): 82–91.

Hammer, M. and Champy, J. (1983) *Reengineering the Corporation – A Manifesto for Business Revolution*. Sydney, Allan & Unwin.

Hampson, K.D. (1993) Technology Strategy and Competitive Performance: A Study of Bridge Construction. PhD, *Civil Engineering*. Stanford, CA, Stanford University.

Hampson, K.D. and Tatum, C.B. (1997) 'Technology Strategy and Competitive Performance in Bridge Construction.' *Journal of Construction Engineering and Management*. **123** (2): 153–161.

Hansen, M.T., Chesbrough, H.W., Nohria, N. and Sull, D.N. (2000) 'Networked Incubators Hothouses of the New Economy.' *Harvard Business Review*. **78** (5): 73–84.

Hansen, M.T., Nohria, N. and Tierney, T. (1998) 'What's Your Strategy for Managing Knowledge?' *Harvard Business Review*. **77** (2): 106–116.

Hargadon, A. and Sutton, R.I. (2000) 'Building an Innovation Factory.' *Harvard Business Review*. **78** (5): 157–166.

Harkola, J. (1994) *Diffusion of Construction Technology in a Japanese Firm*. PhD, Department of Civil Engineering. Stanford, CA, Stanford University.

Hilmer, F.G. and Donaldson, L. (1996) *Management Redeemed – Debunking the Fads that Undermine Corporate Performance*. East Roseville, New South Wales, Australia, Free Press.

Imai, M. (1986) *Kaizen: The Key To Japan's Competitive Success*, New York, McGraw Hill.

Kikawada, K. and Holtshouse, D. (2001) The Knowledge Perspective in the Xerox Group. *Managing Industrial Knowledge – creation, transfer and utilization*, eds Nonaka, I. and Teece, D. London, Sage: 283–314.

Kim, W.C. and Mauborgne, R. (1999) 'Creating New Market Space.' *Harvard Business Review*. **77** (1): 83–93.

Kluge, J., Stein, W. and Licht, T. (2001) *Knowledge Unplugged – The McKinsey & Company Global Survey on Knowledge Management*. New York, PALGRAVE.

Kulkki, S. and Kosonen, M. (2001) How Tacit Knowledge Explains Organizational Renewal and Growth: the Case of Nokia. *Managing Industrial Knowledge – creation, transfer and utilization*, eds Nonaka, I. and Teece, D. London, Sage: 244–269.

Leadbeater, C. (2001) How Should Knowledge be Owned? *Managing Industrial Knowledge – creation, transfer and utilization*, eds Nonaka, I. and Teece, D. London, Sage: 170–181.

Lenard, D.J. and Bowen-James, A. (1996) *Innovation: The Key to Competitive Advantage*. Adelaide, Australia, Construction Industry Institute Australia, University of South Australia.

Leonard, D. and Rayport, J.F. (1997) 'Spark Innovation Through Empathic Design.' *Harvard Business Review*. **75** (6): 102–113.

Leonard-Barton, D. (1995) *Wellsprings of Knowledge – Building and Sustaining the Sources of Innovation*. Boston, MA, Harvard Business School Press.

MacMillan, I. and McGrath, R.G. (1997) 'Discovering New Points of Differentiation.' *Harvard Business Review*. **75** (4): 133–145.

Mintzberg, H. (1994) *The Rise and Fall of Strategic Management*. London, Prentice-Hall International (UK).

Mitropoulos, P. (1996) Technology Adoption Decisions in Construction Organisations. PhD, *Department of Civil Engineering*. Stanford, CA, Stanford University.

Morris, P.W.G. (1994) *The Management of Projects: A New Model*. London, Thomas Telford.

Nam, C.H. and Tatum, C.B. (1989) 'Towards Understanding of Product Innovation Process in Construction.' *Journal of Construction Engineering and Management*. **115** (4): 517–534.

Nam, C.H. and Tatum, C.B. (1992) 'Strategies for Technology Push: Lessons from Construction Innovation.' *Journal of Construction Engineering and Management*. **118** (3): 507–524.

Nonaka, I. and Takeuchi, H. (1995) *The Knowledge-Creating Company*. Oxford, Oxford University Press.

Nonaka, I., Toyama, R. and Konno, N. (2001) SECI, *Ba* and Leadership: A Unified Model of Dynamic Knowledge Creation. *Managing Industrial Knowledge – creation, transfer and utilization*, eds Nonaka, I. and Teece, D. London, Sage: 13–43.

Porter, M.E. (1990) *The Competitive Advantage of Nations*. New York, Free Press.

Porter, M.E. (2001) 'Strategy and the Internet.' *Harvard Business Review*. **79** (3): 63–78.

PricewaterhouseCoopers (2001) Innovation in the Australian Building and Construction

Industry Survey Report, Canberra, Australia, Australian Construction Industry Forum and the Department of Industry Science and Resources: 75.

Prusak, L. (1997) *Knowledge in Organizations – Resources for the Knowledge-based Economy*. Oxford, Butterworth-Heinemann.

Rogers, E.M. (1995) *Diffusion of Innovation*. New York, Free Press.

Shen, Y.J. and Walker, D.H.T. (2001) 'Integrating OHS, EMS and QM with Constructability Principles when Construction Planning – A Design & Construct Project Case Study.' *TQM*, **13** (4): 247–259.

Tatum, C.B. (1989) 'Organizing to Increase Innovation in Construction Firms.' *Journal of Construction Engineering and Management*. **115** (4): 602–617.

Thomke, S. (2001) 'Enlightened Experimentation – The New Imperative for Innovation.' *Harvard Business Review*. **79** (2): 66–75.

Tulley, S. (1998) 'How Cisco Mastered the Net.' *Fortune*. **4** (138): 207–209.

Turner, J.R. (1999) *The Handbook of Project-based Management: Improving the Processes for Achieving Strategic Objectives*. London, UK, McGraw Hill.

Van de Ven, A.H. (1986) 'Central Problems in the Management of Innovation.' *Management Science*, **32** (5): 590–607.

Von Hipple, E., Thomke, S. and Sonnack, M. (1999) 'Creating Breakthrough at 3M.' *Harvard Business Review*. **77** (5): 47–57.

Walker, D.H.T. and Lloyd-Walker, B.M. (1999) Organisational Learning as a Vehicle for Improved Building Procurement. *Procurement Systems: A guide to best practice in construction*, eds Rowlinson, S. and McDermott, P. London, E&FN Spon, London, UK. **1**: 119–137.

Walker, D.H.T. and Shen, Y.J. (2002) 'Project Understanding, Planning, Flexibility of Management Action and Construction Time Performance – Two Australian Case Studies.' *Construction Management and Economics*, **21** (in print).

Walker, D.H.T. and Sidwell, A.C. (1996) *Benchmarking Engineering and Construction: A Manual For Benchmarking Construction Time Performance*. Adelaide, Australia, Construction Industry Institute Australia.

Wenger, E.C. and Snyder, W.M. (2000) 'Communities of Practice: The Organizational Frontier.' *Harvard Business Review*, **78** (1): 139–145.

Womack, J.P. and Jones, D.T. (2000) From Lean Production to Lean Enterprise. *Harvard Business Review on Managing the Value Chain*. Boston, MA, Harvard Business School Press: 221–250.

Womack, J.P., Jones, D.T. and Roos, D. (1990) *The Machine that Changed the World – The Story of Lean Production*. New York, Harper Collins.

Yukl, G. (1988) *Leadership in Organisations*. Sydney, Prentice-Hall.

Chapter 10
Implications of Human Capital Issues

Derek Walker

We discussed in Chapter 3 the development of cross-team relationships including the role of power and influence, encouraging performance improvement through innovation, problem resolution, trust and commitment and teams and leaders. This provided a sound foundation for this chapter in which theory presented in Chapter 3 is extended with specific reference to relationship-based procurement examples. An important source of data that helps us expand upon Chapter 3 was data gathered and analysed in the National Museum of Australia project research study.

This chapter focuses on human capital and ways in which its effectiveness can be leveraged through relationship-based procurement systems. Human capital consists of the skills, competences and knowledge and creative energy that employees bring an organisation.

10.1 Introduction

In attempting to understand how winning organisations can be developed, Lawler identified human capital, organisational capabilities and core competencies as principal sources of competitive advantage (Lawler III 2001, p. 14). Porter's early work identified cost competition as one aspect of competitive advantage but argues that providing a quality advantage through differentiation of product/services offered provides a more sustainable advantage (Porter 1990, 1998, 2001). The contribution that human capital can make in adding value is based upon unique perspectives that each individual develops through the way in which the neural connections of their brain have been wired by experience (Greenfield 2000). Hence by maximising the use of human perspective in designing products, systems and services, greater customisation and differentiation is possible.

Organisational capabilities refer to the abilities of an organisation to perform in ways that deliver products and services that provide distinctive value to the customer. Core competences are described as the technical capabilities and services that allow organisations to win in the marketplace (Lawler III 2001, p. 15). Lawler highlights three pivotal competitive advantage implications of organisational capabilities. First, he identifies the issue of the type of human capital that organisations need to attract, second the question of how long they want to retain their human capital and third, the nature of reward systems required to attract and retain their human capital (Lawler III 2001, p. 17). This issue of working with

258

talented human capital and its central place in sustained competitive advantage of organisations is critical and has occupied the minds of an increasing number of management thinkers. Charles Handy is perhaps one of the more popular proponents of a change to the way in which organisations deal with this issue (Handy 1997; Handy 2001). Some organisations are highly proactive in developing systems to attract and retain the best talent. Cisco Systems, for example, maintained a database of over 60 000 knowledge workers that they considered as potential employees with details to readily identify and recruit the bright talent (Davenport 2001, p. 52). Clearly there is a need for motivated knowledgeable and talented people who have the capacity to develop high quality cross-team relationships discussed in Chapter 3 when dealing with partners in an alliance or other relationship-based procurement system.

We recognise that most people involved in project management bring with them a wealth of knowledge and so we concentrate in this chapter upon the individual knowledge workers and the environment in which they can best contribute to relationship-based procurement systems. In doing so we draw upon data gathered from a number of case studies and more specifically on the National Museum of Australia project. The issues and questions that we have decided to focus upon are:

- How can ethical positions and beliefs affect negotiating styles when negotiating the resolution disputes and day-to-day interaction in relationship-based procurement teams?
- Which organisational environment factors were indicated as significant in affecting trust and commitment on the National Museum of Australia case study project?
- What human capital skills are required to be developed to best operate an alliance and how may these identified skill gaps be met?
    - In what way and to what extent are these skills being developed at present?
    - What reward systems may be appropriate to alliancing projects?

In addressing these issues and important questions we hope to develop a best practice model for relationship-based procurement system teams.

10.2   Ethical positions and beliefs in negotiating the resolution of disputes and day-to-day interaction

Negotiation takes place in a wide variety of forms. Resolution of sensitive conflicts concerning formal issues such as contractual arrangements and the rights and obligations of each party to the negotiation process is a more easily recognisable form. Mutual adjustment by people simply working out acceptable ways in which they get on with their job is a less formal but it could be argued, equally important form. Both can have a significant impact upon the relationship between parties and the building and maintenance of commitment to work together to further and achieve project goals. The issue of what 'ethical' means is interesting. 'Being ethical' is a construct that needs to be explored so that we can fully appreciate where it logically fits into this section.

Ethics has been generally described as 'doing the right thing' or the 'morally justifiable thing' or conforming to the 'golden rule'. All these are vague and potentially unhelpful terms because they are not specific enough to well be defined. Moreover, ethics is culturally filtered (workplace, national or social group). It is an intensely interesting subject and invariably sparks much debate and passion. Referring to morals is equally difficult. Morality has been defined as 'the standards that an individual or a group has about what is right or wrong, or good or evil. Moral standards are also said to include the *norms* we have about the kinds of actions we believe are morally right as well as the *values* we place on the kinds of objects we believe are morally good and morally bad.' (Velasquez 1998, p. 8).

Ethical positions are therefore arrived at from analysing a situation in terms of a framework of norms and values held to be justifiable, valid and relevant. Thus, two negotiators can believe that they are acting morally yet use quite different values and norms to justify a particular course of action. Take, for example, the kind of dilemma that arises out of a corporate representative negotiating in regard to a proposed infrastructure road construction development with an environmentalist representing a local community. Corporate representatives are obliged by law and their business ethics to uphold the interest of their shareholders. These representatives may well hold other subsidiary interests and positions but they have an obligation to protect the shareholder interests. Such negotiations are very complex because while, for the corporate representative, the fiduciary interest can include a financial or pecuniary interest there are issues of short-term versus long-term sustainability to consider.

The environmentalist would be defending the interests of the local community, the general community and the 'environment'. For the environmentalist, the situation is complex because the various communities may have conflicting interests. The neighbourhood community might gain advantage from an enhanced road system – perhaps through raised property values and other elements of utility. However, offsetting against any advantages may be disadvantages from potential discomfort caused by noise, fumes and potential danger to children from traffic on the new roadway. Consequently, this local community may be divided in their opinion and justification for any negotiating positions being considered. At a wider community level, the road may best serve the public 'good'. Issues of employment growth, potential enhanced business activity, freedom of movement and a host of other defendable positions could be cited as support for the road proposal. From the environmental perspective, arguments could be made regarding advantages from reduced fuel consumption and pollution due to elimination of vehicle engines wasting fuel while idling. Such arguments may be contrasted with the encouragement of private rather than public transport and other ancillary arguments about which transport system is more environmentally beneficial. The arguments and counter-arguments can easily be imagined and potentially conflicting nature understood.

Mutual adjustment between team members or between teams provides a more mundane but relevant concept of negotiation. Examples include negotiations about who should do what, when and in what sequence. The same paradoxes, the same fundamental complexities, and the same moral dilemmas apply but at a different scale. This book is primarily concerned with relationship-based procurement systems and in doing so the issue of stakeholders is important because negotiators represent stakeholder interests. Moral justification is often shaped by the wants and needs of stakeholders and the assumption that they have a right to consideration and that their representatives are morally obliged to protect their interests.

Figure 10.1 indicates the typical model of stakeholders for a project. The client organisation is perhaps a more obvious stakeholder, however, this may not be the project sponsor and is unlikely to be the end-user. The community and external independent groups have been intimated in the earlier discussion regarding the environmentalist. Team members from the supply chain hold a stake in the success of the project. The core project team members are an obvious group with a visible stake in project success. Shadow team members are an interesting group whose interests are also often in the minds of other stakeholders but often not recognised as such. For example the family, friends and close associates of core team members have a poorly recognised stake in how projects are managed. These are also referred to as invisible team members (Briner *et al.* 1997, p. 83). Increasingly there are calls for a more family-friendly workplace and this is extending to a concept of a balanced life being promoted, even with talk of a corporate athlete requiring balance between rest and work and play and work (Loehr and Schwartz 2000). Invisible team members form important shadow networks that can have considerable but nebulous influence that must be recognised and are often in the form of norms and values that underpin an individual's moral and ethical framework of justice and fairness and a sense of what is correct or good.

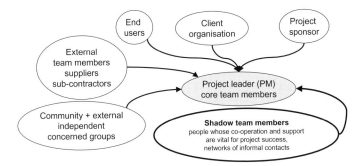

Figure 10.1:   Stakeholder model

In developing the mental picture of how to represent the complex interests of an identified stakeholder a subconscious or explicit map may be developed in which stakeholders can be set in one column with all identifiable stakeholders in other columns. A symbol representing the level or intensity of vested interest can then be placed in each cell, such as traffic lights, to provide an influence map. Cleland (1999, p. 175) provides a simplified version of this concept.

Before concluding this part of the discussion, it is worth briefly touching on the concept of culture and how this may affect negotiated positions. Hofstede identified from a famous long-term and substantial study of IBM employees four dimensions of culture: small versus large power distance; collectivism versus individualism; femininity versus masculinity; weak versus strong uncertainty avoidance (Hofstede 1991). He argues that the way that people behave, the values that they hold and the symbols, rituals and heroes that govern their worldviews are based upon positions along the continuum of those four dimensions. He attempts to generalise across national boundaries to explain what might be understood as a 'western' or 'oriental' cultural bias. For example, in developing his power distance index (PDI) he maintains that in small PDI cultures subordinates readily approach and contradict organisational superiors. The perception being that the power distance is small whereas in large PDI cultures, the distance is perceived to be large and thus subordinates prefer, or are more comfortable with, an autocratic or authoritarian 'boss'. On the collectivist versus individualist cultures dimension he argues that relationships between people at various levels of an organisational hierarchy are seen in moral terms of mutual obligation. In masculine cultures roles are generally seen to be distinct whereas in feminine cultures they overlap. In terms of the uncertainty avoidance dimension, high uncertainty avoidance cultures shun ambiguous situations and need rules and structure (Hofstede 1991). These dimensions are partially contested or at least redefined by others interesting in measuring culture.

Charles Hampden-Turner and Fons Trompenaars identify six dimensions; universalism versus particularism; individualism versus communitarianism; specificity versus diffusion; achieved status versus ascribed status; inner direction versus outer direction; sequential time versus synchronous time. These are briefly explained in Figure 10.2.

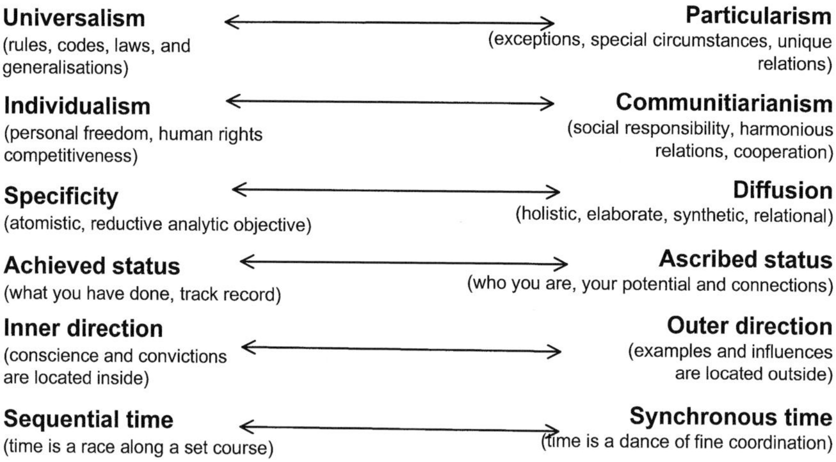

Universalism
(rules, codes, laws, and generalisations)
⟵⟶
Particularism
(exceptions, special circumstances, unique relations)

Individualism
(personal freedom, human rights competitiveness)
⟵⟶
Communitiarianism
(social responsibility, harmonious relations, cooperation)

Specificity
(atomistic, reductive analytic objective)
⟵⟶
Diffusion
(holistic, elaborate, synthetic, relational)

Achieved status
(what you have done, track record)
⟵⟶
Ascribed status
(who you are, your potential and connections)

Inner direction
(conscience and convictions are located inside)
⟵⟶
Outer direction
(examples and influences are located outside)

Sequential time
(time is a race along a set course)
⟵⟶
Synchronous time
(time is a dance of fine coordination)

Figure 10.2: Six cultural dimensions
Source: Hampden-Turner and Trompenaars (2000, p. 11) copyright © John Wiley & Sons.

The illustrations from Hofstede and Hampden-Turner and Trompenaars show us the complexity facing negotiators at both the formal and informal level as each party may or may not share the same ethical frameworks, yet each may be convinced of their morality. Take, for example, the sixth dimension of sequential versus synchronous time. A sequential time-oriented person may believe that a synchronous time-oriented person is needlessly dragging out a negotiated decision while the latter may believe that the best solution will naturally emerge at the naturally 'correct' time. Frustration is both an understandable and foreseeable outcome if both parties fail to understand each other's perspective. This empathic quality has been variously described as 'good communication' skills through to good 'cross-cultural understanding' skills and has been more widely advocated. A major government report of the 1990s sponsored by the then Australian Labour Federal Government advocated cross-cultural 'literacy' as vital for a multicultural and globally trade-oriented nation (Karpin 1995).

Much has been written about the relationship-based procurement approach and its impact upon conflict resolution, which we say in Chapter 3 is a pillar of the partnering and alliancing framework. Bresnen and Marshall provide much insight in their assessment of the dilemmas and tensions that may occur in partnering arrangements that reaches beyond the 'hype' delivered by many proponents of this cooperative form of relationship. They stress that behavioural change does not necessarily follow the adoption of a partnering charter and that a set of espoused standards of behaviours may well vary considerably from that which is practised (Bresnen and Marshall 2000). This is also a matter of concern noted in a Construction Industry Institute Australia (CIIA) survey in which 100 per cent of respondents agreed that issues and problems were allowed to escalate and that 86 per cent of respondents agreed that continuity of open and honest communication was not achieved (Lenard *et al.* 1996). This clearly indicates that all team members engaged in negotiation of any kind need to develop negotiating skills that provide the necessary foundation of trust and commitment for parties to genuinely seek ethical and fair solutions. One of the essential features of partnering has been the way in which problem resolution is managed (Bennett and Jayes 1995). Problem resolution requires open discussion and high levels of trust and commitment to gaining a fair and ethical solution – given the expectation that problem resolution will be undertaken in a non-adversarial manner.

Problem resolution also benefits from a number of perspectives being applied to the surrounding information and perceptions of what may be described as 'facts' (Senge 1990). In a major treatise on leadership, Burns stresses the link between leader and follower and the role of conflict. He argues that conflict is necessary for change as it challenges the *status quo* or 'facts' in the mind of those engaged in negotiating a changed state. In order for change to take place, one party in a change negotiation must create a conflict in the mind of the other party. This challenge results in a mental conflict taking place about what was perceived to be the 'right way' to do something based on new arguments, persuasive views expressed, presentation of new evidence, or perhaps existing evidence presented from a different perspective (Burns 1978). Thus conflict at least expressed in this way may be constructive and valid without being coercive. The need for diversity and an environment where alternative perspectives can be shared is well recognised (Senge 1990; Cope and Kalantzis 1997). As Loosemore *et al.* (2000b) argue, '... the managerial challenge is to harness the potential good in conflict rather than to develop methods of minimising it'.

The important link between trust, commitment and conflict is that it is necessary for open communication to take place for a clearer understanding of mutual positions to be developed. This means that one party must feel it has the right to argue from perhaps a self-centred position but must respect the other party's right to do the same. Moreover, in a more mature negotiation in resolving problems or seeking solutions to problems, this openness and preparedness to accept and debate alternative perspectives on issues helps to build trust and commitment which are essential ingredients of alliancing (Kwok 1998). Thus conflict management from an alliancing or partnering perspective requires open communication to facilitate equal-power negotiation between parties and a mechanism that allows problems to be resolved at the lowest level of management appropriate to the knowledge of the problem circumstances and context. Proponents of partnering and strategic alliances later stressed this aspect (Bennett and Jayes 1995; Lenard *et al.* 1996; Lendrum 1998; ACA 1999). We saw earlier in this section however, that openness to express conflicting perspectives on what the 'truth' might be is also governed by cultural influences (Hofstede 1991; Hampden-Turner and Trompenaars 2000).

Negotiation takes place continuously on projects as a part of the process of mutual adjustment. It also takes place between individuals resolving technical problems as well as trading energy and favours to make life easier and reduce friction involved in generally getting things done. Negotiation can be formal involving clear contractual or other obligation issues or merely the kind of mutual adjustment discussed above. Negotiation is a cooperative exercise, independent people communicate to explore ways to achieve mutual objectives (Thompson 1998). The way that negotiation takes place reveals attitudes, behavioural tendencies and value systems of parties to these negotiations. Lewicki has developed a model of ethical negotiation that idealises the process for mutual gain arising out of a negotiation process undertaken between individuals or groups (Lewicki 1983, p. 80). Selection of strategies and tactics to be employed by negotiators can be categorised as external and internal influences. Negotiators monitor these for acceptability by the internal and external motivational forces that influence them and consequential impact.

The influencing situation can be viewed as the perceived need for integrity, openness and truthfulness that should characterise the negotiating interaction. For example, in negotiating as part of a team with the expectation of a lengthy continuing relationship would require parties to contemplate the likely downside of a tactical victory (win–lose) on future negotiations. This may be a totally different situation to a once-off transaction such as procuring a stand-alone item X with very short continuing requirement for cooperation. Lewicki's work on studying negotiation was undertaken well before the literature of culture was developed and it is interesting to note the judgement of there being 'liars' whereas from a multicultural or post-modernist perspective we could also see this applying to different interpretations of truth based upon cultural dimension biases.

The degree of truthfulness can only be measured on a continuum. Adopting the extremes of this continuum, pure truth and pure deception, is rare. Truth in any event can be judged as a perception of perspective – it depends upon the context. Between these continuum extremes are various degrees of lying or deception where lying can be accepted as a conscious act and deception an unconscious act. These take several forms such as misrepresenting the position to an opponent, bluffing, falsification, leading an opponent to draw an incorrect conclusion or deduction, or being selective in what is disclosed or hidden as well as leaving false clues or trails. Such strategies or tactics have their cost as well as perceived benefits.

Negotiations are undertaken with the currency being information about the interests, motivation and expectation of outcomes of the parties. As Lewicki (1983, p. 74) puts it, 'Lies distort the quality of information that people receive and their ability to make accurate and informed choices about how to behave'. The interesting aspect of this is that it is mindful of the parties' feelings and attitudes and that long-term consequences are seriously considered. Lewicki argues that from the liar's perspective the primary motivation is to increase power and control. Liars are likely to grasp for justification or excuses about the act of misrepresentation to legitimise any actions that may be at internal conflict with their sense of integrity. Internal influences such as demographic factors may come into play such as the negotiator's sense of religious obligation to others. They may have personality traits that predispose them to believe themselves as being liked, admired or being trusted. The situation may motivate them towards a win–lose or win–win desired outcome. They may have a personal moral code that dictates their preferred negotiation style. All these are within the person's psychological make-up.

External factors include the type of interaction and its implication upon the relationship with the negotiating party – once-off versus continuing, the importance of reputation being maintained and other such influences. The relative power that a negotiating party may have could also influence the style of interaction and openness of information exchange. This is where natural power imbalances need to be considered. If one party has more power and wishes the other party to be open then the more powerful party must find ways of reducing the impact and influence of this power differential. Group norms of organisational culture also provide peer pressure to act in a certain way. This may be manifest in a highly 'macho' approach or a highly appreciative approach to choose extremes.

Both the internal and external influences exert pressure to adopt a style or negotiating stance that will affect the interaction. The parties will be constantly monitoring cues, overt and subliminal, to check both their own actions and those of the people they are negotiating with. They will be checking action against the internal models that represent their ethical standards and notions of integrity. They will also be asking questions of themselves of the likely impact of any given tactic and they will be pragmatically assessing the effectiveness of the tactic to check if it results in a desired outcome. Ethical negotiators purposefully sense their own comfort levels from their internal ethical standards and sense feedback from their negotiation partner. Much of this will be taking place at the subconscious level. As Figure 10.3 indicates, this leads to explanation and justification to either or both of the negotiating parties. The unconscious discourse with one's inner being can provide interesting physical manifestations that can be read as relief or pressure often in body language despite attempts to disguise them – being aware of these has its advantages.

Relationship marketing (RM) developed as a form of aligning customer and producer needs more closely. RM provides the converse to the transactional model: client focus, benefits oriented toward project product success, long time-scales that assure high levels of service emphasis in turn guaranteeing high levels of client commitment and a holistic quality view (Gronröos 1990; Gronröos 1994). A highly insightful model of ethics with respect to relationship marketing (see Figure 10.3 below) provides useful indicators of ethical conflicts. This model maps the ethical factors that trigger ethical problems, which in turn spark ethical conflicts that RM and other forms of relationship-based procurement seek to address.

Figure 10.3 goes some way to illustrate the complexity involved in addressing the ethical problems and conflicts and the source of those conflicts. It can be appreciated that the feedback required justifying ethical positions held by parties needs also to be rich and sophisticated in order that both parties can understand the pressures and influences to which their negotiating partners are subject to. Feedback communication in these cases includes not only the more tangible senses but also intangible sense making (Weick 2001) – meaning conveyed through body language, hidden cues and empathic transmission of understanding undertaken in a flexible context-based often heuristic or subconscious manner.

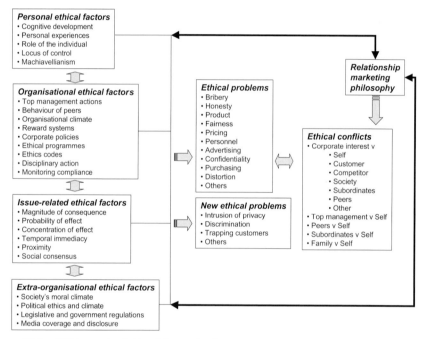

Figure 10.3:   Ethics and relationship marketing
Source: Kavali *et al.* (1999, p. 575), used with permission from MCB University Press

Given the above discussion of an ethical and empathic approach to negotiation, short-term win–win negotiation strategies appear somewhat crude and unsophisticated. To provide a quick win may suffice in highly transactional situations where neither party is likely to repeat a negotiation or influence others who may negotiate with either party – the transactional cost[1] in terms of time and energy expended to achieve the result may be small but the result may be only satisficing, that is satisfying the base-line requirement but not seeking an optimal solution. When undertaking negotiations with longer-term impact, particularly when a

---

[1] For a wider discussion of transaction cost see Walker (1993) in which he cites the work of Coase (1937) on the theory of the firm.

continuing relationship is likely, it is important to explore options and understand how each partner's needs and desires may be satisfied (Fisher and Ury 1983; Thompson 1998).

Win–win solutions as well as the more protracted 'I win–you lose' negotiations require a detailed process of pre-negotiation intelligence gathering, in which, as much as is practically possible, knowledge is gained about the negotiating partner's wants, needs, desires, and background. If this phase is successful the partner's underlying interests (what he/she/they want and why) and positions (what the partner may state that he/she/they want) are. Also the problem to be solved needs to be fully researched to determine its scope, scale and boundaries as well as potential for linking it to other issues so as to expand the solution possibilities. Additionally, internal research needs to be undertaken to know what authority the negotiator has or can claim as well as the scope, scale and flexibility of options that may be investigated or traded (Fritz *et al.* 1998). Once this groundwork has been completed it is possible through dialogue and sharing mental models to articulate that which is hidden so that undisclosed or unstated wants can be tested as realistic needs or possible positions. Often one party has the key to a win–win solution but does not know this because the unstated want or need of the negotiating partner may seem irrelevant to the negotiating context or beyond the immediate scope of the solution envelope. Only by unearthing these hidden agendas can feasible options be explored. In negotiations where trust may be conditional, the purposeful withholding of positions and hiding interests in a bargaining mode (a cat and mouse game) may be played but this requires high levels of transaction cost in terms of second-guessing rather than open discussion and energy expended in hiding and extracting information rather than openly sharing perspectives. It should be remembered that pure transaction-based negotiation where a result means an end to the parties' relationship is rare in many project-based enterprises of today. It is more likely to be the case that a continuing relationship will be the reality and so trust, commitment, integrity and reputation are more valuable assets than those achieved through short-term opportunism.

Fisher and Ury identify three commonly identified styles of negotiation. These are soft, hard and principled negotiation styles. Soft negotiation involves creating situations where the relationship between the parties is all-important, compromises are made and single answers or perspectives are focused upon for an agreed outcome that may well be highly sub-optimal. Soft negotiation often results in harsh realities being ignored. Hard negotiation involves a self-centred and sometimes harsh or brutal approach. The negotiation can easily become a battle or contest where the victory is perceived as being more important than a solution being generated to a problem. Hard bargaining may confront harsh realities but third-way solutions may be ignored that could produce a more productive outcome. Principled negotiation is argued to produce the best negotiation outcome, because parties are focusing on the problem rather than individual interests, participants act as problem solvers rather than resource optimisers addressing issues rather than winning battles or doing *quid-pro-quo* favours (Fisher and Ury 1983). If the nature of the problem is better understood from multiple perspectives then multiple potential solutions may be unearthed. If each party's interests, preferences and positions are known then these may be recast, reframed and reshaped to arrive at hitherto unconsidered solutions. Problem solving in a volatile environment where numerous possible solutions exist can offer a sensible approach to joint solution finding rather than the more dogmatic approach of bargaining (Weick 2001). The idea is to build solutions incrementally rather than trading off set-piece demands that could well be sub-optimal.

Principled negotiation is clearly a more sophisticated approach in terms of the need for better knowledge about each party to the negotiation, the environment in which the problem exists and the range of possible solutions both within the more obvious solutions envelope and those outside the immediately obvious, which may be linked. The demands for this approach in individual cases may seem excessive and so a satisficing or sub-optimal solution may be reached, however, if the parties build a relationship then there is already a reservoir of knowledge and under-standing building within the depth of the relationship.

To illustrate how relationship-based procurement systems affect negotiating tactics it is worth reflecting upon results from research undertaken on the National Museum of Australia project in January 2001 in which the project alliance team (the Acton Peninsula Project alliance team) were surveyed. The initial negotiation survey was pilot tested in September/October 1999. Subsequently three surveys were responded to over the life of the project. Three structured questionnaire surveys of relationship characteristics comprising over 120 questions were admi-nistered over the construction period. The Acton Peninsula Project alliance team were multi-disciplinary and co-located on site in one building (on the Acton Peninsula in Canberra). The results presented comprise a small sample size, 14 out of the approximately 60 available people who were part of the team at that time, but a significant one.

Respondents were asked to compare their experience of negotiation in the following three situations using a scale of 1 = strongly disagree and 7 = strongly agree for:

(1) Average to normal business-as-usual (BAU) – the most common situation – usually high/constant conflict.
(2) Best BAU – the occasional project where all parties to the project work exceptionally well together as a team.
(3) Project alliancing – the project delivery strategy that the parties are currently using on the Acton Peninsula Project for the National Museum of Australia – aim is to force collaboration as the only means to achieve the best outcome for the project and hence all teams involved.

A relative agreement index (RAI) was then constructed using the formula where:

$$RAI = \frac{\sum w}{A \times N}$$

where
w = weight given to each statement by the respondents from the 1 to 7 range described previously
A = 7 (the highest weight)
N = the total number of respondents

The closer the RAI is to 100 the higher the level of agreement is with the statement proposed and conversely the closer the RAI is to 0 the lower the level of agreement is with the statement proposed.

Table 10.1 illustrates some interesting behavioural traits associated with the project and its organisational culture. It indicates many respondents believed their negotiation styles fell significantly between average to normal BAU responses and project alliancing. There is a significant difference between BAU and alliancing for hard negotiation reduced from 29 per cent for BAU to 7 per cent for alliancing indicating a more hitherto uncharacteristic empathic approach to negotiating and a similarly significant increase in principled negotiation from 29 per cent for BAU to 86 per cent for alliancing. This indicates a phenomenal change in behaviours between the negotiation approaches for BAU versus alliancing. This conclusion is reinforced with data relating to the way in which National Museum of Australia alliance team members felt that they had been treated.

Table 10.1: Negotiation styles for the National Museum of Australia project

Negotiation styles (*Adapted from Fisher and Ury, 1983, p. 13*) Please tick the box that best describes your negotiation style – not what you think your negotiation style should be	**Average to normal BAU**	**Best BAU**	**Project alliancing**
Soft negotiation: Involves avoidance of any personal conflict and the making of many concessions	0%	14%	14%
Hard negotiation: Involves treating negotiation as a contest between stronger and weaker where 'hanging tough' and 'holding out' are treated as virtues	**29%**	21%	**7%**
Principled negotiation: Involves deciding issues on their merits rather than through a 'haggling' process.	**29%**	71%	**86%**

Source: Adapted from Peters *et al.* (2001, p. 98)

Table 10.2 raises three interesting points:

(1) Fourteen per cent of respondents in the project alliancing environment as opposed to 21 per cent for BAU believed that they had been exploited and compromised.
(2) Seven per cent of respondents believed they had damaged relationships negotiating in the project alliance environment as opposed to 21 per cent for average to normal BAU.
(3) Seventy-one per cent of respondents under project alliancing believed they focused on issues and respected people, whereas only 21 per cent in average to normal BAU believed they behaved that way.

Table 10.2 indicates that the style and impact of negotiation tactics has significantly changed from BAU. The questions presented do not ask the relevant question whether constructive conflict was encouraged to promote diverse

Table 10.2: Behavioural perceptions relating to negotiation outcomes

Negotiation outcomes Please tick the box that best describes **how you feel at the end of negotiations**: not how you would like to feel. At the end of negotiations do you feel:	Average to normal BAU	Best BAU	Project alliancing
You have been exploited and compromised	21%	7%	14%
You have damaged relationships	21%	7%	7%
You have dealt with issues harshly but people have been respected	21%	57%	71%

Source: Adapted from Peters *et al.* (2001, p. 198)

worldviews to be presented in problem solving. However, in another part of the survey the response to the statement 'We trust our partners' integrity to be able to discuss sensitive issues with them in order to resolve disagreements over such issues without fear of appearing to be a 'non-team player' if these issues are important' was responded to with a 36 per cent average BAU value and 84 per cent for the alliancing value. This clearly indicates on this particular project that negotiation style and tactics had changed dramatically.

It is important to acknowledge the small survey size and the limited conclusions that can be made from such a survey. However, if the survey is viewed as an indication of a potential trend, then it is clear that in relation to negotiation styles, negotiation outcomes and negotiation tactics, respondents indicated a significant and encouraging difference between average to normal BAU negotiation and project alliance negotiation in terms of a more sophisticated and empathic response. Respondents also believed that this change in negotiation style had reduced the negative impact of conflict. Data presented from the study of this project clearly indicates a potential for a sea change in attitudes towards negotiation tactics and the way in which people approach problem solving, task allocation and resource distribution. The project was highly successful from a cost (value for money), time delivery, quality, and relationship point of view (Hampson *et al.* 2001). The existing form of dispute resolution allows disputes to grow until they become a major issue (Loosemore 1999; Loosemore *et al.* 2000a). Many hours may be wasted taking different approaches to finding a solution until eventually either a solution is found or the claim moves to arbitration or litigation. This approach leads to large amounts of expenditure, a huge wastage of time and bad feeling being developed between the participants. Eventually relationships completely break down making future business unlikely.

Project alliancing is a delivery strategy that offers construction professionals an opportunity to play different roles. Instead of being cast in an adversarial role determined by a traditional contract, business imperatives encourage parties to form a team and make decisions based on 'what's best for project' (a focus on the problem at hand) rather then individual concerns. The National Museum of Australia offered an exceptional opportunity to study the nuances of behavioural change with the implementation of project alliancing. All team members were chosen for their expertise and personal skills – price was not part of the selection criteria. The project manager and the rest of the ALT believed that the best people available were selected for the specific project to make decisions based on 'what's best for project' rather than pursuing their own or employer's interests at the expense of the project. This supports our contention that ethical or principled negotiation can deliver superior results when negotiating resource sharing/task distribution.

The following three broad areas that were identified to have an impact on negotiation styles:

- Aim of negotiation – relationship or contract.
- Cultural setting of negotiation – individual or collaborative.
- Emotional awareness of negotiator – awareness of their own and others' wishes, fears and beliefs.

The structure of project alliancing addresses all three of the above:

- The **aim** of negotiation should be relationship-based – as there is an ongoing relationship for the length of the project.
- The **cultural setting** was collaborative.
- The individuals that make up a project alliance were well trained in communication and the underlying psychology behind the negotiation process and therefore have a high level of **emotional awareness.**

10.3   Organisational environment factors affecting trust and commitment

The change management literature provides interesting insights for understanding how an organisational environment may be designed or may evolve that facilitates trust and commitment. In Chapter 3 we introduced the concepts of trust and commitment and its place in developing productive cross-team relationships. In this chapter we will explore the organisational factors that facilitate trust and commitment that, in turn, drives superior performance through intrinsic motivation. We will provide examples of this using the National Museum of Australia case study data.

Etzioni (1961, p. 12) constructed a matrix of commitment and compliance relationship types based upon three kinds of kinds of power applied and three types of involvement. This provides a useful map for understanding how relationships function at an active, reactive or proactive level. **Coercive power** is a similar term to that discussed in Chapter 3 of this book – relying upon fear by the use of punishment. **Remunerative** power is synonymous with reward power. **Normative** power relies upon allocation of symbolic rewards such as esteem and thus is akin to referent, legitimate and connection power and he also describes this as the power to persuade, suggest or manipulate. Etzioni describes three natures of involvement. **Alienative** involvement is an intensely negative state described as the involvement of prostitutes, slaves or prison inmates. **Calculative** involvement is one in which effort of compliance is based on the extent and nature of the rewards offered. **Moral** involvement designates a positive orientation of high intensity and suggests intrinsic motivation. Of the nine possible interactions between the kinds of power applied and kind of involvement only three of these predominate.

Applying coercive power tends to overwhelm resulting in grudging acceptance or compliance as does, to a less grudging extent, remunerative power in generating a calculative response. This is a misleading response because the organisation may mistake it for commitment yet when the reward system is withdrawn, reduced or replaced with a less attractive one, the result is mere compliance. The normative power base when skilfully applied predominantly leads to a moral response, which often drives intrinsic motivation.

One of the striking characteristics of this analysis is that it resonates in the literature on how to treat knowledge workers (those whose principal tools of their trade are their intelligence, skills, knowledge base and commitment) is that organisational influence or power base is increasingly one of being an enabler rather than director. To gain commitment and get the best out of this new kind of workforce the manager becomes more of a coordinator of organisational infrastructure to clear obstructions from the path forward and to provide the technical and knowledge infrastructure that best facilitates the task at hand. The age of the knowledge worker is upon us and the competition for talent is fierce (Davenport and Prusak 2000; Bernick 2001; Davenport 2001; Handy 2001). Organisations must now use normative power bases described by Etzioni over 40 years ago to persuade knowledge workers of the value in making a commitment to a project and then reduce, or better still, eliminate barriers to achieve success. Table 10.3 illustrates seven identified key factors for creating and sustaining commitment.

The following briefly explains key commitment factors (Burgess and Turner 2000):

Table 10.3: Creating and sustaining commitment

Key commitment factors	Before the project	During the project	After the project
Free will to join or leave	Free will to join	Free will to leave, but not rejoin	Freedom to become committed after the fact
Role of uncertainty		Increasing certainty	
Start small and build up		Increasing levels of buy-in	
Joining requires an individual effort	Creation of elitism based upon individual input	Management of the potential 'them and us' syndrome	System to reintegrate team members after the project
Public acts of commitment	Demonstrated commitment from others, especially senior executives	Demonstrated commitment from team members and those that will be affected by change	Recorded commitment and appreciation
Active involvement		Increasing scale and scope of involvement	
Clear messages and lines of communication	Communication of expectations and goals	Open and free communication of ideas, problems and feedback	Feedback and corporate learning

Source: Burgess and Turner (2000, p. 230)

- Individuals are more likely to be committed if they volunteer or are given a choice in whether or not to join a project team;
- Uncertainty provides the chance to question standard approaches and commit to fresh approaches. Uncertainty weakens the status quo and allows room for experimentation and discovery of the new. This should be tempered by earlier comments made about cultural dimensions of uncertainty avoidance by some groups and individuals (Hampden-Turner and Trompenaars 2000);
- Starting small and building the scale of requested commitment up when asking for commitment to large-scope changes is less confronting and frightening as they will appear more manageable and able to be coped with rather than expecting large once-off changes;
- People tend to be tribal by nature so enlisting commitment requires symbols, ceremonies and rituals that bind individuals to the project or change proposal;
- Public acts of commitment need to be made to ensure that those engaged on the project know that it is special and deserves special attention and commitment;
- People need to have a sense of belonging, whether to a group or idea. Commitment is developed through action rather than passive submission;
- Clear messages and clear lines of communication ensure that the purpose of the project or change is unambiguous and that its value to making a positive difference can be accepted and internalised.

It is interesting to link the concepts of compliance versus commitment with commitment building steps to see how this affects effective collaboration. In a paper on power and the project manager, Lovell (1993) investigates the tension between the needs of satisfying one's own requirements and these of others. He cites a well-known source from the late 1970s (Thomas 1976) in which he proposes that being unassertive in attempting to satisfy one's own needs and being uncooperative in satisfying the needs of others results in avoidance behaviour. At the other end of the scale is collaborative behaviour where the satisfaction of one's own needs are assertively attempted while at the same time attempting to satisfy the needs of others. These ideas have been further developed in Figure 10.4.

Figure 10.4 illustrates how the concepts of assertiveness and compliance and commitment can be linked. Compliance behaviour can be assertively resisted, that is with pretence at accommodation of the needs with possible covert or overt sabotage. Conversely it may be unassertively resisted where the concerns of others are grudgingly accepted, perhaps with bureaucratic behaviour patterns emerging or undermined through work-to-rule approaches. Neither of these produces satisfactory results for focusing on project goals or the urgency of the task at hand. Commitment can be unassertively delivered through self-sacrificial behaviours though these would probably not be sustained for very long. The most mutually beneficial situation is to have assertive behaviour to ensure that avenues are vigorously explored in finding solutions to problems combined with intense commitment in which intrinsic motivation such as responding to challenges, is evident. This produces not only collaboration but the opportunity for innovation and creative solutions being discovered (Leonard-Barton 1995). The middle ground, compromise, provides only limited means of getting high levels of commitment.

We have explored trust and commitment in some depth but we still need to

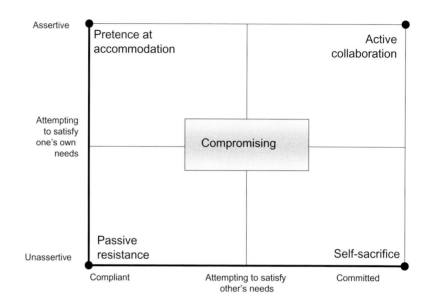

Figure 10.4: Linking assertiveness to compliance/commitment

explore how this may be internalised in change behaviours that maintain commitment. Project management has been also defined as a change process and is relevant to relationship-based procurement systems for projects. Peter Senge is one of the more high-profile writers on change management. He provides some useful insights into drivers and inhibitors of change that are useful for helping us understand how an environment for improving project performance can be created and maintained. These will be briefly reviewed.

Three reinforcing or driver cycles were identified for change (Senge *et al.* 1999) and these have significant implications for designing a work environment that supports commitment within and between teams. All three cycles begin with enthusiasm and commitment (Figure 10.5); this is assumed to be a pre-condition for change. Cycles R1 and R2 lead to an investment in change initiatives by both the organisation and individuals – primarily the investment is in education and training including mentoring and both official and unofficial support. Cycle R1 then moves after a delay in absorbing the investment into an increase in learning capabilities. These generally lead to personal results such as expansion of competencies, finding things easier to do and other outcomes that build confidence and intrinsic motivation including greater feeling of self-worth. Reward systems may provide concrete benefits. These build enthusiasm and commitment because the initiative is proved to be of value. Cycle R2 leads from the change investment to greater involvement with people, which leads to networking and diffusion of the change initiative. This positive socialisation of tacit knowledge helps to make knowledge explicit (Nonaka and Takeuchi 1995; Nonaka *et al.* 2001) and this further reinforces a sense of worth and value for knowledge workers. Tacit knowledge is described as the kind of intuitive and complex knowledge that is held in people's heads or through their bodies as automatically 'knowing' how to do something (Polanyi 1997). Socialisation builds enthusiasm and commitment. Cycle

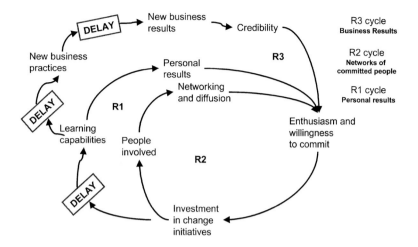

Figure 10.5:   Driving cycles for enthusiasm and willingness to commit
Source: The Dance of Change (copyright © 1999) by Peter Senge, Art Kleiner, Charlotte Roberts, and Richard Ross. Used with permission of Doubleday, a division of Random House Inc., adapted from pp. 46, and 51.

R3 builds upon the learning capabilities because after some time delay while the benefit of the changes are realised and absorbed into the workplace culture it translates into changed work practices and after a further delay this drives improved business performance. This boost to productivity and other positive outcomes drives organisational confidence in the change initiative, which in turn builds upon the enthusiasm and commitment of team members. If the cycle is not subject to atrophy then success will build upon success. This would be an ideal condition that is rarely experienced because of restraining cycles that adversely impact upon this virtuous cycle.

The National Museum of Australia research case study provides interesting insights into the way that the above cycle may have worked in practice on a successful alliancing project. Table 10.4 illustrates a sample of RAI findings from the last of three relationship surveys undertaken during the project's construction. The results were consistent across the three surveys (Peters *et al.* 2001).

The table provides a sample of results from that project that indicate some of the reinforcing cycle elements at work on that project. We selected a number of issues to illustrate – commitment (first four), learning capabilities (next three), organisational empowerment and support for new business practices (last but one) and new business results (last one). The difference between BAU and alliancing is significantly large to indicate that on that particular project, where approximately 25 per cent of the management team was interviewed and responded to the surveys, project alliancing made a big difference to the way in which a changed attitude in team relationships was handled.

Senge *et al.* (1999) identify a number of restraining vicious cycles that inhibit effective change initiatives building momentum. One of these is the problem of making sufficient time available for staff to take advantage of investment, for example in training and development or the energy cost of building effective

Table 10.4: Drivers of commitment and positive change behaviours

Question/statement (with page reference, our bolding of text) Relative Agreement Index (RAI) scores?	Average to Normal BAU	Project alliancing
1. 'We abide by the **spirit of agreements** with our partners rather than concern ourselves about the detail' (S3 page 202)	50%	86%
2. 'We **actively** attempt to **build trust** with our partners through mutual moral and other types of support' (S6 page 202)	51%	93%
3. '**Our actions towards others** (in the supply chain who are not our direct partners) reflects **how we would like them to act toward us**' (O7 page 202)	58%	91%
4. 'We **volunteer help** and support to our partners when they need help and we are happy to provide resources in a crisis' (P1 page 205)	45%	87%
5. 'We respond to disagreements by **rationally debating** and discussing ways to resolve conflicts rather than withdrawing or seeking formal remedies' (C11 page 205)	43%	94%
6. 'I have been stimulated to learn more **technical** knowledge as a result of working in an alliance than I would have expected to have in a non-alliance work situation' (B4 page 211)	34%	71%
7. 'I have been stimulated to learn more **people handling** knowledge as a result of working in an alliance than I would have expected to have in a non-alliance work situation' (B5 page 211)	34%	76%
8. 'The hierarchy provides a lot of **support and encouragement**' (E29 page 219)	50%	88%
9. 'We are **continually exploring options** outside the immediately obvious' (V21 page 219)	48%	85%

Source: Peters *et al.* (2001, page numbers indicated within the table

within and cross-team relationships required in most project work. This is illustrated in Figure 10.6.

A time trap may develop through a combination of flexibility of time and availability of time that leads to frustration dampening enthusiasm and commitment (barrier B1) and reduction in the effectiveness of learning capabilities (barrier B2). Evidence from the National Museum of Australia project reveals some mixed organisational support behaviour. Some examples are illustrated in Table 10.5 below.

Results from Table 10.5 are probably typical of many organisations. There was a greater commitment to enhance skills on project alliancing than BAU projects but participants felt that the 'time' cost was born by them. While there is intent and perhaps policies in place to support learning in tangible ways including providing organisational slack in terms of time for learning and work relief for those participating, the reality is somewhat different.

One of the ways in which the time gap may be obviated is by creating 'virtual time' to at least in part help solve the problems caused by the time gap. Others,

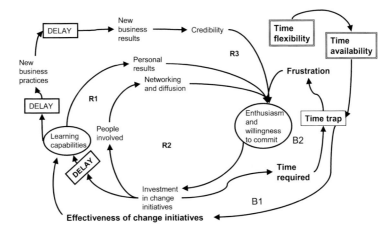

Figure 10.6:   Time trap
Source: The Dance of Change (copyright © 1999) by Peter Senge, Art Kleiner, Charlotte Roberts, and Richard Ross. Used with permission of Doubleday, a division of Random House Inc., adapted from p. 69.

internal to and external from the work teams can provide this virtual time. If this help is withheld for any reason then not only is the time trap exacerbated but additional pressures contributed by this 'help gap' slow the momentum of the reinforcing virtuous cycle driving successful change. Figure 10.7 illustrates the help gap and shows a barrier to the virtuous cycle being developed through insufficient quality or quantity of help being provided. This is hypothesised as influencing commitment, which results in a help gap that in turn affects the effectiveness of the change initiative.

Existing or temporarily available resources can be positively used to reduce the help trap by potentially developing compensating forces that reduce the impact of the time gap. The help gap, however, can (and usually does) negatively influence commitment. Committed people tend to crowd out resources provided to them to help cope with training and development because they invest high levels of personal emotional investment in project success that they are loath to relinquish.

Table 10.5: Time trap factors acting as barriers to positive change		
Question/statement (with page reference, our bolding of text) **Relative Agreement Index (RAI) scores**	**Average to normal BAU**	**Project alliancing**
1. 'I feel that I am expected to **work too many hours**' (I27 page 218)	46%	51%
2. 'My immediate supervisor firm **encourages** me to **develop** my **skills** through structured learning activities (courses, training, etc.)' (F11 page 211)	44%	64%
3. 'My employer expects me to develop my skills but this is **on my own time** so I have to **catch up** on work activities' (F12 page 211)	42%	52%

Source: Peters *et al.* (2001, p. 202, response, page numbers indicated within the table)

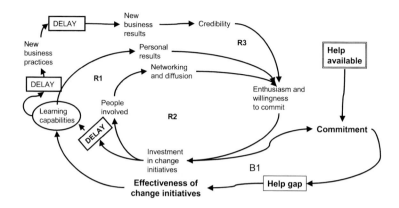

Figure 10.7:   Help trap
Source: The Dance of Change (copyright © 1999) by Peter Senge, Art kleiner, Charlotte
Roberts, and Richard Ross. Used with permission of Doubleday, a division of Random
House Inc., adapted from pp. 46, 48, and 105

This crowding out may take the form of being so over-committed that opportunity
to use help in the form of temporary assistance or offers of additional training,
even in a more flexible mode, may be met with resistance by those targeted for
help. Such people may rationalise this behaviour because much of the knowledge
employed that is of most value is tacit not explicit. The need to socialise knowledge, to get the temporary helpers 'up-to-speed' can be more disruptive and add
more pressure than the benefits of having that help available. However, this
behaviour provides even greater time-trap pressure. The link between the time gap
and help gap is subtle. It is essential when designing systems to provide resolution
to the help gap problem so that the time gap is not exacerbated. Frustration and
burnout are serious consequences of the time trap that may be caused by positive
efforts to address the time trap through providing inappropriately or poorly
thought through help. Mutual adjustment (providing help) through informal
mechanisms based upon effective negotiation techniques may help resolve some of
the time trap problems without raising additional burdens associated with real-
locating resources. Such mutual adjustment may also be assisted by an ability to
call on additional help or resources to be applied seeming outside the help-trap
identified area. The mindset of a project 'pool' of resources rather than individual
team pools of resources may help because self-adjustment and help can be
switched in more creative ways. This is very difficult in non-partnering or non-
alliancing environments. The question of who directly benefits and who directly
pays for such help is generally a highly defended part of an accountability sub-
system of project or business control. On the National Museum of Australia
project some interesting findings reveal the changed mindset that operates in a
BAU and project alliancing environment.

The first four findings illustrated in Table 10.6 clearly indicate an environment
where the time gap has some chance of being obviated by providing help in a self-
adjusting informal manner. The difference between BAU and project alliancing
averages at close to or over 100 per cent, which is a remarkable if not stunning
result. Additionally, close team proximity and a willingness to help other teams in

Table 10.6: Help trap factors acting as barriers to positive change

Question/statement (with page reference, our bolding of text) Relative Agreement Index (RAI) scores	Average to normal BAU	Project alliancing
1. 'We believe that by **cooperating with our partners** openly we reduce the likelihood of opportunistic behaviour' (S5 page 202)	42%	87%
2. 'We actively attempt to build trust with our partners through **mutual moral and other types of support**' (S6 page 202)	51%	93%
3. 'We **share resources** through contributing to the general resource pool available to us and our partners' (P4 page 205)	39%	89%
4. 'We believe that **close physical proximity** to our partner organisations for extended periods of time on site is **of vital importance** in maintaining a good team relationship' (R6 page 208)	44%	93%
5. 'We generally like to **help other** teams when possible' (I14 page 218)	51%	90%
6. Managers feel that level and frequency of **training is below average** (page 240)	N/A	
7. User **utility measures** – which tracked the support as well as technical support of the **supporting information technology** system ProjectWeb (page 242) • For designers • For managers • For administrators	65% 70% 55%	

Source: Peters *et al.* (2001, page numbers indicated within the table

this environment reinforce self-adjustment help that can be provided. This requires considerable empowerment of the teams. Newcombe, for example, in a study of co-located project teams observed the positive effect co-location may have on team empowerment and processes of mutual-adjustment (Newcombe 1996).

Table 10.6 also draws upon the National Museum of Australia study relating to the role and implementation of information and communications technology (ICT). This part of the study investigated the web-based browser technology (ProjectWeb) used for integrated project management and control on the project. Bovis Lend Lease, the lead contractor in the project alliance, developed ProjectWeb and used the National Museum of Australia as a demonstration project for its use. The project manager stated at several seminars[2] that over a million dollars was spent on providing the infrastructure (hardware and training and support) for the project. The alliance partners committed this considerable investment as part of a deliberate policy to demonstrate and learn how ICT could

---

[2] These formed part of the dissemination process for fulfilling the performance criteria of innovation dissemination and were held in Melbourne (30 April 2001), Canberra (8 May 2001), Sydney (1 May 2001) and Brisbane (9 May 2001). They were open to the general public, though of the 400 people in total who attended these were mostly professional, educational and industry people.

more effectively be used on construction projects. Despite this considerable investment the survey conducted by the Commonwealth and Science Research Organisation (CSIRO) representatives of the National Museum of Australia project research team revealed that the project team managers felt that insufficient training was undertaken (the result shown in the sixth row in Table 10.6 was not based on a question comparing BAU with project alliances, hence the not applicable (N/A) result indicated). The CSIRO part of the study on ICT implementation also revealed some very positive aspects of the way in which ICT helped project team members carry out their work. The last result row in Table 10.6 indicates high utility measures for the design team and managers with a somewhat more ambivalent though positive response by administration staff.

Senge *et al.* identify a further trap that restrains the reinforcing virtuous cycle, namely a commitment gap emerging through perceived lack of relevance of the commitment either to the team participants or more commonly by those in leadership positions on projects (Figure 10.8). Relevance may be perceived as diminished if either personal or business benefits appear marginal or negative. This is the 'what is in it for me or my organisation' syndrome. Commitment to change or changed approaches requires continual positive feedback that either leads to improvement through identified ways to improve or confirms the nature and/or extent of benefits derived.

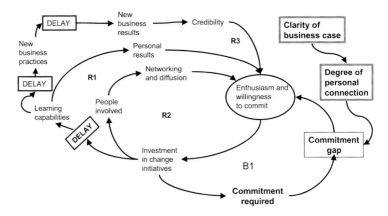

Figure 10.8: Relevance and commitment trap
Source: The Dance of Change (copyright © 1999) by Peter Senge, Art Kleiner, Charlotte Roberts, and Richard Ross. Used with permission of Doubleday, a division of Random House Inc., adapted from p. 161

From the business driver perspective, clear motives for the change need to be understood and probably articulated through a sound business case. The reason for enduring any 'possible pain' needs to be clear for management to support the initiative. Feedback on what works or does not appear to work is important.

From a personal perspective, benefits need to be demonstrated, as much of the motivation to continue putting energy into the change initiative is both intrinsic and extrinsic. Tangible benefits and rewards will help to satisfy extrinsic motivational factors including praise and celebration of success. Intangible benefits also needs to be addressed. These may include the job being easier to do, that additional useful skill sets are developed, or that individual interests are satisfied with aspects of the work being treated as if it were a hobby.

Table 10.7 results strongly indicate that for the National Museum of Australia project, the factors that might impact upon relevance are highly supportive for project alliancing as opposed to BAU. Again many of the RAI scores are double for project alliancing than BAU. The first five measures indicate that there seems to be high goal alignment between individuals, their immediate team organisations, and the project. The project alliance establishment of a clear charter of behavioural expectations was not only developed but clearly expressed and broadcast throughout the site office. The sixth measure, which was expressed as a negative statement, indicates high integrity for both BAU and project alliancing with the alliancing being very low. The seventh measure indicates very strong commitment to the project with very high levels of personal and project sense of identity. This indicated that for the National Museum of Australia project there was a very high degree of relevance both at the organisational and personal level for alignment of goals and behaviours to support the project alliance concept. The stark difference between BAU and project alliance helps to explain how the virtuous cycle of improvement was less adversely affected on the National Museum of Australia project than on BAU projects. The relevance gap and commitment gap naturally leads to discussion of the results/performance gap.

Figure 10.9 illustrates two restraining cycles B1 and B2 to the reinforcing cycles that build barriers to commitment and enthusiasm. The investment in change such as training and development naturally leads to expectations of results at both the personal and organisational level. Cycle B1 focuses upon the organisational results expectations within the implicit time horizon. The problem that occurs is that this

Table 10.7: Relevance trap factors acting as barriers to positive change

Question/statement (with page reference, our bolding of text) Relative Agreement Index (RAI) scores	Average to normal BAU	Project alliancing
1. 'Our goals are those of our partners do not conflict' (P8 page 205)	36%	80%
2. 'Our goals and those of our partners are in harmony with project goals' (P9 page 205)	33%	73%
3. 'I am confident the majority of time that I understand what is expected of me' (C1 page 217)	56%	86%
4. 'Achieving what is expected of me for work-related objectives for the project is very important to me' (C4 page 217)	66%	92%
5. 'There are few conflicts between project and company objectives' (C6 page 217)	47%	73%
6. 'I believe that there is large gap in our team between what we say/commit to do and what we actually intend to do' (I9 page 218) NOTE negative question	32%	28%
7. 'I feel part of the project community' (I35 page 218)	47%	92%

Source: Peters *et al.* (2001, page numbers indicated within the table

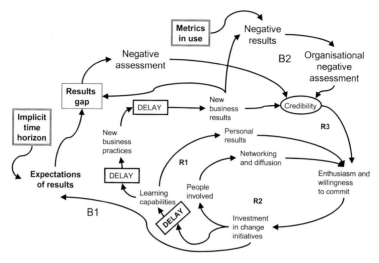

Figure 10.9: Results trap
Source: The Dance of Change (copyright © 1999) by Peter Senge, Art Kleiner, Charlotte Roberts, and Richard Ross. Used by permission of Doubleday, a division of Random House Inc., adapted from pp. 46, 48 and 286

time horizon is often too optimistic. In change initiatives concerning project work such as the engagement of various teams and their joint operation, perhaps from a communication and decision-making perspective, the level of trust and cooperation necessary for this to happen takes considerable effort over a long period of time while team members build a trust bank (see Chapter 3, Figure 3.7). Evidence of success may be expected far too soon when change initiative investments are introduced. These might include training programmes to align project objectives with that of the various teams and individuals or ICT systems for shared communication and decision making. The time trap, help trap, or one of the other traps identified later in this chapter may well exacerbate this. This will lead to a results gap between what was expected of the new business results and what was delivered. When this occurs within a blame-oriented organisational culture the natural reaction will be a search for a scapegoat, defensive routines and systems of deception aimed to mislead and obfuscate. These outcomes have been well described in some of the texts on organisational behaviour and organisational learning, such as the Type I and II behaviours in which the espoused theory (or that which is put forward as the operating paradigm) can be contrasted to the theory-in-use (Argyris and Schön 1996). When this occurs the defensive routines quickly establish and undermine the virtuous cycle for building commitment. Figure 10.9 indicates how negative assessment reduces credibility. This in turn dampens enthusiasm and commitment.

Restraining cycle B2 indicates that even if expectations were realistic, credibility of business results can be undermined by the use of inappropriate performance metrics for measuring business results. This frequently happens when small business units or teams have their performance assessment based upon short-term output or efficiency outcomes relating to a small part of the system rather than their contribution to the whole system or their long-term impact of the change initiative. The change initiative may be likely to produce a temporary drop in perceived efficiency, perhaps it is part of a learning curve effect or because the cycle time of the output or outcome lengthens. If the metrics used to assess results are wrongly aligned then the reward/punitive system may actually penalise effective adherence to the changed process undermining the change initiative. BAU behaviours that encourage defensive routines such as developing paper trails used for making claims or counter-claims exemplify the B2 cycle. In a project alliance scenario the expressed ethos is one of cross-team help and support to overcome problems rather than reverting to defensive blame laying and associated administrative effort to document the support arguments for laying blame or deflecting it. On the National Museum of Australia project, like many other alliancing projects, profits were isolated and placed at risk along with the rewards for success and these tied to project success rather than individual team success. The separate teams and organisations taking part in the alliance had an incentive to ensure that if any teams or team members were experiencing difficulty, others would step in and help them for the good of the project. The metric 'project success' was deemed more appropriate than an individual team 'efficiency' metric under the project alliancing ethos prevailing on this project.

Table 10.8 starkly indicates the difference between BAU and project alliancing on how the respondents viewed success. A constant theme that emerged from the report of the project (Peters *et al.* 2001) was a best-for-project ethos that underpinned decision making from the top management down to operatives on site. The first two rows of results confirm this theme. The third and fourth rows indicate the

Table 10.8: Results trap factors acting as barriers to positive change

Question/statement (with page reference, our bolding of text) Relative Agreement Index (RAI) scores	Average to normal BAU	Project alliancing
1. 'We recognise our **partners' contribution** as of equal **importance** in achieving **project goals**' (A7 page 205)	43%	85%
2. 'We maintain **open lines of general communication** with our partners in order **to prevent** hesitation, reservation or other **defensive behaviour**' (A1 page 208)	49%	92%
3. 'We take considerable effort to **learn from our partners'** experience' (B2 page 211)	39%	76%
4. 'One of the reasons I was attracted to this project was to be **mentally stimulated to learn new things**' (B15 page 211)	45%	71%
5. 'I believe that I get **good recognition** for my contribution' (I34 page 218)	44%	85%
6. 'I feel that this project is **good for my career** plans' (I35 page 218)	52%	77%
7. 'I get **fair reward** for my contribution for the work I do relative to others' (E32 page 219)	52%	78%
8. 'We get **little feedback** on our general performance' (E31 page 219)	52%	52%

Source: Peters *et al.* (2001, page numbers indicated within the table

intrinsic mental stimulation that provided high satisfaction feedback for personal results. Rows 5 to 7 indicate that individuals felt that they had achieved positive results for project alliancing, well over that of BAU. It is interesting that they felt that they had the same mediocre level of feedback for their general performance and this indicates some room for improvement, however, there was plenty of evidence on site for celebration of significant team culture events such as individuals joining or leaving the project team, births of team members' children and even one funeral of a team member's partner. Letters of thanks and appreciation, which were part of a team member's exit process, provided emotional feedback. There was, however, a perception indicated in the eighth row of Table 10.8, that personal performance feedback was not significantly evident.

Figure 10.10 illustrates the fear and anxiety trap. It stems from the learning capabilities part of R1 being restricted in its effectiveness by a lack of candour and openness, which leads to an openness gap. Psychological safety and trust has a direct impact upon the individual and team capacity for openness. This in turn influences the extent of the openness gap. When this gap is wide there is an atmosphere of hidden action, of saying one thing but meaning another, of hidden agendas and a swamp of murkiness that engulfs the ability to rationally and openly discuss difficulties and to offer praise when appropriate. The hidden nature of vital communication about what is really valued and appreciated results in a no-man's land of second guessing what might really be happening. This saps enthusiasm and commitment that often leads to people aspiring to mediocrity and failing to achieve even that modest level of performance.

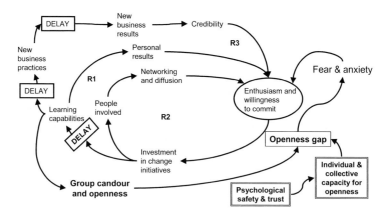

Figure 10.10:   Fear and anxiety trap
Source: The Dance of Change (copyright © 1999) by Peter Senge, Art Kleiner, Charlotte Roberts, and Richard Ross. Used with permission of Doubleday, a division of Random House Inc., adapted from pp. 46, 48, and 245

Table 10.9 illustrates a number of supporting data for the model presented in Figure 10.10. On the National Museum of Australia project there was clearly a high level of openness that contrasts sharply with BAU. Results presented in row 1 indicate a high level of recognition of conflict as inevitable for both BAU and

Table 10.9: Fear and anxiety trap factors acting as barriers to positive change

Question/statement (with page reference, our bolding of text) Relative Agreement Index (RAI) scores	Average to normal BAU	Project alliancing
1. 'I believe that some level of **conflict**, disagreement and **different ways of seeing** issues is **inevitable**' (C10 page 205)	66%	86%
2. 'We respond to disagreements by **rationally** debating and discussing ways to resolve conflicts **rather than withdrawing** or seeking **formal remedies**' (C11 page 205)	43%	94%
3. 'When **problems** arise we concentrate on **solving** them **rather than** trying to find somebody to **blame**' (C12 page 205)	43%	93%
4. 'We see our partners and us **sharing risk** on the basis of mutual **competence**, whoever can **best control risk** volunteers to **accept** and **manage** it' (C13 page 205)	36%	89%
5. 'We communicate **openly** with our partners when problem solving and are not afraid to **own up to mistakes** made' (A2 page 208)	39%	88%
6. 'We **know** what is an **acceptable** behaviour **to our partners**' (A8 page 205)	49%	80%
7. '**Our partners know** what is an **acceptable behaviour to us**' (A9 page 205)	43%	76%
8. 'I feel that the working **atmosphere** between groups is mainly characterised by conflict and **point scoring**' (I40 page 218). NOTE negative question	39%	28%
9. 'I feel a lot like I am **manipulated**' (I41 page 218) NOTE negative question	28%	18%

Source: Peters et al. (2001, page numbers indicated within the table

project alliancing. This is healthy as conflict leads to exposure and exploration of numerous points of view and this enhances innovation and organisational learning (Nonaka and Takeuchi 1995; Pedler *et al.* 1996; Hampden-Turner and Trompenaars 2000). The presence of conflict is not a problem, it is the way that conflict is handled that is the main issue that requires careful handling (Loosemore *et al.* 2000b). Results on rows 2, 3 and 4 illustrate differences from BAU to project alliancing of over 100 per cent for aspects of problem solving in which many contrasting and conflicting views are sought and thought through. The way that risk sharing was handled on the National Museum of Australia project is a particularly good example. The openness exemplified in the results shown in row 5 is also supportive of the model illustrated in Figure 10.10. Openness can also be measured in terms of behaviours between partners – understanding each partner's expectation of treatment is an important element of building the capacity for psychological safety and trust. Results indicated in rows 6 and 7 again starkly show the difference between BAU and project alliancing on the National Museum of Australia project. Results in the last two rows posit a negative question, the responses revealed low values for a toxic atmosphere experienced by teams. The project alliancing values were very small indicating a healthy environment that reduces fear and anxiety. The main thrust of Table 10.9 results indicates how the fear and anxiety trap may be developed and maintained and how, in the case of the National Museum of Australia project, this restraining force on the virtuous cycle of enthusiasm and commitment was dampened.

The final model in this series (Figure 10.11) illustrates the trust trap. Two barrier cycles B1 and B2 illustrate how this gap emerges. In B1, clarity and credibility of management values and aims define the level of trust management required. If this is poorly handled then it adversely affects credibility, this in turn impacts upon enthusiasm and willingness to commit. The trust gap also impacts upon B2, the reflection gap. In this cycle, if the investment in the change initiative does not encourage reflection, and this may be exacerbated through a management value system that fails to support reflection, then the reflection gap emerges. Reflection is vital as it allows us to re-evaluate the systemic drivers and inhibitors of action. The literature indicates that reflection is a key core professional competency and the ability to frame and reframe questions, problems, and issues are the hallmark of intelligence (Schön 1983; McNiff and Whitehead 2000). This reflection gap clouds the clarity of personal values and aims, which dampens enthusiasm and willingness to commit.

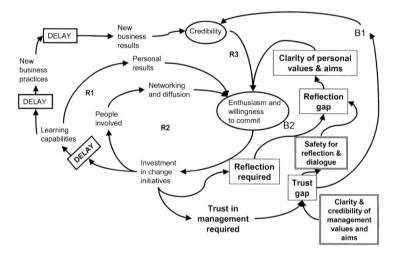

Figure 10.11:   Trust trap
Source: The Dance of Change (copyright © 1999) by Peter Senge, Art Kleiner, Charlotte Roberts, and Richard Ross. Used by permission of Doubleday, a division of Random House Inc., adapted from p. 197

Results illustrated in Table 10.10 provide interesting insights. Results in rows 1, 2 and 3 relate to integrity and an ability to discuss sensitive issues, which is a real measure of trust. These provide amazing results of a difference of over 100 per cent between BAU and project alliancing. The result in row 4 illustrates how this is manifested through a consultative framework. The next two rows contain results relating to sharing ideas facilitating reflection and these indicate a willingness to share ideas across teams as well as there being a large increase in general idea and perception sharing for project alliancing. The last result in row 7 indicates a high level of empowerment and

Table 10.10: Trust trap factors acting as barriers to positive change		
Question/statement (with page reference, our bolding of text) **Relative Agreement Index (RAI) scores**	**Average to normal BAU**	**Project alliancing**
1. 'We **trust our** partners' **integrity** to be able to discuss sensitive issues with them in order to resolve disagreements over such issues without fear of appearing a 'non-team' player if these issues are important' (O8 page 202)	36%	84%
2. 'We **trust our partners' integrity** to be able to discuss sensitive issues with us in order to resolve disagreements over such issues without fear of appearing a 'non-team' player if these issues are important' (O9 page 202)	35%	84%
3. 'We are **alert** to issues that our **partners** may find **sensitive**' A4 page 208)	39%	83%
4. 'We **consult** with our partners before making **key decisions affecting them**' (A5 page 208)	42%	87%
5. 'I **regularly share ideas** from my colleagues from **my firm**' (C9 page 211)	57%	82%
6. 'I **regularly share ideas** from my colleagues from **different companies**' (C10 page 211)	41%	77%
7. 'I feel that I have **insufficient authority** to make contractual or ethical obligations' (I39 page 218) NOTE this is a negative question	26%	19%

Source: Peters *et al.* (2001, page numbers indicated within the table)

trust in individuals to be given the authority to make contractual or ethical commitments.

In this section we have discussed in some depth, the organisational factors that affect trust and commitment. The Senge *et al.* (1999) models, while based on change management issues, are relevant to the study of how best organisations may be framed, structured and developed to sustain the driving virtuous cycles indicated and inhibit the restraining impact of the vicious cycles that provide barriers to enthusiasm and commitment. The evidence for the National Museum of Australia project illustrates an overwhelming and pertinent contrast between BAU and project alliance behaviours that helps to explain how the Senge *et al.* (1999) models can be applied to the study of organisational and individual trust and commitment.

10.4   Human capital skills required to facilitate trust and commitment

Earlier we saw that relationship-based procurement systems require a different mind set from business-as-usual. We also saw that relationship-based procurement systems need closer attention to a wider range of stakeholders and appreciation of their expressed and unvoiced needs. It would be advantageous for team members to be empathic insofar that they are aware of the holistic and long-term impact that their actions and interactions have upon all stakeholders on projects (Leonard-

Barton 1995). It is unlikely, however, that this will ever fully occur. Even if it were to do so it could result in paralysis by analysis, unless the cultural settings within organisations imbued a sense of concerns for outcomes that has been internalised. We know that internalised values are very quickly accessed when action is needed and that the basis of nimble reaction to environments is that these behaviours are hard-wired into the brain (Greenfield 2000). In a relationship-based procurement system all team members need to be attuned to the needs, interest and culture of their negotiating partners to achieve win–win outcomes. We have argued earlier in the first section of this chapter that by imbuing team members with a desire for adopting a principled approach to mutual adjustment and other more formal negotiations, these better outcomes for the project (focus of the work) will be more likely achieved. We saw from the evidence presented on the National Museum of Australia project that the culture on that project supported principled negotiation. The advantages of that were presented. We saw that there was a reduction in team members feeling exploited and damaged by their day-to-day negotiations from BAU to project alliancing on that project. We also saw that team members concentrated vigorously on issues rather than people and that they felt that they were respected by their negotiating partners because they adopted this approach. Benefits of principled negotiation appear clear.

In the section relating to the organisational environment factors that affect trust and commitment we saw that an adaptation of a change model (Senge *et al.* 1999) provided a sound framework for understanding how creating and maintaining commitment can be achieved through active collaboration.

The extensive data presented in Tables 10.4 to 10.10 provide a compelling case that project alliancing on the National Museum of Australia project provided an organisational environment in which behaviour supporting a virtuous cycle of enthusiasm and willingness to commit was realised. We also saw on the National Museum of Australia project that team behaviour also minimised the impact of vicious cycles that provide barriers to enthusiasm and willingness to commit.

We could consider the model for innovation offered by Leonard-Barton (1995) when attempting to summarise the skills and behaviours that need to be encouraged (and were evident on the National Museum of Australia project). At the core of these competences lies the physical systems infrastructure. This would include, for example, the facilities for co-location of teams and the ICT infrastructure to facilitate virtual co-location as well as rapid transfer of necessary documentation to assist decision making and monitoring for project control. This implies that project teams will need ICT support personnel and a workforce that can apply ICT collaboratively to discharge their work effectively. This requires a level of computer literacy not generally widespread in project teams in the construction industry.

The next level in the model indicates management systems. These are the formal (explicit knowledge) and informal (tacit knowledge) routines, procedures and rules that teams work by. These form real intangible assets that are hard to evaluate (Sveiby 1997, p. 12). However, when companies relocate an office or when they form a new organisation from scratch, the value of company policies, manuals and the framework of experience is immediately understood as the cost of acquiring this resource becomes evident. The need for team members to have a full grasp of the organisation's operating paradigm and systems is manifest.

Skills and knowledge relate to the tasks required to be undertaken and the knowledge required to gain an holistic appreciation of project goals. These combine technical skills relating to professional or trade competence with human interaction skills – how to understand people and what 'makes them tick', how to communicate effectively and how to read the political environment especially with respect to shadow or invisible team members/stakeholders as illustrated in Figure 10.1. These skills are often poorly appreciated, not tested or sought at recruitment time, or not subsequently developed as part of a team skills enhancement strategy.

Leonard-Barton's model (1995, p. 19) illustrates the cultural competence 'values' as an the outer layer of her model. This may range from the development of a vision statement that makes organisational values explicit to the development of an organisational culture through shared history of overcoming hurdles and barriers to organisational objectives. Organisational cultures are often highly embedded and difficult to change. In many organisations the culture of 'not informing upon a colleague' or conforming to a non-representative dominant minority or elite has been highlighted as a problem in the literature (Cope and Kalantzis 1997). The need for culturally sensitive skills become evident when developing an environment where trust and commitment is built.

Leonard-Barton's model (1995, p. 19) also indicates the need for openness to new ideas within teams. Teams hone their skills when problem solving. They continually need to import knowledge from the more recognisable sources such as learning institutions, professional bodies or industry representative associations. There needs to be an implementing and integration of knowledge and skills so that these can be internalised. Finally, experimentation is also an important facet of skills development. Framing problems to be solved is an important if not vital skill and the development of appropriate ways in which theory can be tested against observed conditions and situations (Schön 1983). Reflection forms part of this skill base as it is needed for analysis and for figuring out what is relevant about lessons learned and what is not. Experimentation helps unearth new possibilities from existing knowledge re-packaged in novel ways. These four interconnecting elements shape the organisation's core competencies.

A discussion of skills or organisational maturity building would not be complete without some explanation of how tacit knowledge may be diffused and made available to a broader base of individuals within organisations. Nonaka and Takeuchi (1995, p. 62) illustrate this process as a knowledge process that integrates knowing with existing knowledge to create new knowledge. This process involves four phases of moving from tacit to explicit knowledge. These are firstly through **socialisation** by sharing experience, mental models and skills and this transfers tacit knowledge from one person or group of people to another. The next phase **externalisation** transforms tacit to explicit knowledge through articulating tacit knowledge in explicit concepts such as metaphors, models, concepts and equations. This can then absorbed into an organisation through **combination** – systemising different bodies of explicit knowledge by analysing, categorising, and reconfiguring information into knowledge that can be explicitly stored for retrieval. Finally this explicit knowledge can be transformed into new tacit knowledge through **internalisation** – absorption of explicit knowledge into tacit knowledge through verbalisation, systems document processes, simulations.

The process of sharing knowledge is highly complex. Management systems, skills and knowledge and values all encompass elements of tacit and explicit knowledge, which need to be transformed in order that worldviews may be shared to clarify and modify. Developing an organisation and its team members can be seen as a complex task. Leadership skills are also a critical skill that needs to be developed in teams to be able to effectively operate in relationship-based procurement settings. One of these skills is to have the capacity to form a shared will that enables enthusiasm and commitment to flourish.

Scharmer (2001, p. 85) illustrates how a shared will may be developed. He developed a model of four quadrants of a circle segregated by a vertical and horizontal axis. The horizontal axis represents a continuum from primacy of the parts (a highly devolved and detailed approach) to primacy of the whole (a more 'big-picture' and holistic approach). The vertical axis represents a continuum with non-reflection dimension (attempt to think about what has happened and its impact and consequences) and a reflective dimension (carrying out a conversation with one's self to investigate what is happening and why and to place this in a learning context for the future). Within this framework he presents four behaviours. Behaviour 1 (primacy for the whole and non-self reflective) where 'talking nice' refers to a primacy for being polite and not 'making waves' in a politically correct manner in which rules and protocols determine what may be said and what must remain unsaid. The outcome is not to say what one thinks most of the time. This is called Type I behaviour (Argyris and Schön 1996). As knowledge is considered more specifically with a technical or detail focus, the conversational tone is characterised as Behaviour 2 or 'talking tough' that is being explicit about what needs to be known about at the detail level, i.e. the what, how, why and who specifics.

Behaviour 3 indicates a reflective dialogue in which knowledge in use is tested against prevailing theories, belief systems and ethical frameworks that are accepted as guides for action. This represents a more thoughtful operating level and requires skills of those operating at this level to be researchers into their own actions and practices. This is often referred to as engaging in an action learning perspective (McNiff and Whitehead 2000).

Behaviour 4 is related to generative learning in which skills are developed in reflection in action so that these team members are constantly self-evaluating what they do and how that fits the operating environment. It also leads to imagination, inspiration and intuition in action when skills are so hardwired into the consciousness that intelligence guides action at a rapid and subliminal level.

To develop a shared will, leaders must know when to apply the appropriate behaviours illustrated in Behaviours 1 to 4 to articulate the project vision and to impart the true complexity of the task ahead in a manner that inspires confidence rather than dismay. Moving from Behaviour 1 to 4 allows a more informed and sophisticated outcome. Shared will needs to be linked to educating team members about broader implications and consequences of the quality of their interactions. Behaviours 3 and 4 relate to the concept of principled negotiation.

What, then are the implications for human capital in being involved in relationship-based procurement systems?

We understand through mapping (see Table 10.11) desired characteristics of human capital for effectively undertaking relationship-based procurement projects that the input into this system is a supportive environment, an appropriate individual skill base for team members to be quietly proficient at their job, and

Table 10.11: Elements of successful human capital

Work environment	Skill set	Attitudes/rewards

Cultural:
• Conflict management versus conflict aversion
• Encouraging diversity of views
• Encouraging questioning, probing and reframing issues
• Providing time, flexibility and space to reflect
• Providing systemic help mechanisms, personal and professional development that support the culture
 	
Leadership:
• Clarity of goals, vision and ethical standards
• Supporting the culture and re-inventing when necessary
• Supporting inter- and intra-team interaction
• Style that facilitates enthusiasm and commitment
 	
Performance appraisal:
• Clear success KPIs
• Rewards cooperation and knowledge sharing
• Success metrics supportive of building skills, learning and cooperative attitude
	Experiential (shared through common experiences):
	• Technical skills, know-how
	• Caring, trust, attachment and security
	• Energy passion, tension and commitment
	
	Routine (embedded in actions and practice):
	• Technical skills, know-how
	• Organisational routines
	• Organisational culture
	
	Systemic (systemised and packaged knowledge):
	• Documents specification and manuals
	• Databases and ICT systems
	• Patents, manuals procedures (adapted from (Nonaka *et al.* 2001, p. 29)
	• Broad education that facilitates wider view
	• Cultural literacy
	• Good listening skills
	• Good communication skills
	• Good empathy skills
		Internal orientation:
		• Ethical integrity to act by doing 'the right thing'
		• Strong sense of self value
		• Assertive but not aggressive
		• Open minded
		• Stickability and endurance
		
		External to others:
		• Willing to help others
		• Willing to share knowledge and ideas
		• Supportive of organisational goals particularly if that means lobbying and arguing for changes and reframing culture
		• Best for project
		
		Rewards
		
		• Evidence of sharing knowledge
		  ○ Participating in a community of practice
		  ○ Mentoring
		  ○ Knowledge 'editor'
		  ○ Catalyst and communicator
		  ○ Internal consultant in lesson learned
		• Project teams having a risk–reward system that ensures that they sink or swim together to encourage supportive behaviours
		• Career kudos for working on relationship-based procurement systems

requisite attitudinal characteristics that maximise the synergy of interaction between individuals in achieving project goals and aspirations.

Figure 10.12 provides a model that illustrates the three elements required. Attitudes are difficult to develop without a reward system that encourages the required outcome. Many of the leading knowledge management authorities argue that appropriate reward systems must be put in place to encourage knowledge sharing (Leonard-Barton 1995; Nonaka and Takeuchi 1995; Garvin 1998; Hansen *et al.* 1998; Davenport and Prusak 2000; Leadbeater 2001). There is a debate, still unresolved, between those who see performance rewards for individuals as an important driver for improvement while others recommend that rewards should be based upon team, group or project performance. From an individual point of view, rewards need to encourage sharing knowledge. This can include mentoring and/or actively contributing to a community of practice (Wenger and Snyder 2000) perhaps through being part of an editorial panel screening lessons learned to ensure that duplication and swamping of suggestions does not occur (Leadbeater 2001; Robinson *et al.*

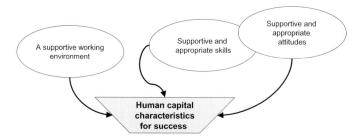

Figure 10.12:   Successful human capital characteristics

2001). Rewards should be given to those who are a catalyst for positive change (Kim and Fisher 2001). From an organisational point of view, the risk–reward system used on alliancing projects such as the National Museum of Australia project promoted project success rather than team success. Risks and rewards were based upon project rather than individual organisation or team success so that it was in the interest of all alliance partners to help and assist when necessary to ensure project success. Finally, involvement in relationship-based procurement projects needs to be valued and rewarded in terms of career prospects to encourage the most talented people to engage in this arrangement.

10.5   Chapter conclusions

We have seen in this chapter that highly sensitive versatile leaders and team members need to be developed and that their skills need to be broad ranging. Technical skills are simply insufficient to marshal the required level of empathy and insight. The National Museum of Australia project study example cited in this chapter provides useful indicators of the characteristics of individuals needed to participate effectively in relationship-based procurement systems. While much of this may seem somewhat idealistic or unrealistic in a fiercely competitive world, the need for collaboration and cooperation has been manifestly demonstrated throughout this book. If we accept this view then we must accept the need for a more sophisticated workforce. The participants in the National Museum of Australia project were not super-beings and were not given any unusual training to act in the way that they have done. The project organisation took careful steps to create an environment where the required behaviours could be nurtured and the alliancing workshops did serve to reinforce the skills needed. Results from the research into the National Museum of Australia project case study indicates that people seemed to naturally want to do 'the right thing', think in terms of 'best for project' and share knowledge. While this was not an easy mindset to develop, it was a preferred paradigm for many people – simply because it requires less energy than constantly fighting shadow wars to prepare for a worse case scenario. People seem to dislike working in a confrontational manner. This bodes well for us preparing human capital to perform in a relationship-based procurement context.

10.6 References

ACA (1999) *Relationship Contracting – Optimising Project Outcomes*, Sydney, Australian Constructors Association.

Argyris, C. and Schön, D. (1996) *Organizational Learning II: Theory, method, and practice*. Reading, MA, Addison-Wesley.

Bennett, J. and Jayes, S. (1995) *Trusting the Team*. Reading, UK, Centre for Strategic Studies in Construction, The University of Reading.

Bernick, C.L. (2001) 'When Your Culture Needs a Makeover.' *Harvard Business Review*. **79** (6): 53–61.

Bresnen, M. and Marshall, N. (2000) 'Partnering in construction: a critical review of the issues, problems and dilemmas.' *Construction Management and Economics*. **18** (2): 229–238.

Briner, W., Hastings, C. and Geddes, M. (1997) *Project Leadership*. Aldershot, UK, Gower.

Burgess, R. and Turner, S. (2000) 'Seven Key Features for Creating and Sustaining Commitment.' *International Journal of Project Management*. **18** (4): 225–233.

Burns, J.M. (1978) *Leadership*. New York, Harper & Row.

Cleland, D.I. (1999) *Project Management Strategic Design and Implementation*. Singapore, McGraw Hill.

Coase, R.H. (1937) 'The Nature of the Firm.' *Economica*. **4**: 386–405.

Cope, W. and Kalantzis, M. (1997) *Productive Diversity – A New, Australian Model for Work and Management*. Annandale, NSW, Pluto Press Australia Limited.

Davenport, T.H. (2001) Knowledge Workers and the Future of Management. *The Future of Leadership – Today's Top Leadership Thinkers Speak to Tomorrow's Leaders*, eds Bennis, W., Spreitzer, G.M. and Cummings, T.G. San Francisco, Jossey-Bass: 41–58.

Davenport, T.H. and Prusak, L. (2000) *Working Knowledge – How Organizations Manage What They Know*. Boston, Harvard Business School Press.

Etzioni, A. (1961) *A Comparative Analysis of Complex Organizations: On Power, Involvement, and Their Correlates*. New York, The Free Press.

Fisher, R. and Ury, W. (1983) *Getting to Yes: Negotiating an Agreement Without Giving In*. London, Penguin Books.

Fritz, P., Parker, A. and Stumm, S. (1998) *Beyond Yes – Negotiating and Networking the Twin Elements for Improved People Performance*. Sydney, Harper Collins.

Garvin, D.A. (1998) Building a Learning Organisation. *Harvard Business Review on Knowledge Management*. Boston, MA, Harvard Business School Publishing: 47–80.

Greenfield, S. (2000) Professor Susan Greenfield. *The Science Show*, 10th June. Williams R. Sydney, Australia, ABC, http://www.abc.net.au/rn/science/ss/stories/s137294.htm.

Gronröos, C. (1990) 'Relationship Approach to Marketing in Service Contexts: The Marketing and Organisational Behavior Interface.' *Journal of Business Research*. **20**: 3–11.

Gronröos, C. (1994) 'From Marketing Mix to Relationship Marketing: Towards a Paradigm Shift in Marketing.' *Management Decision*, **32** (2): 4–20.

Hampden-Turner, C. and Trompenaars, F. (2000) *Building Cross-Cultural Competence – How to create wealth from conflicting values*. New York, John Wiley & Sons.

Handy, C. (1997) *The Hungry Spirit*. London, Random House.

Handy, C. (2001) A World of Fleas and Elephants. *The Future of Leadership – Today's Top Leadership Thinkers Speak to Tomorrow's Leaders*, eds Bennis, W., Spreitzer, G.M. and Cummings, T.G. San Francisco, Jossey-Bass: 29–40.

Hansen, M.T., Nohria, N. and Tierney, T. (1998) 'What's Your Strategy for Managing Knowledge?' *Harvard Business Review*. **77** (2): 106–116.

Hofstede, G. (1991) *Culture and Organizations: Software of the Mind*. New York, McGraw Hill.

Karpin, D. (1995) *Enterprising Nation – Renewing Australia's Managers to Meet the Challenge of the Asia-Pacific Century, Taskforce Report*. Canberra, Industry Taskforce on

Leadership and Management Skills – Ministry for Employment, Education and Training, Australian Federal Government.

Kavali, S.G., Tzokas, N.X. and Saren, M.J. (1999) 'Relationship Marketing as an Ethical Approach; Philosophical and Managerial Considerations.' *Management Decision*, **37** (7): 573–581.

Kim, J.H. and Fisher, N. (2001) *Knowledge Management for Small Occasional Construction Industry Clients: A Theoretical Framework*. ARCOM Seventeenth Annual Conference, Salford, UK, ARCOM.

Kwok, T. (1998) *Strategic Alliances in Construction: A Study of Contracting Relationships and Competitive Advantage in Public Sector Building Works*. PhD, Faculty of the Built Environment. Brisbane, Queensland University of Technology.

Lawler III, E.E. (2001) The Era of Human Capital Has Finally Arrived. *The Future of Leadership – Today's Top Leadership Thinkers Speak to Tomorrow's Leaders*, eds Bennis, W., Spreitzer, G.M. and Cummings, T.G. San Francisco, Jossey-Bass: 14–25.

Leadbeater, C. (2001) How Should Knowledge be Owned? *Managing Industrial Knowledge – creation, transfer and utilization*, eds Nonaka, I. and Teece, D. London, Sage: 170–181.

Lenard, D.J., Bowen-James, A., Thompson, M. and Anderson, L. (1996) *Partnering – Models for Success*. Adelaide, Australia, Construction Industry Institute Australia.

Lendrum, T. (1998) *The Strategic Partnering Handbook*. Sydney, McGraw Hill.

Leonard-Barton, D. (1995) *Wellsprings of Knowledge – Building and Sustaining the Sources of Innovation*. Boston, MA, Harvard Business School Press.

Lewicki, R.J. (1983) Lying and Deception – A Behaviour Model. *Negotiating in Organizations*, eds Bazerman, M.H. and Lewicki, R.J. London, Sage: 68–90.

Loehr, J. and Schwartz, T. (2000) 'The Making of a Corporate Athelete.' *Harvard Business Review*. **79** (1): 120–128.

Loosemore, M. (1999) 'Responsibility, Power and Construction Conflict.' *Construction Management and Economics*. **17** (6): 699–709.

Loosemore, M., Nguyen, B.T. and Denis, N. (2000a) 'Conflict in the Construction Industry.' *Construction Management and Economics*. **18** (4): 447–456.

Loosemore, M., Nguyen, B.T. and Denis, N. (2000b) 'An Investigation into the Merits of Encouraging Conflict in the Construction Industry.' *Construction Management and Economics*. **18** (4): 447–456.

Lovell, R.J. (1993) 'Power and the Project Manager.' *International Journal of Project Management*. **11** (2): 73–78.

McNiff, J. and Whitehead, J. (2000) *Action Research in Organisations*. London, Routledge.

Newcombe, R. (1996) 'Empowering the Construction Project Team.' *International Journal of Project Management*. **14** (2): 75–80.

Nonaka, I. and Takeuchi, H. (1995) *The Knowledge-Creating Company*. Oxford, Oxford University Press.

Nonaka, I., Toyama, R. and Konno, N. (2001) SECI, *Ba* and Leadership: A Unified Model of Dynamic Knowledge Creation. *Managing Industrial Knowledge – creation, transfer and utilization*, eds Nonaka, I. and Teece, D. London, Sage: 13–43.

Pedler, M., Burgoyne, J. and Boydell, T. (1996) *The Learning Company: A Strategy for Sustainable Development*. London, McGraw Hill.

Peters, R.J., Walker, D.H.T., Tucker, S., Mohamed, S., Ambrose, M., Johnston, D. and Hampson, K.D. (2001) *Case Study of the Acton Peninsula Development*, Government Research Report. Canberra, Department of Industry, Science and Resources, Commonwealth of Australia Government: 515.

Polanyi, M. (1997) Tacit Knowledge. *Knowledge in Organizations – Resources for the knowledge-based economy*. Prusak, L. Oxford, Butterworth-Heinemann: 135–146.

Porter, M.E. (1990) *The Competitive Advantage of Nations*. New York, Free Press.

Porter, M.E. (1998) 'Clusters and the New Economics of Competition.' *Harvard Business Review*. **76** (6): 77–90.

Porter, M.E. (2001) 'Strategy and the Internet.' *Harvard Business Review*. **79** (3): 63–78.

Robinson, H.S., Carrillo, P.M., Anumba, C.J. and Al-Ghassani, A.M. (2001) *Linking Knowledge Management Strategy to Business Performance in Construction Organizations.* ARCOM Seventeenth Annual Conference, Salford, UK, ARCOM.

Scharmer, C.O. (2001) Self-transcending Knowledge: Organizing Around Emerging Realities. *Managing Industrial Knowledge – creation, transfer and utilization*, eds Nonaka, I. and Teece, D. London, Sage: 69–90.

Schön, D.A. (1983) *The Reflective Practitioner – How Professionals Think in Action*. New York, Basic Books.

Senge, P., Kleiner, A., Roberts, C., Roth, G. and Smith, B. (1999) *The Dance of Change: The Challenges of Sustaining Momentum in Learning Organisations*. New York, USA, Doubleday.

Senge, P.M. (1990) *The Fifth Discipline – The Art & Practice of the Learning Organization*. Sydney, Australia, Random House.

Sveiby, K.E. (1997) *The New Organizational Wealth: Managing and Measuring Knowledge-based Assets*. San Francisco, Berrett-Koehler Publishers, Inc.

Thomas, K. (1976) Handbook of Industrial and Organisational Psychology. USA, Rand-McNally.

Thompson, L. (1998) *The Mind and Heart of the Negotiator*. London, Prentice-Hall International.

Velasquez, M.G. (1998) *Business Ethics Concepts and Cases*. Upper Saddle River, New Jersey, Prentice-Hall.

Walker, A. (1993) *Project Management in Construction*. Oxford, Blackwell Science.

Weick, K.E. (2001) Leadership as the Legitimation of Doubt. *The Future of Leadership – Today's Top Leadership Thinkers Speak to Tomorrow's Leaders*, eds Bennis, W., Spreitzer, G.M. and Cummings, T.G. San Francisco, Jossey-Bass: 91–102.

Wenger, E.C. and Snyder, W.M. (2000) 'Communities of Practice: The Organizational Frontier.' *Harvard Business Review*, **78** (1): 139–145.

Index